DIGITAL COMMUNICATIONS WITH CHAOS

DIGITAL COMMUNICATIONS WITH CHAOS

Multiple Access Techniques and Performance

Wai M. Tam, Francis C. M. Lau, Chi K. Tse

Department of Electronic and Information Engineering, Hong Kong Polytechnic University, Hunghom, Kowloon, Hong Kong

ELSEVIER

Amsterdam • Boston • Heidelberg • London • New York • Oxford
Paris • San Diego • San Francisco • Singapore • Sydney • Tokyo

ELSEVIER Ltd.
Radarweg 29
P.O. Box 211
1000 AE Amsterdam
The Netherlands

ELSEVIER Inc.
525 B Street, Suite 1900
San Diego, CA 92101-4495
USA

ELSEVIER Ltd.
The Boulevard, Langford Lane
Kidlington, Oxford OX5 1GB
UK

ELSEVIER Ltd.
84 Theobalds Road
London WC1X 8RR
UK

First edition 2007

ISBN-13: 978 0080 451510
ISBN-10: 0080 451519

Printed and bound in Great Britain

07 08 09 10   10 9 8 7 6 5 4 3 2 1

To our families

Tung-Ying, Kin-Pan and Kin-Ho

Karen, Valerie and Michelle

Belinda and Eugene

PREFACE

Since the 1970s, there has been a great deal of research effort spent on studying chaotic systems and the properties of the chaotic signals generated. Characterized by their wideband, impulse-like autocorrelation and low cross-correlation properties, chaotic signals are considered to be useful spread-spectrum signals for carrying digital information. Spectrum spreading has become one of the most popular modulation techniques for high-speed wireless communications. It makes use of signals of very wide bandwidth to carry information at relatively low data rates, and possesses advantages such as low probability of interception, resistance to jamming, multiple-access capability and mitigation to multipath effect, which are particularly important in a wireless scenario. In addition to enjoying the aforementioned benefits, chaotic signals can be generated using simple circuitries, thus lowering the cost of transceivers.

Early study of chaos-based communication systems was focused on a single-user case. In the past few years, more effort has been put on investigating systems with multiple-access capability, which is a key feature of spread-spectrum communication systems. In this book, we aim to present a detailed study of some multiple-access schemes used for chaos-based communications. At the same time, we will evaluate the performance of such systems. In addition, the effectiveness of the multiuser detection techniques, whose primary objective is to reduce interference between users and hence improve performance, will be evaluated in the context of multiple-access digital communication systems.

While this book intends to serve as a reference for the graduate students and researchers who are involved in the study of chaos communications, it also provides useful information on some key features of chaos-based communications for communication engineers and technology developers. Although much theoretical work has already been performed on understanding chaos-based communication systems, there are outstanding implementing issues such as robust bit synchronization and channel equalization, which have to be resolved before practical chaos-based communication systems can become fully available.

This book begins with an overview of spread-spectrum communication systems and application of chaos to spread-spectrum communications in Chapter 1. An overview of a few basic modulation and demodulation schemes used in chaos-based communication systems is then presented in Chapter 2, along with a brief description of several multiple-access systems. Starting from Chapter 3, we investigate in detail some selected chaos-based digital communication systems with multiple-access capability. Simulation results of the bit error rates of these systems are presented and compared with analytical ones. The results are also compared with those of some well-known chaos-based communication systems. Assuming that the chaos generators at the transmitter and receiver are synchronized, in Chapter 3, we study the performance of a coherent chaos-based communication system with multiple users. We also present some mathematical techniques which can be used to estimate or to exactly calculate the bit error rates. In Chapter 4, we explore the use of multiuser detection techniques, which aim at reducing the interference between users and improving the overall performance of the multiple-access systems. While robust chaos synchronization is yet to be achieved, noncoherent chaos-based communication systems, which do not require chaos synchronization, are more practical and have been widely studied. Several multiple-access techniques for noncoherent chaos-based systems, together with their operational details, are presented in Chapters 5 to 8.

We would like to thank Prof. Géza Kolumbán and Prof. A.J. Lawrance for their invaluable comments and suggestions to our research work. We are also grateful to the colleagues at the Hong Kong Polytechnic University for all the fruitful discussions we have had on the subject of chaos-based communication systems. Further, we are indebted to the University and the Hong Kong Research Grants Council (CERG Project PolyU5137/02E) for funding our work. We also thank the University for providing us with a splendid environment to perform quality research work. Last but not least, we sincerely thank our families for their enduring support and patience in all these years.

Wai M. Tam
Francis C. M. Lau
Chi K. Tse

CONTENTS

Preface vii

1 Introduction 1

1.1 Chaos 1
1.2 Spread-Spectrum Communication Systems 2
1.3 Application of Chaos to Spread-Spectrum Communication Systems 5
1.3.1 Analog Modulation 5
1.3.2 Digital Modulation 6
1.3.3 Direct-Sequence Spread-Spectrum 8
1.4 Motivation 8
1.5 Book Organization 9

2 Chaos-Based Digital Communication Systems 11

2.1 Chaos-Based Digital Modulation and Demodulation Schemes 12
2.1.1 Chaos-Shift-Keying (CSK) Modulation 12
2.1.2 Differential Chaos-Shift-Keying (DCSK) Modulation 20
2.1.3 Frequency-Modulated DCSK (FM-DCSK) Modulation 22
2.1.4 Correlation-Delay-Shift-Keying (CDSK) Modulation 24
2.2 Application to the Multiple-Access Systems 26
2.2.1 Two-User DCSK/FM-DCSK System 26
2.2.2 Multiuser FM-DCSK System Based on Walsh Codes 27
2.2.3 Multi-User System Based on Chaotic Frequency Modulation (CFM) 29
2.3 Summary 31

3 Multiple-Access Coherent Antipodal Chaos-Shift-Keying Systems 33

3.1 Introduction 33
3.2 Multiple-Access Antipodal Chaos-Shift-Keying (MA-ACSK) System 34
3.3 Analysis of the MA-ACSK System Based on Gaussian Approximation 36
3.4 Exact Bit Error Rate Analysis of the MA-ACSK System 38
3.5 Results and Discussions 42

3.6 Summary 48
Appendix 3A. Derivation of the Statistical Properties for the Chaotic Sequences Generated by the Chebyshev Map 49

4 Multiuser Detection Techniques for Multiple-Access Coherent Antipodal Chaos-Shift-Keying Systems 55

4.1 Introduction 55
4.2 Linear Multiuser Detection Techniques 56
4.2.1 System Description 56
4.2.2 Performance Analysis 61
4.2.3 Results and Discussions 73
4.3 Nonlinear Multiuser Detection Techniques 83
4.3.1 Parallel Interference Cancellation (PIC) Detectors 83
4.3.2 Performance Analysis of the PIC Detectors 84
4.3.3 Results and Discussions 88
4.4 Summary 93
Appendix 4A. Derivation of Means and Variances Relevant to the Analyses of the Decorrelating Detector and MMSE Detector 93
Appendix 4B. Derivation of Means, Variances and Covariances Relevant to the Analysis of the Decorrelating Detector 100
Appendix 4C. Derivation of Covariances and Mean Relevant to the Analysis of the MMSE Detector 103
Appendix 4D. Calculation of the Probability of Occurrence of $N_{e_{(n-1)}}^{(j)}$ for the PIC Detector 106

5 Multiple-Access Generalized Correlation Delay-Shift-Keying Systems 111

5.1 Review of Correlation-Delay-Shift-Keying (CDSK) Scheme 112
5.2 Generalized Correlation-Delay-Shift-Keying (GCDSK) Scheme 115
5.2.1 Transmitter Structure of the GCDSK system 115
5.2.2 Receiver Structure of the GCDSK system 116
5.2.3 Gaussian-Approximated Bit Error Rates 117
5.2.4 Results and Discussions 119
5.3 Multiple-Access GCDSK Scheme 126
5.3.1 Transmitter Structure of the MA-GCDSK System 126
5.3.2 Receiver Structure of the MA-GCDSK System 127
5.3.3 Gaussian-Approximated Bit Error Rates 129
5.3.4 Results and Discussions 130
5.4 Summary 135
Appendix 5A. Expressions for the Variables in the GCDSK System 138

Appendix 5B. Derivation of Bit Error Rates for GCDSK System Based on Simple Gaussian Approximation 141
Appendix 5C. Derivation of Bit Error Rates for CDSK System Based on Simple Gaussian Approximation 148
Appendix 5D. Expressions for the Variables in the MA-GCDSK System 149
Appendix 5E. Derivation of Bit Error Rates for MA-GCDSK System 154

6 Enhancements of Correlation-Delay-Shift-Keying Scheme 157

6.1 Enhanced Correlation-Delay-Shift-Keying System with Method One 158
6.1.1 Transmitter Structure 158
6.1.2 Receiver Structure 159
6.1.3 Gaussian-Approximated Bit Error Rates for ECDSK-M1 System 162
6.2 Enhanced Correlation-Delay-Shift-Keying System with Method Two 164
6.2.1 Transmitter Structure 164
6.2.2 Receiver Structure 165
6.2.3 Gaussian-Approximated Bit Error Rates for ECDSK-M2 System 167
6.3 Permutation-Based Multiple-Access CDSK Systems 169
6.3.1 Transmitter Structure 169
6.3.2 Receiver Structure 171
6.3.3 Gaussian-Approximated Bit Error Rates 172
6.4 Results and Discussions 176
6.4.1 Effect of Gain *G* 176
6.4.2 Effect of Spreading Factor 176
6.4.3 Comparison of the Simulated and Gaussian-Approximated Bit Error Rates 179
6.4.4 Comparison with PMA-CDSK and PMA-DCSK Systems 179
6.5 Summary 183

7 Multiple-Access Scheme Utilizing Transmitted References And Training Sequences 187

7.1 Transmitter Structure 187
7.2 Receiver Structures 189
7.2.1 Adaptive Transversal Filter 190
7.2.2 Inverse-and-Average Receiver 195
7.3 Results and Discussions 197
7.4 Summary 208

8 Near-Optimum Communication Schemes 209

8.1 Near-Optimum Coherent (NOC) and Noncoherent (NON) Systems 210
8.1.1 Single-User System 210
8.1.2 Multiple-User System 212
8.2 Near-Optimum Noncoherent System with Transmitted Reference 214
8.2.1 Single-User System 214
8.2.2 Multiple-User System 216
8.3 Results and Discussions 217
8.3.1 Effects of the Choice of Transformation Function $F(\cdot)$ in the NOC System 217
8.3.2 Using Chaotic Maps with Symmetrical Probability Density Function 220
8.3.3 Using Chaotic Maps with Asymmetrical Probability Density Function 224
8.3.4 Effect of the Choice of Spreading Factor 226
8.4 Summary 228

Bibliography 229

Index 235

1

INTRODUCTION

1.1 CHAOS

Chaotic signals are non-periodic, random-like and bounded signals that are generated in a deterministic manner [Alligood *et al.*, 1997; Brown, 1996; Elaydi, 2000; Hilborn, 2000]. Chaotic systems, i.e., systems that produce chaotic signals, form a special category in deterministic dynamical systems. The chaotic dynamical systems can be broadly classified into continuous-time and discrete-time types. In a continuous-time system, the chaotic signal is derived from a set of differential equations, i.e.,

$$\boldsymbol{g}(\boldsymbol{x},t)=\dot{\boldsymbol{x}}, \quad \boldsymbol{x}(t_0)=\boldsymbol{x}_0, \tag{1.1.1}$$

where $\boldsymbol{x}$ is a vector representing the state of the system at time t and $\boldsymbol{g}$ denotes the set of differential equations governing the dynamical system. In a discrete-time chaotic system, the chaotic sample at the kth iteration is generated from a set of difference equations, i.e.,

$$\boldsymbol{x}_k=\boldsymbol{g}(\boldsymbol{x}_{k-1})=\boldsymbol{g}^{(k)}(\boldsymbol{x}_0), \tag{1.1.2}$$

where $\boldsymbol{x}$ is the state vector and $\boldsymbol{g}(\cdot)$ denotes the iterative function, which is usually called a *chaotic map*. Figure 1.1 plots a chaotic signal, generated by a one-dimensional chaotic map, against normalized time. It can be observed that the signal never repeats itself, looks random-like and is bounded in the interval $[-1,+1]$.

In addition, the chaotic system exhibits *sensitive dependence on initial conditions*, meaning that the chaotic signals produced are very different even with a very small difference in the initial conditions. This phenomenon is illustrated in Fig. 1.1 which plots two chaotic signals with a difference of only 0.0001 in their initial conditions. Even though a small difference is introduced between the initial values, the two chaotic signals separate rapidly from each other after a short time period. Therefore, by using different initial values, it is possible to produce a large number of chaotic signals. Further, chaotic signals are characterized by their impulse-like autocorrelation and very low cross-correlation properties. Figure 1.2 plots the typical autocorrelation and cross-correlation functions of the chaotic signals. Because of the impulse-like nature of the autocorrelation function, the power spectrum of the chaotic signal exhibits a wideband feature, which is shown in Fig. 1.3.

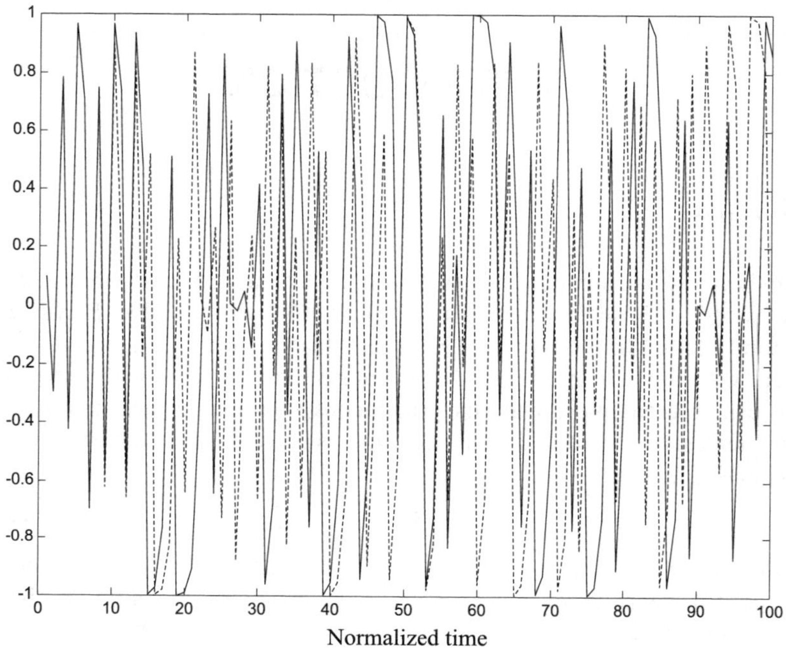

Figure 1.1. Waveforms of chaotic signals with slightly different initial values plotted against normalized time. The chaotic signal is generated by using the cubic map $x_k = 4x_{k-1}^3 - 3x_{k-1}$.

1.2 SPREAD-SPECTRUM COMMUNICATION SYSTEMS

When a narrowband signal modulates a wideband carrier, the signal bandwidth will be increased substantially. Also, the power spectral density (psd) can be lowered accordingly without affecting the bit error performance. Because of the low psd, the signal can be hidden under the background noise, thus guaranteeing a low probability of detection by unintended parties. In addition, the wide signal bandwidth can provide anti-jamming capabilities when appropriate demodulation techniques are applied. Such features, namely low probability of detection and anti-jamming, are typical characteristics possessed by spread-spectrum communication systems [Peterson *et al.*, 1995; Simon *et al.*, 2002; Skaug and Hjelmstad, 1985]. Further, spread-spectrum communication systems can accommodate multiple users if spreading codes with good correlation properties are selected.

Another advantage possessed by broadband signals is the mitigation of multipath fadings. Multipath effect occurs when a signal propagates through differ-

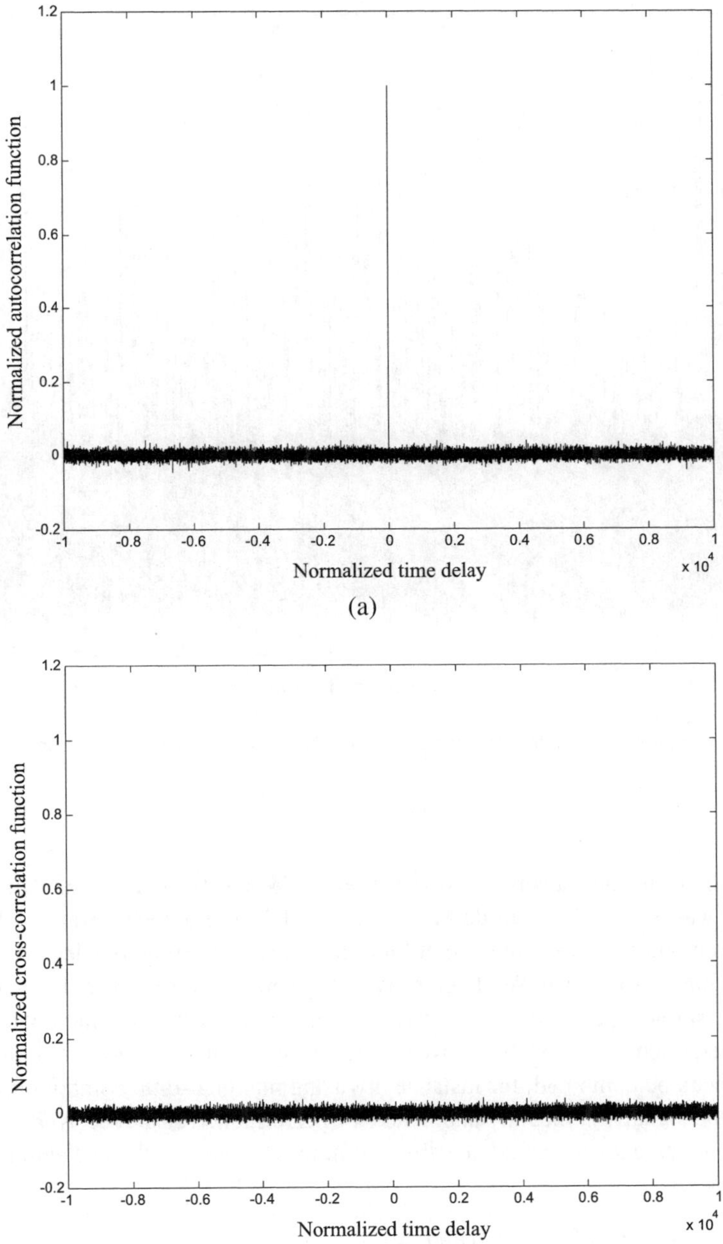

(b)

Figure 1.2. Normalized correlation functions of chaotic signals against normalized time delay. The generating system is $x_k = 4x_{k-1}^3 - 3x_{k-1}$. (a) Autocorrelation; (b) cross-correlation.

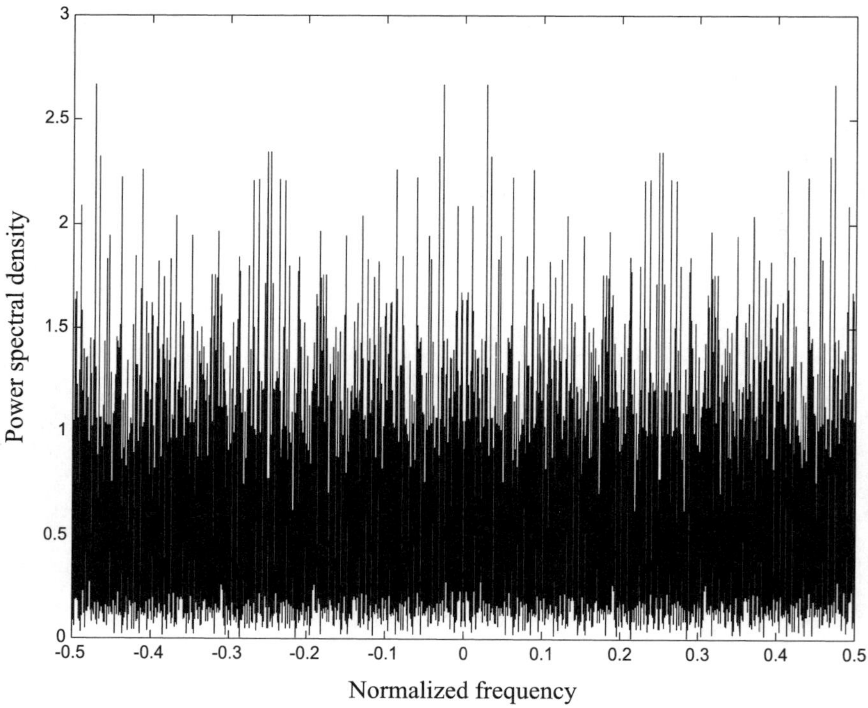

Figure 1.3. Power spectrum of a chaotic signal against normalized frequency. The generating system is $x_k = 4x_{k-1}^3 - 3x_{k-1}$.

ent paths from the transmitter to the receiver. When it occurs, a number of reflected waves with different delays are received. These reflected waves interfere with each other, causing intersymbol interference and leading to a degradation of the system performance. While narrowband systems cannot resolve the multipath signals, spread-spectrum systems can differentiate these paths using appropriate receivers, such as a RAKE receiver. After resolving these paths, the individual signals can be combined, for instance, by using maximal-ratio combining, to enhance the signal significantly and hence the performance of the receiver.

The advantages of spread-spectrum communications can thus be summarized as follows:

1. Low probability of interception.
2. Resistant to jamming.
3. Multiple-access capability.
4. Multipath protection.

1.3 APPLICATION OF CHAOS TO SPREAD-SPECTRUM COMMUNICATION SYSTEMS

Recently, chaotic signals, having a wideband nature, have been proposed as carriers for spread-spectrum communications. This is because chaos-based spread-spectrum systems can offer several advantages over conventional spread-spectrum systems. For example, wideband chaotic signals can be generated using very simple circuitry. Thus, the cost of hardware can be much cheaper. Also, due to the nonperiodic nature of the chaotic signal, the message cannot be easily intercepted and hence the security is enhanced. In addition, chaotic signals have very good auto- and cross-correlation properties. These are important features in a multiple-access environment because they can produce low co-channel interference and hence better system performance. Last but not least, a large number of chaotic signals can be easily generated by using different initial conditions and these signals can be very useful in differentiating users under a multi-user environment.

A number of chaos-based communication systems have been proposed in the last decade. They can be broadly categorized into three main types, namely, chaos-based analog modulation [Kocarev *et al.*, 1992], chaos-based digital modulation [Dedieu *et al.*, 1993; Kolumbán *et al.*, 1998; Parlitz *et al.*, 1992] and direct-sequence spread-spectrum [Heidari-Bateni and McGillem, 1994; Mazzini *et al.*, 1997, 1998; Parlitz and Ergezinger, 1994]. In the following, we will briefly introduce these communication systems.

1.3.1 Analog Modulation

For chaos-based analog modulation systems, two main techniques have been suggested — chaotic masking [Cuomo and Oppenheim, 1993; Kocarev *et al.*, 1992] and chaotic modulation [Itoh and Murakami, 1995; Itoh *et al.*, 1994]. One approach for the chaotic masking is by adding the information signal to the noise-like chaotic signal. The detection is accomplished by regenerating and subtracting the chaotic signal from the received signal. This method is simple to be implemented, but a robust synchronization circuit is required to reproduce the chaotic signal at the receiver.

In chaotic modulation, the information is injected into the parameters of the chaotic map. In consequence, the chaotic map changes from time to time. Synchronization is not required for the demodulation of the analog signal. There are three main demodulation techniques, namely inversion approach [Elmirghani and Cryan, 1994; Miller *et al.*, 1995], linear demodulation approach using the adaptive filter with least-mean-square, recursive least-squares and Kalman-filter-based algorithms [Leung and Lam, 1997], and nonlinear demodulation approach using the radial-basis function (RBF) neural network [Chow *et al.*, 2000; Feng and Tse, 2001]. Among these demodulation techniques, the first approach is the simplest.

But it is difficult to recover the information when the chaotic signal is corrupted by noise. The second technique mitigates the effect of noise by making use of an adaptive filter to demodulate the information. In the third approach, an RBF neural network at the receiver is trained to continuously estimate the state of the nonlinear dynamical system at the transmitting side. Based on the estimation results, the analog information is recovered. Unfortunately, existing analog modulation schemes still do not possess sufficient noise immunity, hence limiting their practical use in communication systems.

1.3.2 Digital Modulation

Compared with analog modulation schemes, digital modulation schemes are more practical and quite a number of chaos-based digital modulation and demodulation techniques have been proposed [Schimming and Hasler, 2000; Jáko, 1998; Kennedy *et al.*, 1998; Kennedy and Kolumbán, 2000; Kis *et al.*, 1998; Kolumbán, 1998; Kolumbán, 2002; Sushchik *et al.*, 2000]. Typically, in a chaos-based digital communication system, the digital symbols are mapped to nonperiodic chaotic basis functions. For instance, in *chaos shift keying* (CSK), different symbols are mapped onto different chaotic attractors which are produced either by a dynamical system for different values of a bifurcation parameter or by a set of completely different dynamical systems [Kennedy, 1998; Kennedy and Kolumbán, 2000; Kis *et al.*, 1998].

A number of coherent systems have been suggested and studied [Dedieu *et al.*, 1993; Kocarev *et al.*, 1992; Kolumbán, 2000a; Kolumbán *et al.*, 1998]. For a coherent system, an exact replica of the chaotic signal needs to be reproduced at the receiving end. Because robust synchronization techniques are not yet available, coherent systems are still not realizable in a practical environment.

Noncoherent communication schemes, which do not require the reproduction of the chaotic signals at the receiving end, are more feasible in practice. The first noncoherent chaos-based digital communication scheme, namely *differential chaos-shift-keying* (DCSK) scheme, was proposed by Kolumbán *et al.* [1996]. In DCSK, a reference chaotic signal is sent, followed by the same signal modulated by a binary symbol. At the receiver, these two pieces of chaotic signals are correlated. The binary symbol is then decoded based on the sign of the correlator output [Kolumbán and Kennedy 2000; Kolumbán *et al.*, 1998; Kolumbán *et al.*, 2002b]. Correlation delay shift keying (CDSK) is another noncoherent modulation scheme in which a reference signal is transmitted to the receiver [Sushchik *et al.*, 2000]. Unlike DCSK, the reference chaotic signal is added to the information-bearing signal with a certain time delay. In consequence, the bandwidth requirement is lower but the performance is degraded compared with the DCSK scheme.

Another noncoherent detection technique, which is applicable to CSK modulation scheme, has been proposed by Hasler and Schimming [2000]. The technique is based on an optimal classifier which optimizes the bit error rate by selecting the symbol that minimizes the *a posteriori* probability. The computational complexity of the classifier was further studied by Lau and Tse [2003b], and an approximate-optimal detection scheme was also proposed [Lau and Tse, 2003c]. In addition, Tse *et al.* reported another noncoherent detector for CSK [Tse *et al.*, 2001, 2002]. The detection principles are formulated on the reconstruction of the return map of the chaotic signals and the regression method.

Besides the aforementioned basic noncoherent detection schemes, a number of DCSK-based derivatives have evolved to further enhance the DCSK scheme. For example, *frequency-modulated DCSK* (FM-DCSK) has been proposed to overcome the varying bit-energy problem in DCSK [Kennedy and Kolumbán, 2000; Kolumbán *et al.*, 1997b]. Instead of feeding a chaotic signal into a DCSK modulator, the chaotic signal is applied to modulate the frequency of a sinusoidal carrier, producing a chaotic FM signal. By sending the chaotic FM signal to a DCSK modulator, an FM-DCSK output is produced which has a constant amplitude, and hence constant power and energy per bit duration. *Quadrature CSK* (QCSK), a multilevel version of DCSK, has been investigated by Galias and Maggio [2001]. Based upon the generation of orthogonal chaotic functions, QCSK allows an increase in data rate with respect to DCSK, with the same bandwidth occupation. In the DCSK scheme, because of the similarity between the reference and information-bearing chaotic samples, the bit rate of the system can be easily derived by viewing the frequency spectrum of the transmitted signal. This may not be very desirable if we want to hide the signal away from unintended parties. In the *permutation-based DCSK* (P-DCSK) scheme [Lau *et al.*, 2002a, 2003], a permutation transformation is introduced in the modulator to shuffle the chaotic samples, destroying the similarity between the reference and information-bearing samples in the DCSK system. By doing so, the bit rate is made undetectable from the frequency spectrum, thereby enhancing the data security.

Recently, another noncoherent communication system, namely chaotic on-off keying (COOK) system, has gained much interest among researchers because of a new category of applications. In February 2003, the Federal Communications Commission (FCC) released 3.1 GHz to 10.6 GHz, a radio bandwidth of 7.5 GHz (called ultra-wideband or UWB), for use by the public. One of the applications of the UWB bandwidth is low-speed and low-power wireless communications (e.g., for sensor networks and wireless personal area networks). Developing a low-cost and low-power UWB device is one of the major challenges faced by researchers. One way to lower the cost of a UWB device is to design simple transmitters and receivers. Because chaotic signals are well known for their inherently wideband nature and their ease of generation, chaotic generators can serve in low-cost transmitters to generate information carriers for UWB communications [Dmitriev *et al.*, 2005]. In fact, it has been demonstrated that a very wide bandwidth chaotic signal can be generated directly into the desired microwave band, e.g., 3.1 to

5.1 GHz, by using a simple chaotic circuit [Chong *et al.*, 2005a]. Also, the power spectrum of the chaotic signal after on-off keying remains of the same shape. At the receiving side, a simple energy detector can be used to detect the presence or absence of the chaotic pulses, thus decoding the symbols being sent. Because of the aforementioned advantages, a consortium led by Samsung Electronics Co. Ltd. has proposed and implemented a simple chaotic on-off keying (COOK) technique to transmit binary symbols in the UWB band with a major application in low-rate wireless personal area networks [Chong *et al.*, 2005b].

1.3.3 Direct-Sequence Spread-Spectrum

The application of chaotic sequences to a direct-sequence spread-spectrum (DS-SS) system was first studied by Heidari-Bateni *et al.* [1992]. In their proposed scheme, the nonperiodic chaotic sequence replaces the binary spreading sequences used in DS-SS systems. Instead of having only two values $\{-1, +1\}$, the amplitude of the spreading sequence now varies within a limited range. Such an arrangement makes the eavesdropper very difficult to estimate the spreading sequence being used. Even if one can decode the spreading sequence used within a certain bit duration, such information will not be very useful in the detection of other bits because of the nonperiodic behavior of the chaotic sequence.

Mazzini *et al.* [1997, 1998] have investigated the use of quantized chaotic sequences to spread the binary symbols. During the process, the quantized signals are transformed into periodic signals. Results indicate that the resultant spread-spectrum systems achieve a better performance than the conventional direct-sequence spread-spectrum systems that use m-sequence or Gold sequence [Peterson *et al.*, 1995].

1.4 MOTIVATION

Chaos-based communication systems have been shown to possess certain advantages over conventional spread-spectrum systems. However, research into chaos-based communication systems began only a decade ago and there are a number of issues to be resolved before chaos-based systems can be put into practical use. Consequently, there is plenty of room for further improvement.

This book is focused on the study of chaos-based digital communication systems, with emphasis on a multiple-access environment. Being spread-spectrum systems occupying bandwidths much wider than what their information contents require, chaos-based communication systems are expected to provide multiple-access service and consequently their performance should be evaluated under a multiple-access environment [Jákó *et al.*, 2000; Lau *et al.*, 2002b; Yang and Chua, 1997, 1998]. In this book, we evaluate the multiple-access capability of two widely studied chaos-based digital modulation schemes, namely coherent chaos shift keying (CSK) and noncoherent correlation delay shift keying (CDSK), in detail. Detectors that improve the performance of multiple-access coherent CSK system are also explored and investigated. Although robust synchronization techniques are not yet available, the study of the coherent CSK system is important in that invaluable performance benchmarks are obtained. Also, new techniques developed for analyzing CSK systems can be applied to analyzing other noncoherent systems. Besides studying the multiple-access capability of existing chaos-based modulation schemes, we will investigate novel noncoherent chaos-based communication systems. The aim is to further improve existing single-user or multi-user noncoherent systems.

1.5 BOOK ORGANIZATION

The book is organized as follows.

Chapter 2 gives an overview on a number of digital modulation and demodulation schemes for chaos-based communication systems. Three multiple-access schemes are also presented. Chapters 3 to 8 report our major work. For the purpose of evaluating the performance of the systems under study, we use an equivalent discrete-time low-pass model of the continuous-time bandpass system [Kolumbán, 1998]. In the equivalent low-pass system, the signal bandwidth is half of that in the bandpass system, i.e., the bandwidth of the low-pass system equals B where $2B$ represents the bandwidth of the bandpass system. According to the sampling theorem, the sampling frequency should not be less than twice the signal bandwidth. In our study, we select the sampling frequency $1/T_s$ (where T_s is the sample duration) as the Nyquist sampling rate, i.e. $1/T_s = 2B$. Therefore, the time duration product, denoted by $2BT_b$, where T_b is the bit duration, is given by

$$2BT_b = \frac{T_b}{T_s} = \mathrm{SF}, \tag{1.5.1}$$

and SF is the spreading factor.

In Chapter 3, we study the multiple-access capability of the coherent CSK system. Detailed derivations on the bit error rates (BERs) are given. Two techniques are developed to calculate the BERs. The first one is based on the Gaussian approximation. The technique is simple and produces approximate BERs that are accurate when the spreading factor is large. The second method is exact but the computational complexity is higher.

Chapter 4 explores the application of multi-user detection techniques to coherent CSK systems. Two linear multi-user detectors, namely decorrelating detector and minimum mean-square-error detector, and one nonlinear multi-user detection scheme, namely parallel interference cancellation scheme, are studied in detail. The corresponding theoretical models are also derived.

Chapter 5 presents a generalized correlation-delay-shift-keying (GCDSK) scheme for noncoherent chaos-based communication systems. The scheme possesses the advantages of the correlation-delay-shift-keying scheme with improved bit error performance. The scheme is also extended to include multiple users by assigning different delays to different users. The effects of the spreading factor, the length of delay and the number of delay units on the BER are also examined.

In Chapter 6, we investigate two enhanced approaches for the CDSK system. Based on the permutation method, a multiple-access scheme is developed for the CDSK and enhanced CDSK schemes. The performances of the enhanced systems are compared with those of the CDSK and DCSK systems.

In Chapter 7, we introduce two types of receiver for a multi-user chaos-based communication system. In the first one, an adaptive filter is employed which aims to reduce the inter-user interference. In the second receiver type, a simple inverse-and-average method is used in an attempt to recover the chaotic reference signal, which is then used to correlate with the information-bearing signals for determining the received symbols. The performance bounds of these two schemes are also derived. The BERs of these two systems are compared with that of the multiple-access DCSK system.

Chapter 8 presents a near-optimal chaotic modulation/demodulation scheme, which can be applied to both coherent and noncoherent systems. Based on the scheme, a communication system with multiple-access capability is also developed and studied. Finally, the BERs of the systems are then compared with those of other chaotic systems.

2

CHAOS-BASED DIGITAL COMMUNICATION SYSTEMS

Chaotic communication techniques have emerged rapidly during the last decade, and much research effort has been devoted to the development of chaos-based digital communication systems that can achieve performance comparable to that of the conventional communication systems.

In chaos-based communication systems, detection schemes can be broadly classified into coherent and noncoherent types. Coherent systems require an exact replica of the chaotic carrier used to carry the information to be reproduced at the receiver, whereas noncoherent systems have no such requirement. The reproduction of replica of chaotic carriers in coherent systems is usually accomplished through a synchronization process, such as in coherent chaos-shift-keying (CSK) and symmetric CSK systems [Kolumbán *et al.*, 1998; Lau and Tse, 2002; Sushchik *et al.*, 2000]. Noncoherent detection can take a variety of forms, but its basic principle is to make use of some distinguishable properties of the transmitted signals, which can be some inherent deterministic properties (e.g., return-map based detection [Tse *et al.*, 2001, 2002] and maximum-likelihood method [Kisel *et al.*, 2001]), or fabricated by a suitable transmission and reception arrangement (e.g., DCSK [Kennedy, 1998; Kis *et al.*, 1998; Kolumbán *et al.*, 1997a] and correlation delay shift keying (CDSK) [Sushchik *et al.*, 2000]).

In recent years, a number of digital modulation and demodulation schemes have been proposed. In the following sections, we will review several chaos-based digital modulation schemes and the corresponding demodulation methods. Chaos-based digital modulation schemes occupy a much wider bandwidth than is required to transmit a stream of binary symbols. To ensure an efficient use of the spectrum, multiple users should be allowed to share the same bandwidth. Therefore, we will also introduce two multiple-access techniques for noncoherent DCSK and FM-DCSK systems. Moreover, the multi-user chaotic frequency modulation system (CFM) will be described.

2.1 CHAOS-BASED DIGITAL MODULATION AND DEMODULATION SCHEMES

2.1.1 Chaos-Shift-Keying (CSK) Modulation

For a binary system, we can make use of one or two chaos generators to accomplish CSK modulation.

2.1.1.1 *CSK with One Chaos Generator* There are two types of CSK that utilizes only one chaos generator, namely antipodal CSK (ACSK) and chaotic on-off-keying (COOK).

2.1.1.1.1 Antipodal CSK Figure 2.1 depicts the coherent antipodal CSK modulator and demodulator [Kolumbán, 2000b]. In a coherent antipodal CSK modulator, the chaotic signal $c(t)$ is first generated. Consider the lth symbol duration $[(l-1)T_b, lT_b)$, where T_b is the bit duration. If a "+1" is transmitted, the chaotic signal will be sent. If a "−1" is transmitted, an inverted copy of the chaotic signal is used as the transmitted signal. Hence, the transmitted signal can be expressed as

$$s(t) = \begin{cases} c(t) & \text{when symbol “+1” is transmitted,} \\ -c(t) & \text{when symbol “−1” is transmitted.} \end{cases} \tag{2.1.1}$$

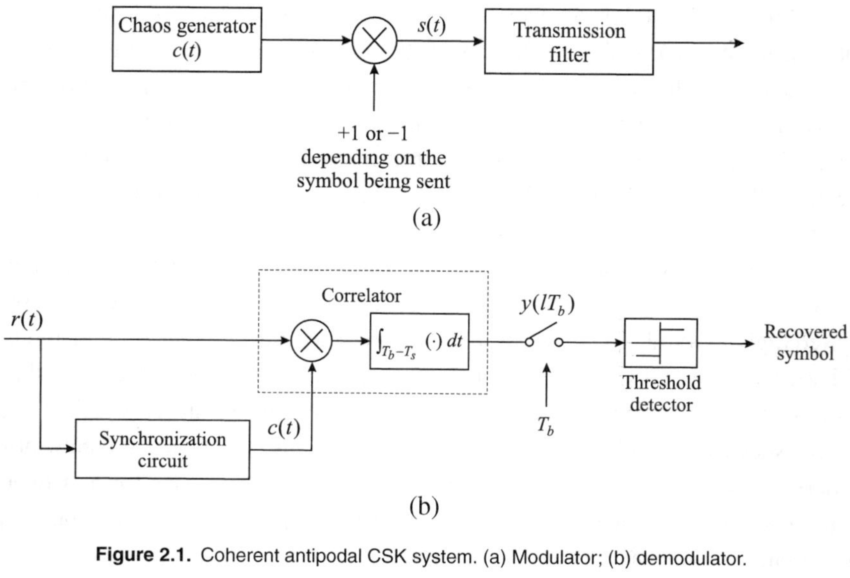

Figure 2.1. Coherent antipodal CSK system. (a) Modulator; (b) demodulator.

Assuming that the transmitted signal is corrupted by additive noise, the received signal is given by

$$r(t) = s(t) + \xi'(t), \tag{2.1.2}$$

where $\xi'(t)$ denotes the noise signal. At the receiver, a self-synchronization circuit will be used to reproduce the chaotic signal. The reproduced signal then correlates with the received signal $r(t)$. The output of the correlator is given by

$$y(lT_b) = \int_{(l-1)T_b+T_s}^{lT_b} r(t)c(t)\,dt, \tag{2.1.3}$$

where T_s is the acquisition time to achieve synchronization. The output of the correlator is compared with the threshold (zero in this case) to determine whether a "+1" or "−1" has been received. If the correlator output is larger than zero, a "+1" is detected. Otherwise, a "−1" is decoded. The probability density functions (pdfs) of the correlator output, $y(lT_b)$, for high and low average bit-energy-to-noise-power-spectral-density ratio values (E_b/N_0) are plotted in Fig. 2.2. When E_b/N_0 is high, it can be observed that $y(lT_b) > 0$ when a "+1" is transmitted and $y(lT_b) < 0$ when a "−1" is transmitted. When E_b/N_0 reduces to a low value, a common region can be found for the cases when "+1" and "−1" are sent. Within this region, an error occurs if $y(lT_b) < 0$ when the transmitted symbol is "+1" or if $y(lT_b) > 0$ when the transmitted symbol is "−1".

2.1.1.1.2 Chaotic On-Off-Keying The block diagram of a chaotic-on-off-keying (COOK) modulator is shown in Fig. 2.3 [Kolumbán *et al.*, 1998]. The COOK modulator is controlled by a switch. If a symbol "+1" is transmitted, the switch is closed and the chaotic signal $c(t)$ is sent. If a symbol "−1" is transmitted, the switch becomes opened and no signal is transmitted. Therefore, the transmitted signal is given by

$$s(t) = \begin{cases} c(t) & \text{when symbol "+1" is transmitted,} \\ 0 & \text{when symbol "−1" is transmitted.} \end{cases} \tag{2.1.4}$$

The demodulation can be accomplished by using a coherent receiver or a noncoherent receiver. The coherent detectionmethod is the same as that used in the antipodal CSK. The noncoherent demodulation process is illustrated in Fig. 2.4. The noncoherent detector is, in fact, a bit-energy estimator whose output for the lth symbol is given by

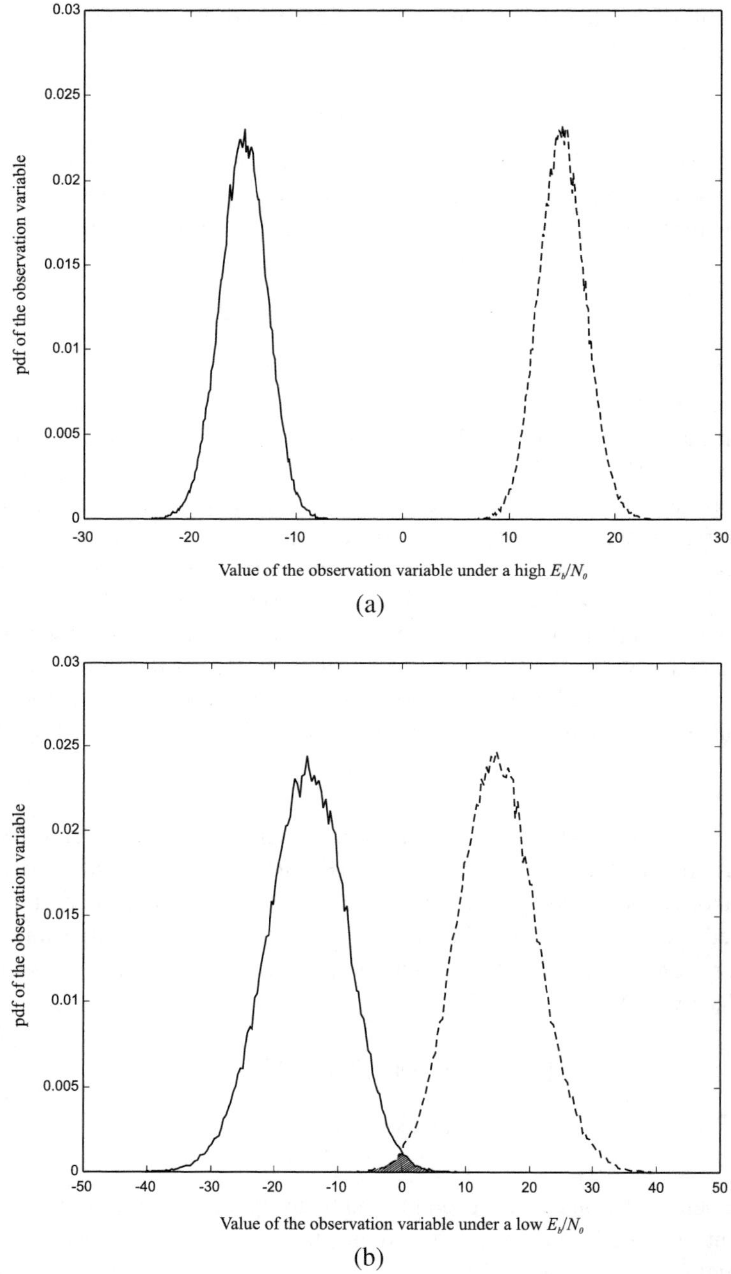

Figure 2.2. Probability density function (pdf) of the observation variable $y(lT_b)$ for a coherent antipodal CSK system for (a) high E_b/N_0; (b) low E_b/N_0.

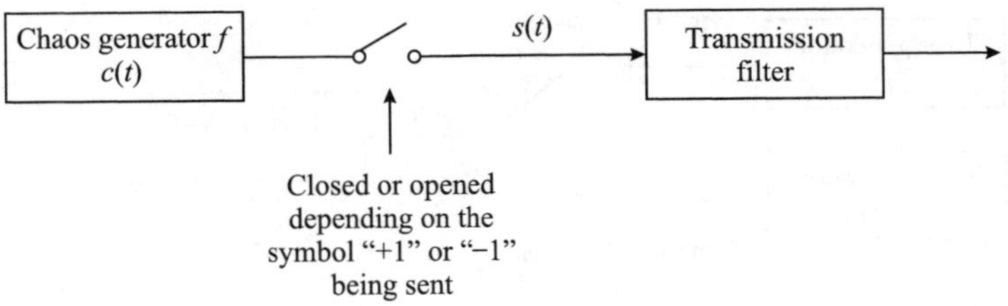

Figure 2.3. COOK modulator.

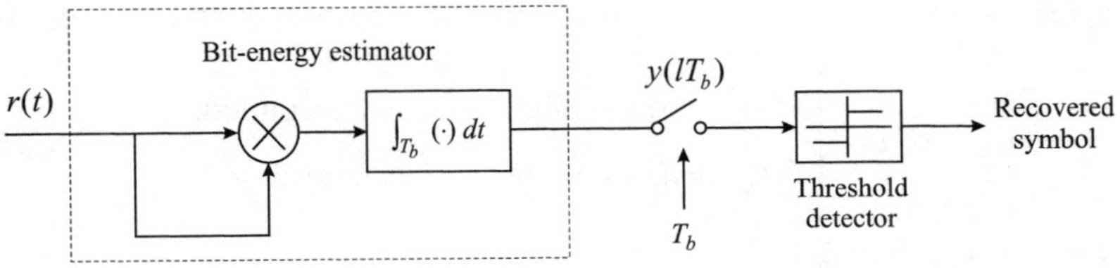

Figure 2.4. Bit-energy-based demodulator.

$$y(lT_b) = \int_{(l-1)T_b}^{lT_b} r^2(t)\,dt = \int_{(l-1)T_b}^{lT_b} s^2(t)\,dt + 2\int_{(l-1)T_b}^{lT_b} s(t)\xi'(t)\,dt + \int_{(l-1)T_b}^{lT_b} \left[\xi'(t)\right]^2 dt. \tag{2.1.5}$$

In the absence of the noise signal $\xi'(t)$, the bit-energy estimator gives

$$y(lT_b) = \begin{cases} \int_{(l-1)T_b}^{lT_b} c^2(t)\,dt & \text{when symbol "+1" is transmitted,} \\ 0 & \text{when symbol "−1" is transmitted.} \end{cases} \tag{2.1.6}$$

The output is then passed to the threshold detector with the threshold set to the mid-value of $\mathrm{E}\left[\int_{(l-1)T_b}^{lT_b} c^2(t)\,dt\right]$ and zero, i.e., $\mathrm{E}\left[\int_{(l-1)T_b}^{lT_b} c^2(t)\,dt\right]/2$ where $\mathrm{E}[\cdot]$ denotes the expectation operator. The decoded symbol is "+1" if the estimated bit energy is larger than the threshold; otherwise, a "−1" is recovered. Note that in the presence of noise, the threshold level needs to be shifted in order to optimize the performance.

2.1.1.2 *CSK with Two Chaos Generators* In this section, we present the modulation and demodulation schemes for chaos-shift-keying systems that make use of two chaos generators.

2.1.1.2.1 Modulation The modulator of CSK with two chaos generators is shown in Fig. 2.5. Assuming that the chaotic signals $\hat{c}(t)$ and $\check{c}(t)$ are sent when

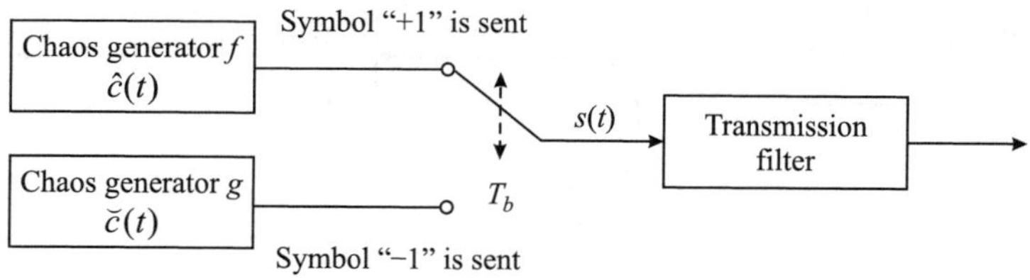

Figure 2.5. CSK modulator with two chaos generators.

the symbols are "+1" and "−1", respectively, the transmitted signal is given by

$$s(t) = \begin{cases} \hat{c}(t) & \text{when symbol "+1" is transmitted,} \\ \breve{c}(t) & \text{when symbol "−1" is transmitted.} \end{cases} \tag{2.1.7}$$

2.1.1.2.2 Demodulation

A. Coherent demodulation

1) Synchronization-error-based detector

The first demodulation method proposed for CSK is based on the synchronization error [Dedieu *et al.*, 1993; Parlitz *et al.*, 1992]. Figure 2.6 depicts a synchronization-error-based demodulator. In this demodulation scheme, self-synchronizing subsystems are used for detection. The subsystems $\hat{f}$ and $\hat{g}$ correspond to the cases when symbols "+1" and "−1" are transmitted, respectively. If a "+1" is sent, the subsystem $\hat{f}$ can synchronize with the received signal, while the subsystem $\hat{g}$ produces a large error. Hence a decision can be made based on the synchronization errors between the received signal and the estimated signals generated by the self-synchronizing subsystems.

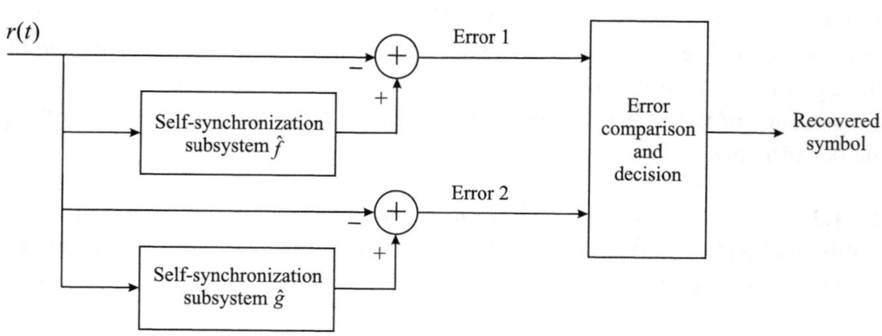

Figure 2.6. Synchronization-error-based demodulator.

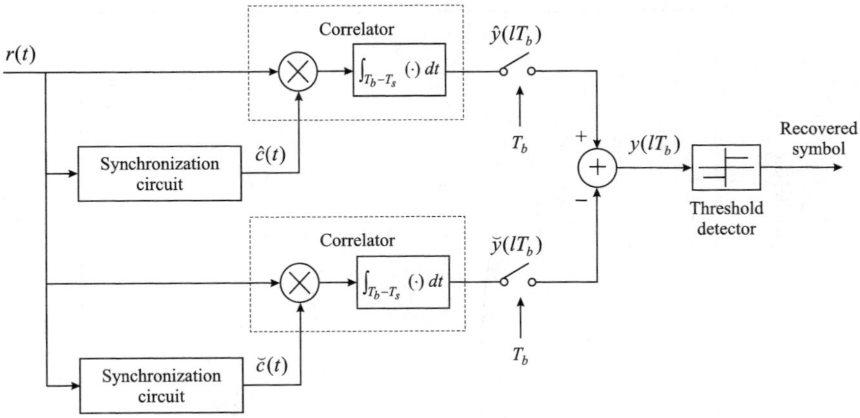

Figure 2.7. Correlator-type coherent demodulator for CSK with two chaos generators.

2) Correlator-type detector

In a correlator-type coherent CSK demodulator, two correlators are employed to evaluate the correlations between the received signal and the two recovered chaotic signals, as shown in Fig. 2.7 [Kolumbán *et al.*, 1998; Lau and Tse, 2002]. The outputs of the correlators for the lth symbol are given by

$$\hat{y}(lT_b) = \int_{(l-1)T_b+T_s}^{lT_b} r(t)\hat{c}(t)\,dt, \tag{2.1.8}$$

$$\check{y}(lT_b) = \int_{(l-1)T_b+T_s}^{lT_b} r(t)\check{c}(t)\,dt. \tag{2.1.9}$$

The input to the threshold detector is obtained using

$$y(lT_b) = \hat{y}(lT_b) - \check{y}(lT_b). \tag{2.1.10}$$

If the output of the first correlator is larger than that of the second one, symbol "+1" is decoded. Otherwise, symbol "−1" is decoded. In Fig. 2.8, we plot the pdf of $y(lT_b)$ for the correlator-type detector. If E_b/N_0 is large, the two output regions separate and the transmitted symbols can be distinguished correctly. If E_b/N_0 is small, the two regions overlap, causing wrong decisions made.

B. Noncoherent demodulation

Coherent detection requires a synchronization process. Under a poor synchronization condition, it is very difficult, if not impossible, to regenerate an exact replica of the chaotic signal. Hence, noncoherent detectors are perceived to be more practical at the present time.

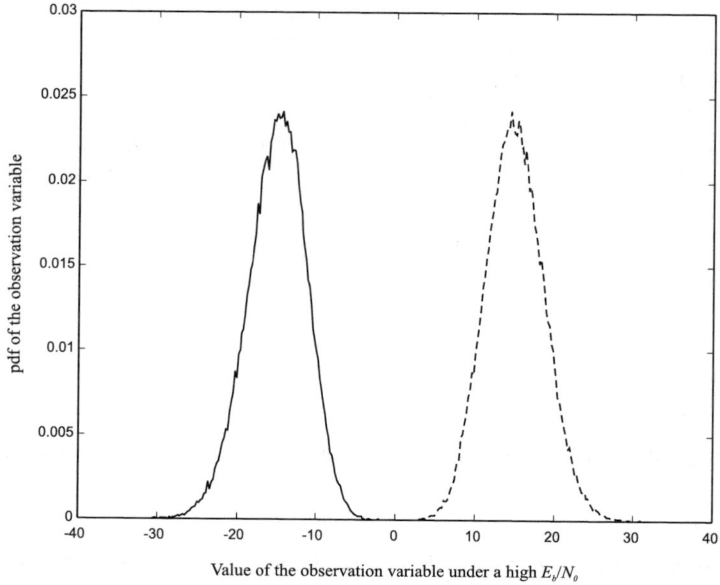

(a)

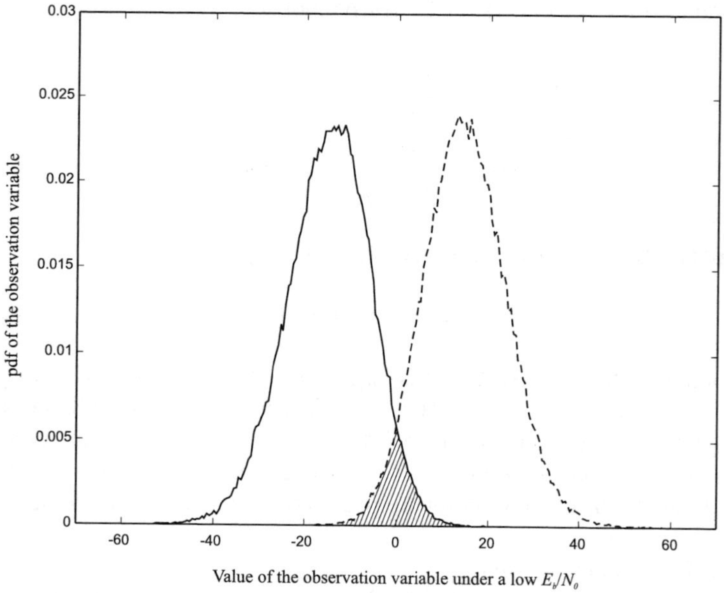

(b)

Figure 2.8. Probability density function of the observation variable $y(lT_b)$ for a correlator-type coherent CSK system for (a) high E_b/N_0; (b) low E_b/N_0.

1) Bit-energy estimator

The simplest noncoherent detection scheme is based on the received bit energy [Kolumbán *et al.*, 1998]. A bit-energy-based demodulator is shown in Fig. 2.4, which is the same one used to decode COOK signals. We assume that the mean bit energies corresponding to $\hat{c}(t)$ and $\check{c}(t)$ are different and denote them by $\hat{E}_b$ and $\check{E}_b$, respectively. In a noiseless environment, the bit-energy estimator gives

$$y(lT_b) = \begin{cases} \int_{(l-1)T_b}^{lT_b} \hat{c}^2(t)\,dt & \text{when symbol "+1" is transmitted,} \\ \int_{(l-1)T_b}^{lT_b} \check{c}^2(t)\,dt & \text{when symbol "}-1\text{" is transmitted,} \end{cases} \tag{2.1.11}$$

for the lth symbol. Similar to the decoding of the COOK, the threshold of the detector is set to the mid-value between $\hat{E}_b$ and $\check{E}_b$, i.e., $(\hat{E}_b + \check{E}_b)/2$. Also, the threshold will shift with the noise level in a noisy environment.

2) Optimal classifier

Assume that each symbol is represented by 2β chaotic samples within each bit duration. Denote the transmitted signal vector, the noise vector and the received signal vector by $\boldsymbol{s}$, $\boldsymbol{\xi}'$ and $\boldsymbol{r}$, respectively, which are given by

$$\boldsymbol{s} = [\, s_1 \quad s_2 \quad \cdots \quad s_{2\beta} \,]^{\mathrm{T}}, \tag{2.1.12}$$

$$\boldsymbol{\xi}' = \left[\, \xi_1' \quad \xi_2' \quad \cdots \quad \xi_{2\beta}' \,\right]^{\mathrm{T}}, \tag{2.1.13}$$

$$\boldsymbol{r} = [\, r_1 \quad r_2 \quad \cdots \quad r_{2\beta} \,]^{\mathrm{T}} = \boldsymbol{s} + \boldsymbol{\xi}'. \tag{2.1.14}$$

where T represents the transpose. The decision criterion of the optimum classifier is to maximize the *a posteriori* probability, defined as $\mathrm{Prob}(\boldsymbol{s}|\boldsymbol{r})$ [Hasler and Schimming, 2000]. According to Bayes' rule, the *a posteriori* probability can be expressed into

$$\mathrm{Prob}(\boldsymbol{s}|\boldsymbol{r}) = \frac{p(\boldsymbol{r}|\boldsymbol{s})\,\mathrm{Prob}(\boldsymbol{s})}{p(\boldsymbol{r})} \tag{2.1.15}$$

where $p(\cdot)$ denotes the probability density function. Also, maximizing (2.1.15) is equivalent to maximizing $p(\boldsymbol{r}|\boldsymbol{s})$ when the symbols are sent with equal probabilities. The computational requirement of the optimal classifier, however, is quite intensive. Further investigation into the optimal classifier has also been conducted by Lau and Tse [2003b, 2003c].

3) Return-map-based demodulation

The transmitter uses two chaos generating functions, f and g, which can be written in a common form $h(\cdot)$ [Tse *et al.*, 2001, 2002], i.e.,

$$\begin{cases} \hat{c}_{k+1} = f(\hat{c}_k) = h(b, \hat{c}_k), \\ \check{c}_{k+1} = g(\check{c}_k) = h(-b, \check{c}_k), \end{cases} \tag{2.1.16}$$

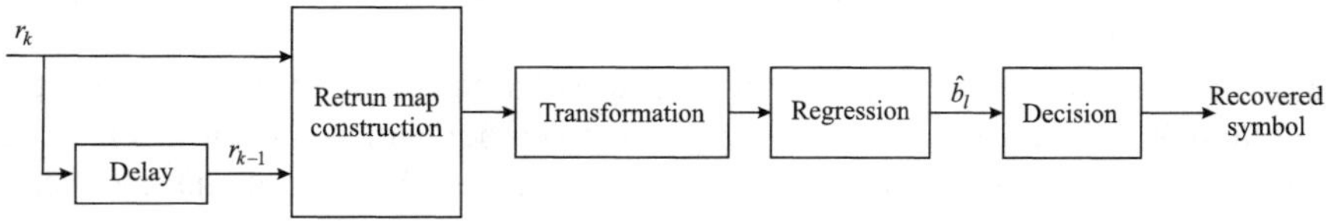

Figure 2.9. Return-map-based demodulator.

where b denotes the distinguishing parameter of the map. Based on the received signals, we reconstruct the return map at the receiver side, from which we estimate the distinguishing parameter b_l for the lth bit period, and decide what the transmitted bit is according to the sign of b_l. Further, when the two chaotic signals are generated using the form

$$\begin{cases} \hat{c}_{k+1} = b\tilde{h}(\hat{c}_k), \\ \check{c}_{k+1} = -b\tilde{h}(\check{c}_k), \end{cases} \tag{2.1.17}$$

the detection step can be simplified by making use of a regression method to estimate the parameter $\hat{b}_l$. The block diagram of the return-map-based demodulator in this case is shown in Fig. 2.9.

2.1.2 Differential Chaos-Shift-Keying (DCSK) Modulation

When the CSK signals are decoded based on the estimation of the bit energy, the threshold of the detector should shift with the noise level. Otherwise, lots of errors would occur. To overcome this issue, differential chaos-shift-keying (DCSK) modulation scheme has been proposed [Kolumbán *et al.*, 1996]. Figure 2.10 shows the block diagram of a DCSK system.

In DCSK modulation, each transmitted symbol duration is divided into two identical time slots. The first one is used to transmit a chaotic reference signal while the second time slot sends an information-bearing signal. During the second time slot, if a symbol "+1" is to be transmitted, the chaotic reference signal is repeated; if the symbol "−1" is to be transmitted, an inverted copy of the reference signal will be sent. Hence, the transmitted signal is given by

$$s(t) = \begin{cases} c(t) & \text{for } (l-1)T_b \leqslant t < (l-1/2)T_b, \\ c(t-T_b/2) & \text{for } (l-1/2)T_b \leqslant t < lT_b, \end{cases} \tag{2.1.18}$$

if the lth symbol is "+1", and

$$s(t) = \begin{cases} c(t) & \text{for } (l-1)T_b \leqslant t < (l-1/2)T_b, \\ -c(t-T_b/2) & \text{for } (l-1/2)T_b \leqslant t < lT_b, \end{cases} \tag{2.1.19}$$

if the lth symbol is "−1", where $c(t)$ denotes the chaotic reference signal.

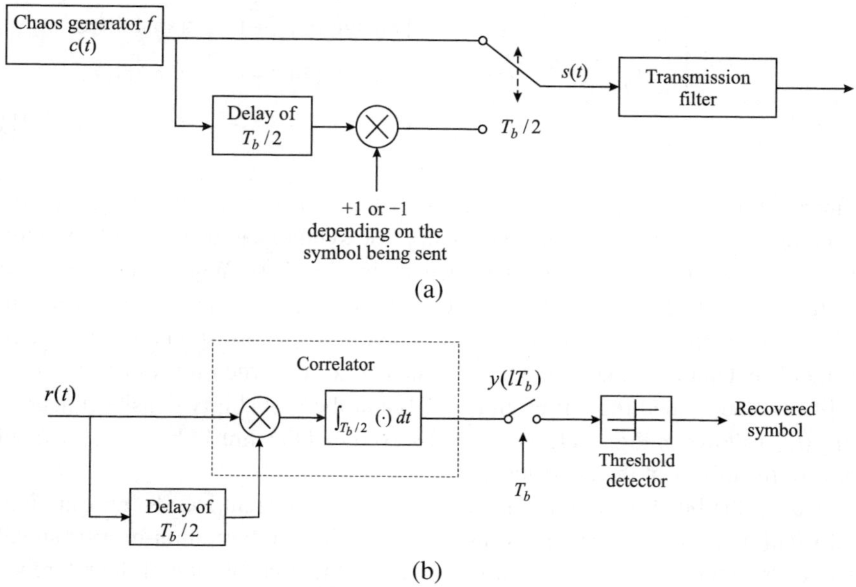

Figure 2.10. DCSK system. (a) Modulator; (b) demodulator.

At the receiver, the received signal in the second half symbol duration will correlate with that in the first half symbol duration. The output of the correlator for the lth symbol duration, $y(lT_b)$, is given by

$$\begin{aligned} y(lT_b) &= \int_{(l-1/2)T_b}^{lT_b} r(t)r(t-T/2)\,dt \\ &= \int_{(l-1/2)T_b}^{lT_b} \left[s(t)+\xi'(t)\right]\left[s(t-T_b/2)+\xi'(t-T_b/2)\right]dt \\ &= \int_{(l-1/2)T_b}^{lT_b} s(t)s(t-T_b/2)\,dt + \int_{(l-1/2)T_b}^{lT_b} s(t)\xi'(t-T_b/2)\,dt \\ &\quad + \int_{(l-1/2)T_b}^{lT_b} \xi'(t)s(t-T_b/2)\,dt + \int_{(l-1/2)T_b}^{lT_b} \xi'(t)\xi'(t-T_b/2)\,dt. \end{aligned} \tag{2.1.20}$$

In a noiseless environment, $y(lT_b)$ equals

$$y(lT_b) = \int_{(l-1/2)T_b}^{lT_b} s(t)s(t-T_b/2)\,dt$$

$$= \begin{cases} \int_{(l-1)T_b}^{(l-1/2)T_b} c^2(t)\,dt & \text{when symbol "+1" is transmitted,} \\ -\int_{(l-1)T_b}^{(l-1/2)T_b} c^2(t)\,dt & \text{when symbol "−1" is transmitted.} \end{cases} \tag{2.1.21}$$

Since $c^2(t) > 0$, when the lth symbol is "+1", $y(lT_b)$ is positive. When the lth symbol is "−1", $y(lT_b)$ is negative. Hence, the symbol can be detected correctly based on the sign of $y(lT_b)$, as shown in Fig. 2.11(a). When noise is added to the transmitted signal, the variance of $y(lT_b)$ becomes larger. As shown in Fig. 2.11(b), for a moderate E_b/N_0, the two regions corresponding to the two symbols extend and mix together. This leads to errors when the value of $y(lT_b)$ falls into the shaded region. From (2.1.20) and the probability density functions of $y(lT_b)$ shown in Fig. 2.11, it can be observed that the threshold should be fixed at zero regardless of the noise level.

Due to the bit structural arrangement in the noncoherent DCSK transmission, only half of the symbol duration is used to transmit the information-bearing signal while the other half symbol duration is used for the transmission of the reference signal. To improve the bandwidth efficiency, an enhanced version of the DCSK scheme has been proposed [Jáko, 1998; Kolumbán *et al.*, 1999]. In the enhanced scheme, one reference signal serves N ($N > 1$) information-bearing signals. Using the enhanced DCSK scheme, the bandwidth is reduced. In addition, there is a reduction of the average bit energy. Further improvement on the enhanced-DCSK scheme has also been achieved by making use of a noise cleaning technique [Kolumbán *et al.*, 1999, 2002a].

2.1.3 Frequency-Modulated DCSK (FM-DCSK) Modulation

Due to the non-periodic nature of the chaotic signal, the DCSK scheme produces time-varying bit energies, which causes the so-called *estimation problem* at the receiver [Kolumbán and Kennedy, 2000]. One way to overcome this problem is to use a chaotic frequency-modulated (FM) signal [Kolumbán *et al.*, 1997b]. An FM-DCSK modulator is shown in Fig. 2.12. Note that except for the generation of the chaotic basis function, the modulation and demodulation processes are the same as those used in the DCSK scheme. As the chaotic FM signal has a constant amplitude, the energy of the FM-DCSK signal remains the same for all bits. Under a noiseless condition, the output of the correlator, $y(lT_b)$ in Fig. 2.10(b), thus gives only two distinct values. If the transmitted signal passes through a noisy channel, these two distinct values will fluctuate with a variance dependent on the noise power. Overall, the FM-DCSK scheme can achieve a better bit error performance than the DCSK scheme.

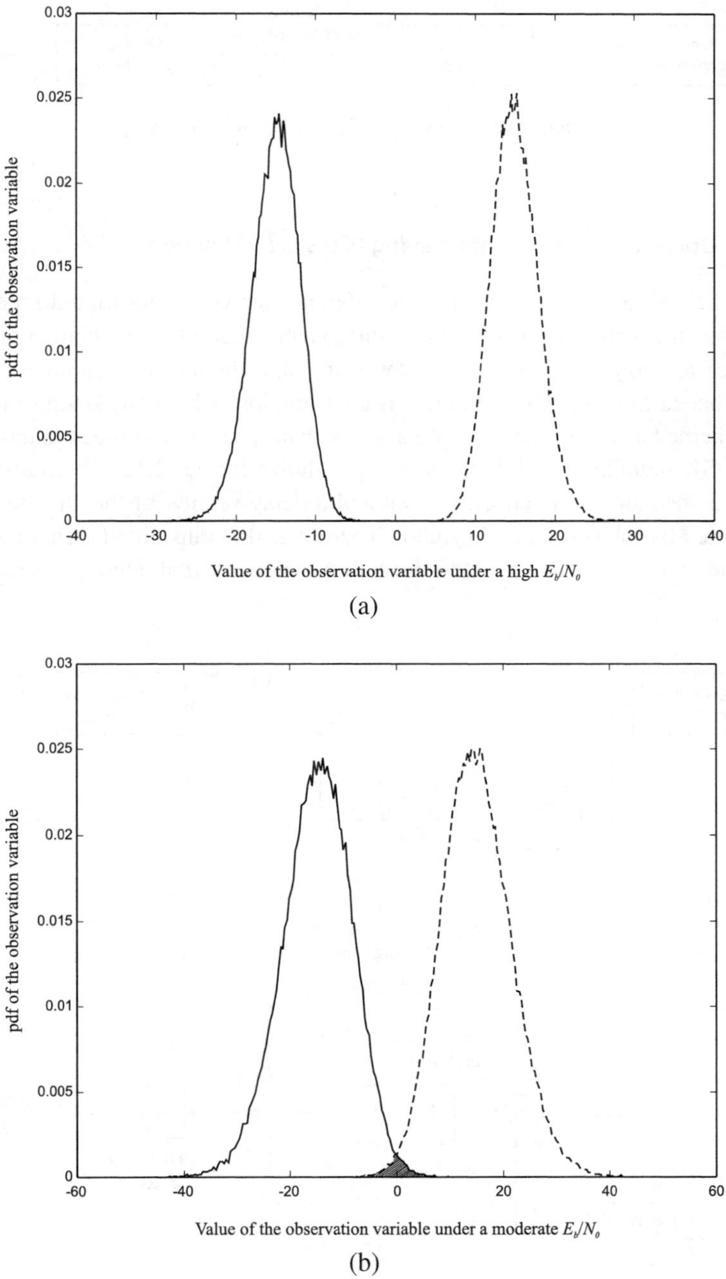

Figure 2.11. Probability density function of the observation variable $y(lT_b)$ for a DCSK system for (a) high E_b/N_0; (b) moderate E_b/N_0.

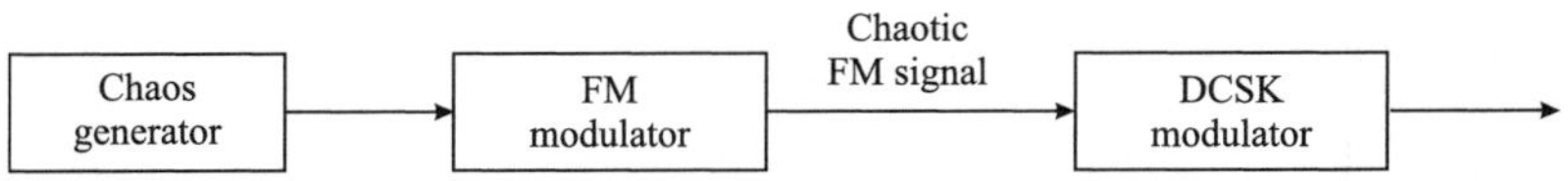

Figure 2.12. Frequency-modulated DCSK modulator.

2.1.4 Correlation-Delay-Shift-Keying (CDSK) Modulation

In the DCSK modulation scheme, the reference signal is transmitted separately from the information-bearing signal. Consequently, half of the symbol duration is not used to carry information. In order to transmit the information continuously and enhance the bandwidth efficiency, a correlation-delay-shift-keying modulation scheme has been proposed by Sushchik *et al.* [2000]. The block diagrams of the CDSK modulator and demodulator are shown in Fig. 2.13. The transmitted signal is the sum of the chaotic signal and a delay version of the chaotic signal modulated by the transmitted symbol. Therefore, the transmitted signal contains both the reference signal and the information-bearing signal. During the lth sym-

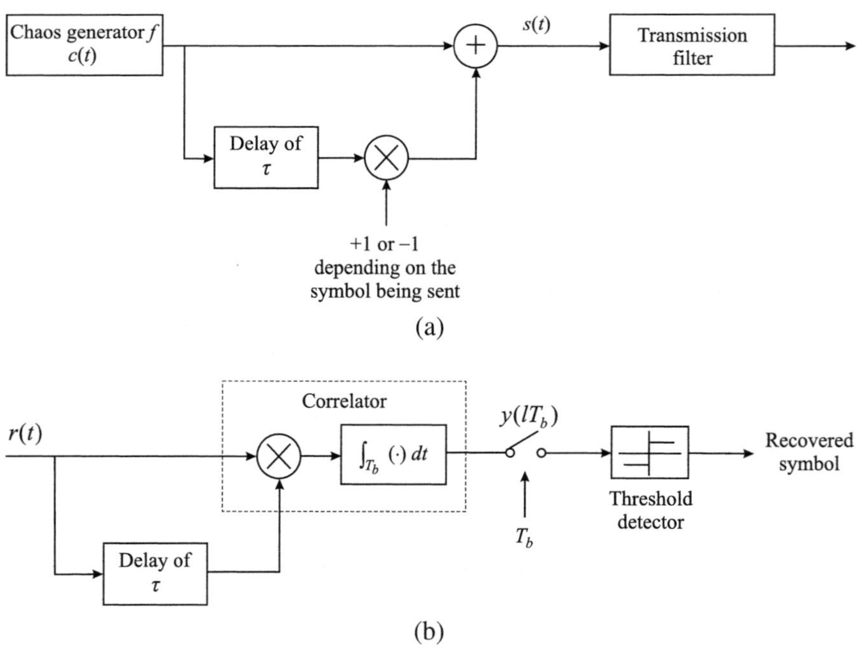

Figure 2.13. Correlation-delay-shift-keying system. (a) Modulator; (b) demodulator.

bol period, the transmitted signal is given by

$$s(t) = \begin{cases} c(t) + c(t-\tau) & \text{when symbol "+1" is transmitted,} \\ c(t) - c(t-\tau) & \text{when symbol "−1" is transmitted,} \end{cases} \tag{2.1.22}$$

where τ is the delay. In (2.1.22), the first term $c(t)$ is the reference signal for the information-bearing signal at time $t+\tau$. The second term $\pm c(t-\tau)$ is used to carry the binary information of the lth symbol.

Similar to the DCSK scheme, a correlator-based receiver is implemented to recover the symbol. The output of the correlator, $y(lT_b)$, at the end of the lth symbol duration is given by

$$y(lT_b) = \int_{(l-1)T_b}^{lT_b} r(t)r(t-\tau)\,dt. \tag{2.1.23}$$

If the transmitted signal passes through a noisy channel, we have

$$\begin{aligned} y(lT_b) &= \int_{(l-1)T_b}^{lT_b} \big[s(t) + \xi'(t)\big]\big[s(t-\tau) + \xi'(t-\tau)\big]\,dt \\ &= \int_{(l-1)T_b}^{lT_b} s(t)s(t-\tau)\,dt + \int_{(l-1)T_b}^{lT_b} s(t)\xi'(t-\tau)\,dt \\ &\quad + \int_{(l-1)T_b}^{lT_b} \xi'(t)s(t-\tau)\,dt + \int_{(l-1)T_b}^{lT_b} \xi'(t)\xi'(t-\tau)\,dt. \end{aligned} \tag{2.1.24}$$

In a noiseless environment, $y(lT_b)$ equals

$$\begin{aligned} y(lT_b) &= \int_{(l-1)T_b}^{lT_b} \big[c(t) + c(t-\tau)\big]\big[c(t-\tau) \pm c(t-2\tau)\big]\,dt \\ &= \int_{(l-1)T_b}^{lT_b} \big[c^2(t-\tau) + c(t)c(t-\tau) \\ &\quad \pm c(t)c(t-2\tau) \pm c(t-\tau)c(t-2\tau)\big]\,dt, \end{aligned} \tag{2.1.25}$$

if a "+1" is transmitted for the lth symbol, and

$$\begin{aligned} y(lT_b) &= \int_{(l-1)T_b}^{lT_b} \big[c(t) - c(t-\tau)\big]\big[c(t-\tau) \pm c(t-2\tau)\big]\,dt \\ &= \int_{(l-1)T_b}^{lT_b} \big[-c^2(t-\tau) + c(t)c(t-\tau) \\ &\quad \pm c(t)c(t-2\tau) \mp c(t-\tau)c(t-2\tau)\big]\,dt, \end{aligned} \tag{2.1.26}$$

if a "-1" is transmitted for the lth symbol. Comparing (2.1.25) and (2.1.26) with the correlator output in the DCSK scheme, i.e., (2.1.21), it is observed that more uncertainty is introduced due to the cross-correlations between the chaotic signal and the delayed versions of it. Hence, the performance of the CDSK scheme is degraded compared to that of the DCSK scheme.

2.2 APPLICATION TO THE MULTIPLE-ACCESS SYSTEMS

Multiple access is an important requirement for spread-spectrum communications [Peterson *et al.*, 1995]. For chaos-based communication systems, only a few multiple-access schemes have been proposed. For instance, a method based on multiplexing chaotic signals has been proposed by Carroll and Pecora [1999], and some chaos-based approaches for generating spreading codes have been applied to conventional code-division-multiple-access (CDMA) systems [Mazzini *et al.*, 1997, 1998; Yang and Chua, 1997, 1998].

Multiple access using DCSK has been introduced by Kolumbán *et al.* [1997a], and the multiple-access capability of FM-DCSK has been studied by Jáko *et al.* [2000]. To minimize the co-channel interference, Kolumbán *et al.* [2002b] has proposed a multi-user FM-DCSK scheme, in which a chaotic signal is combined with two Walsh functions to form the basis functions representing the symbols "$+1$" and "-1" for each user.

In this section, we review the aforementioned multiple-access schemes for use in the DCSK/FM-DCSK systems, namely a two-user DCSK/FM-DCSK system and a multi-user FM-DCSK system based on Walsh codes. Moreover, a multi-user system using the chaotic frequency modulation will be presented.

2.2.1 Two-User DCSK/FM-DCSK System

The transmission scheme of a two-user DCSK/FM-DCSK system is shown in Fig. 2.14 [Kolumbán *et al.*, 1997a; Jáko *et al.*, 2000]. In this scheme, to accommodate two users, two chaotic basis functions are generated. The bit duration T_b is divided into four identical time slots. For the first transmitter, the reference signal is divided into two parts $A_{R,1}$ and $A_{R,2}$, which are sent in the first and third time slots, respectively. The information-bearing signal is also divided into two parts $A_{I,1}$ and $A_{I,2}$, which are transmitted in the second and fourth time slots, respectively. For the second transmitter, the two parts of the reference signal, $B_{R,1}$ and $B_{R,2}$, are sent in the first two time slots. The remaining two time slots are used to transmit the information-bearing parts $B_{I,1}$ and $B_{I,2}$.

The decoding is performed as follows. For the purpose of illustration, we consider the first receiver. The signals in the second and fourth time slots will corre-

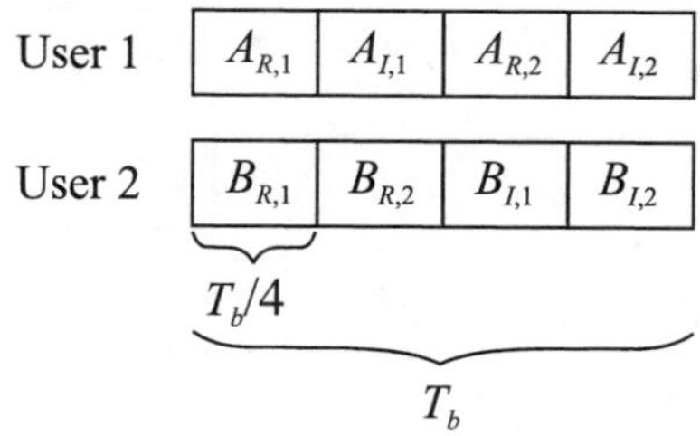

Figure 2.14. Transmission scheme of a two-user DCSK/FM-DCSK system.

late with the signals in the first and third time slots, respectively. Based on the sum of these results, a decision will be made regarding the decoded symbol. Correlation between $A_{I,1}$ and $A_{R,1}$, and that between $A_{I,2}$ and $A_{R,2}$ give the required signals. Correlations between segments like $A_{I,1}$ and $B_{R,1}$, and $B_{R,2}$ and $A_{R,1}$, create co-channel interferences. By producing chaotic signals with low cross-correlations, co-channel interference can be reduced and the system performance can be improved. Similarly, for the second user, the signals in the third and fourth time slots will correlate with the signals in the first and second time slots, respectively. A decision on the recovered symbol is made based on the sum of these results.

While the co-channel interference can be kept low by using chaotic signals with low cross-correlations for different users, it is still finite and increases with the number of users. To achieve eliminating the interference between users, Walsh codes have been proposed to modulate the chaotic signals. In the following section, we will look at the system in more detail.

2.2.2 Multiuser FM-DCSK System Based on Walsh Codes

Consider an N-user FM-DCSK system based on Walsh codes [Kolumbán *et al.*, 2002b]. Two basis functions are used to represent the binary symbols for each user. Hence, a total of $2N$ basis functions are required for the N-user system. Furthermore, each symbol period is divided into 2^M time slots where M is a positive integer and $2^M \geqslant 2N$. The number of time slots equals 2^M which is the length of Walsh codes.

Figure 2.15 illustrates the transmission scheme of a four-user system in which $M = 3$ and $2^M = 8$. Let $m = 1, 2, \ldots, 2^M$ denote the slot number within each symbol period. To construct a pair of basis functions for the ith user, denoted by $\hat{c}^{(i)}(t)$ and $\check{c}^{(i)}(t)$, we multiply two Walsh codes $\{\hat{b}_m^{(i)} \in \{-1, +1\} : m = 1, 2, \ldots, 2^M\}$ and $\{\check{b}_m^{(i)} \in \{-1, +1\} : m = 1, 2, \ldots, 2^M\}$ with a chaotic FM signal $c^{(i)}(t)$ repeatedly. (Details on the generation of Walsh codes can be found in Section 7.2.1.3.) Then, during the mth time slot, the transmitted signal of the ith

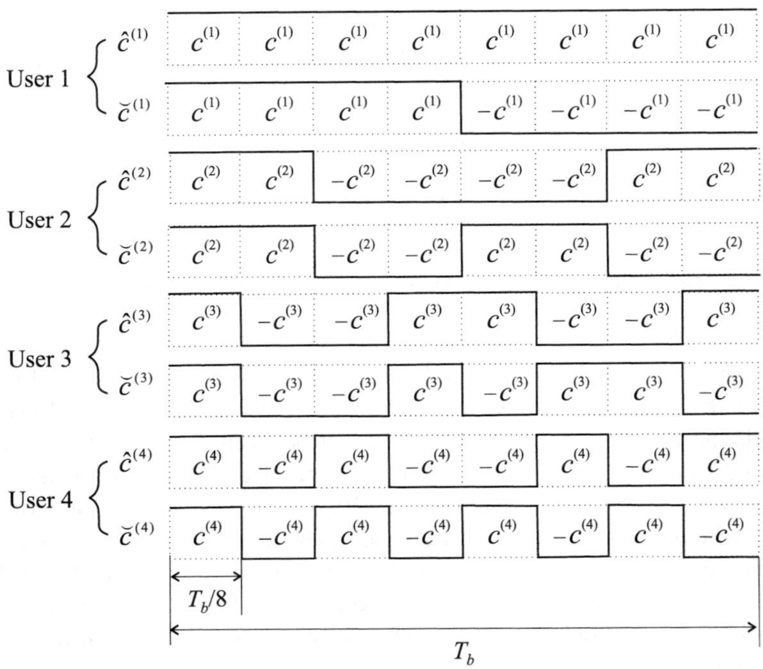

Figure 2.15. Transmission scheme of a multiuser FM-DCSK system.

user is given by ($0 \leqslant t < T_b/2^M$)

$$s_m^{(i)}(t) = \begin{cases} \hat{c}_m^{(i)}(t) = \hat{b}_m^{(i)} c^{(i)}(t) & \text{when symbol “}+1\text{” is transmitted,} \\ \check{c}_m^{(i)}(t) = \check{b}_m^{(i)} c^{(i)}(t) & \text{when symbol “}-1\text{” is transmitted.} \end{cases} \tag{2.2.1}$$

Assume that the transmitted signal is corrupted by noise and denote the noise signal in the mth time slot by $\xi'_m(t)$. The received signal during the mth time slot is given by

$$r_m(t) = \sum_{i=1}^{N} s_m^{(i)}(t) + \xi'_m(t), \quad 0 \leqslant t < T_b/2^M. \tag{2.2.2}$$

For the jth user, the signal in each time slot will multiply into the corresponding element in the Walsh code. Then an averaging process is performed to estimate the chaotic FM signal being sent. This is done for each of the two Walsh codes assigned to each user. For example, suppose a "+1" is transmitted for the jth user. Based on the Walsh code $\{\hat{b}_m^{(j)} \in \{-1, +1\} : m = 1, 2, \ldots, 2^M\}$, the first estimated chaotic signal, denoted by $\tilde{c}^{(j)}_{+1,+1}(t)$, is given by

$$\begin{aligned}
\tilde{c}^{(j)}_{+1,+1}(t) &= \frac{1}{2^M}\sum_{m=1}^{2^M}\hat{b}^{(j)}_m r_m(t) \\
&= \frac{1}{2^M}\sum_{m=1}^{2^M}\hat{b}^{(j)}_m\left[\sum_{i=1}^{N}s^{(i)}_m(t)+\xi'_m(t)\right] \\
&= c^{(j)}(t)+\frac{1}{2^M}\sum_{m=1}^{2^M}\hat{b}^{(j)}_m\xi'_m(t).
\end{aligned} \tag{2.2.3}$$

The last equality holds because of the orthogonality between the Walsh codes. In this case, the chaotic FM signal is estimated with some distortion due to the averaged noise. Using the aforementioned procedure, it is readily shown that using the other Walsh code $\{\check{b}^{(j)}_m \in \{-1,+1\} : m = 1, 2, \ldots, 2^M\}$, the estimated chaotic signal, denoted by $\tilde{c}^{(j)}_{+1,-1}(t)$, equals

$$\begin{aligned}
\tilde{c}^{(j)}_{+1,-1}(t) &= \frac{1}{2^M}\sum_{m=1}^{2^M}\check{b}^{(j)}_m r_m(t) \\
&= \frac{1}{2^M}\sum_{m=1}^{2^M}\check{b}^{(j)}_m\left[\sum_{i=1}^{N}s^{(i)}_m(t)+\xi'_m(t)\right] \\
&= \frac{1}{2^M}\sum_{m=1}^{2^M}\check{b}^{(j)}_m\xi'_m(t).
\end{aligned} \tag{2.2.4}$$

It can be observed that no chaotic signal but noise is obtained.

The estimated chaotic basis functions are then reconstructed using $\{\hat{b}^{(j)}_m\}$ with $\tilde{c}^{(j)}_{+1,+1}(t)$, and $\{\check{b}^{(j)}_m\}$ with $\tilde{c}^{(j)}_{+1,-1}(t)$. Finally, each of the estimated chaotic basis functions correlates with the received signal. If the correlator output corresponding to the estimated chaotic basis function reconstructed using $\{\hat{b}^{(j)}_m\}$ and $\tilde{c}^{(j)}_{+1,+1}(t)$ is larger, a "+1" is decoded. Otherwise, a "−1" will be decoded.

2.2.3 Multi-User System Based on Chaotic Frequency Modulation (CFM)

In the previous two sections, we have briefly described two multiple-access techniques for use in chaos-based communication systems. Moreover, we assume that the systems can be any kind of communication systems. In this section, we present

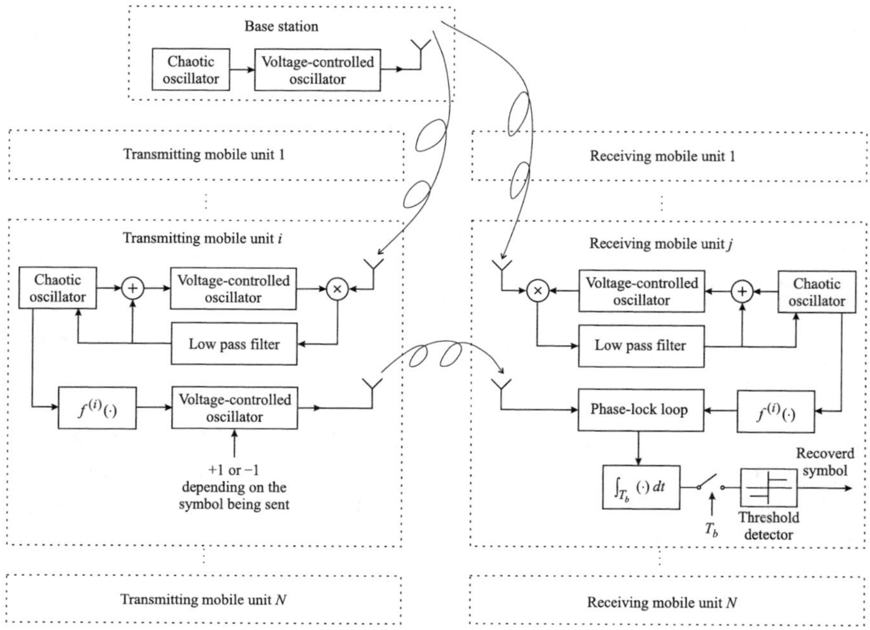

Figure 2.16. A multi-user chaotic frequency modulation (CFM) system.

a multi-user chaotic frequency modulation (CFM) system which is designed to operate in a mobile environment [Tsimring *et al.*, 2001].

In the multi-user CFM system, it is assumed that there is a base station sending out a chaotic signal continuously via the wireless channel. Suppose that there are N transmitting mobile units. Also, there are N corresponding receiving mobile units detecting the transmitted symbols. Figure 2.16 shows the structure of the system.

To establish a communication link successfully between a pair of transmit-receive mobile units, all the transmitting and receiving units must first synchronize their chaotic oscillators with the chaotic signal sent by the base station. After the synchronization process has been completed, the transmitting mobile units will make use of the parameters obtained through the synchronization process to generate their own CFM signals, which are then used to carry information. Consider the ith transmitting mobile unit. The phase of the information-bearing carrier is given by

$$\dot{\phi}^{(i)} = \omega^{(i)} = \begin{cases} \omega^{(1)}(1 + m_1 f^{(i)}(\boldsymbol{x}^{(i)}) + m_2) & \text{when symbol “+1” is transmitted,} \\ \omega^{(1)}(1 + m_1 f^{(i)}(\boldsymbol{x}^{(i)}) - m_2) & \text{when symbol “−1” is transmitted,} \end{cases} \tag{2.2.5}$$

where the parameters m_1 and m_2 have been obtained through the initial synchronization process. The functions $f^{(i)}(\cdot)$ and $\omega^{(i)}$ are the chaotic signal generating function and the frequency of the *voltage controlled oscillator* (VCO), respectively, for the ith transmitting mobile unit. Suppose the jth receiving unit needs to decode the symbols sent from the ith transmitting unit. It will then re-create the same chaotic signal $f^{(i)}(\cdot)$ used by the ith transmitting unit. As shown in Fig. 2.16, the transmitted symbol is then detected in accordance with the sign of the result when the output of the *phase-lock loop* (PLL) for the jth receiving unit is integrated over one bit duration.

Finally, to ensure that the synchronization process between the base station and the mobile units can be accomplished, it is required that the synchronization signal uses a different frequency band from the data communication signal. As for the communication links between the mobile units, co-channel interference occurs because all signals are using the frequency band around ω_1.

2.3 SUMMARY

In this chapter, we have reviewed some common chaos-based digital modulation schemes. Also, the corresponding coherent and/or noncoherent demodulation schemes are briefly explained. Because chaos-based communication systems are spread-spectrum systems, they are expected to possess multiuser capability. Two multiple-access schemes based on DCSK/FM-DCSK and one multiple-access scheme based on chaotic frequency modulation have been introduced and their operation principles have been described.

Based on the fundamental concepts presented in this chapter, we will investigate, in detail, a few other multiple-access chaos-based digital communication systems, which will be described in the following chapters.

3

MULTIPLE-ACCESS COHERENT ANTIPODAL CHAOS-SHIFT-KEYING SYSTEMS

3.1 INTRODUCTION

Chaotic signals, by virtue of their wideband characteristics, can be used to spread narrowband information over a wider frequency spectrum, as in conventional spread-spectrum communications. Moreover, like other forms of spread-spectrum communication systems, chaos-based systems must allow multiple access in order to make efficient use of the available bandwidth. In this chapter, we begin our formal study on the performance of chaos-based digital communication systems under a multiple-access environment. We investigate in depth, using a discrete-time approach, a coherent antipodal chaos-shift-keying (ACSK) system with multiple users. It is assumed that the amplitude of the carrier is modulated by the chaotic signals (amplitude modulation) and we analyze the system using the equivalent baseband model [Hasler and Schimming, 2000; Lau and Tse, 2003a; Proakis and Salehi, 1994]. In theory, correlator-type coherent methods give a better error performance than non-coherent methods under an additive white Gaussian noise (AWGN) channel. Thus, although robust synchronization methods are not yet available, the study of coherent demodulation schemes is important in that it can provide performance benchmarks for evaluating other chaotic modulation/demodulation techniques under similar conditions.

The organization of the chapter is as follows. Section 3.2 describes a multiple-access coherent antipodal chaos-shift-keying communication system. In Section 3.3, we evaluate the bit error rates (BERs) using the traditional Gaussian-approximation-based method. Analytical expressions for the BERs can also be obtained when the chaotic sequences satisfy certain statistical properties. Our analysis covers the cases where each user employs either a distinct or the same chaotic map. In Section 3.4, we apply the exact analytical method, which was proposed by Lawrance and Ohama [2002], to the study of a multiple-access antipodal CSK system [Tam *et al.*, 2004a]. We compare the exact analytical BER results with those obtained from traditional methods based on the assumption of Gaussian distributed decision parameter, the control being provided by brute-force numerical simulations.

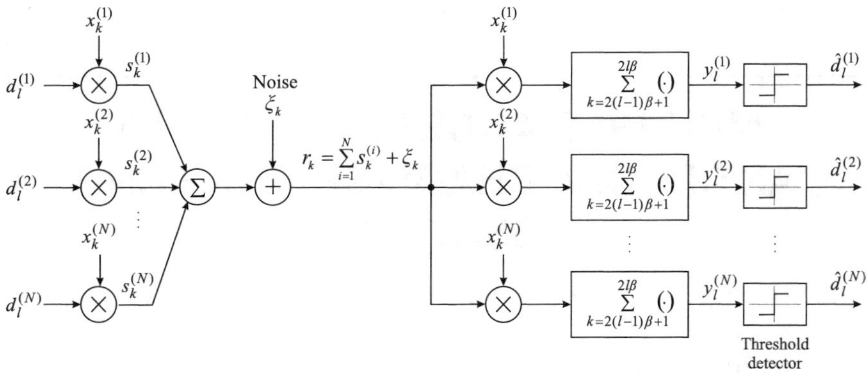

Figure 3.1. A multiple-access antipodal chaos-shift-keying (MA-ACSK) communication system.

3.2 MULTIPLE-ACCESS ANTIPODAL CHAOS-SHIFT-KEYING (MA-ACSK) SYSTEM

We consider a coherent antipodal chaos-shift-keying communication system with N users, as shown in Fig. 3.1. A discrete-time approach will be adopted in the following analysis. Suppose the ith user is transmitting a binary symbol $d_l^{(i)}$ during the lth bit duration, and the symbol $d_l^{(i)}$ is either "+1" or "−1", each with a probability of 1/2. Also, the symbols sent by different users are independent of one another. Essentially, there are N chaos generators corresponding to the N different users, and the ith chaos generator produces the chaotic samples $\{x_k^{(i)}\}$, which is used to spread the binary symbol sequence $\{d_l^{(i)}\}$. We assume that the chaotic sequences are generated independently of one another. Define the spreading factor, 2β, as the number of chaotic samples used to transmit one binary symbol. We also assume that the mean value of each of the chaotic sequences is zero in order to avoid transmitting any dc component which is a waste of power. In other words,

$$\mathrm{E}\left[x_k^{(i)}\right] = 0, \quad i = 1, 2, \ldots, N, \tag{3.2.1}$$

where $\mathrm{E}[\psi]$ denotes the mean value of ψ.

The signal transmitted by the ith user at time $k = 2(l-1)\beta + 1, 2(l-1)\beta + 2, \ldots, 2l\beta$, under the antipodal CSK scheme, can be written as

$$s_k^{(i)} = d_l^{(i)} x_k^{(i)}. \tag{3.2.2}$$

The overall transmitted signal of the system at time k is thus given by

$$s_k = \sum_{i=1}^{N} s_k^{(i)}. \tag{3.2.3}$$

As the signal goes through a communication channel, noise is being added. For an *additive white Gaussian noise* (AWGN) channel, the received signal is given by

$$r_k = \sum_{i=1}^{N} s_k^{(i)} + \xi_k, \tag{3.2.4}$$

where ξ_k is an additive white Gaussian noise sample with zero mean and variance (power spectral density) $N_0/2$. Here, we make the usual assumption for coherent systems that the chaotic spreading signals are reproduced exactly at the receiver. As shown in Fig. 3.1, the receiver basically employs a direct correlation process for demodulation. For the jth user, the output of the correlator, denoted by $y_l^{(j)}$, is given by

$$\begin{aligned} y_l^{(j)} &= \sum_{k=2(l-1)\beta+1}^{2l\beta} r_k x_k^{(j)} = \sum_{k=2(l-1)\beta+1}^{2l\beta} \left(\sum_{i=1}^{N} s_k^{(i)} + \xi_k \right) x_k^{(j)} \\ &= \sum_{i=1}^{N} d_l^{(i)} \sum_{k=2(l-1)\beta+1}^{2l\beta} x_k^{(i)} x_k^{(j)} + \sum_{k=2(l-1)\beta+1}^{2l\beta} \xi_k x_k^{(j)}. \end{aligned} \tag{3.2.5}$$

The decision parameter is the correlator output $y_l^{(j)}$, and the recovered symbol, denoted by $\hat{d}_l$, is given by

$$\hat{d}_l^{(j)} = \mathrm{sgn}\big[y_l^{(j)}\big], \tag{3.2.6}$$

where sgn[·] is the sign function. The above decoding rule simply says that for the jth user, the recovered symbol is a "+1" if $y_l^{(j)} > 0$, and is a "−1" otherwise.

In the following sections, we will present two approaches for calculating the bit error rates for the afore-described multiple-access antipodal chaos-shift-keying (MA-ACSK) system. The conventional method, which assumes Gaussian distribution of the conditional decision parameter $y_l^{(j)}$, is first discussed. We will highlight the main assumption of the method, which may lead to inaccurate performance prediction in some scenarios. An alternative approach which considers the dynamics of the decision parameter is then described. While the second approach produces exact analytical formulae for the required BERs, the computational complexity is much higher.

3.3 ANALYSIS OF THE MA-ACSK SYSTEM BASED ON GAUSSIAN APPROXIMATION

With no loss of generality, let us consider the jth user and derive the probability of error for the first symbol. For brevity, we omit the subscript l (which is 1 in this case) in the variables $d_l^{(j)}$, $\hat{d}_l^{(j)}$ and $y_l^{(j)}$. The decision parameter for the jth user is given by

$$y^{(j)} = d^{(j)} \sum_{k=1}^{2\beta} (x_k^{(j)})^2 + \sum_{i=1, i \neq j}^{N} d^{(i)} \sum_{k=1}^{2\beta} x_k^{(i)} x_k^{(j)} + \sum_{k=1}^{2\beta} \xi_k x_k^{(j)}, \qquad (3.3.1)$$

where the conditioning is only on $d^{(j)}$ and not the others. The BER for the jth user can then be evaluated by

$$\mathrm{BER}_{\mathrm{GA}}^{(j)} = \frac{1}{2} \operatorname{Prob}\big(y^{(j)} \leqslant 0 \mid (d^{(j)} = +1)\big) + \frac{1}{2} \operatorname{Prob}\big(y^{(j)} > 0 \mid (d^{(j)} = -1)\big). \qquad (3.3.2)$$

Moreover, with a Gaussian distribution assumed for the conditional decision parameter, the two conditional error probabilities are approximated by

$$\operatorname{Prob}\big(y^{(j)} \leqslant 0 \mid (d^{(j)} = +1)\big) = \frac{1}{2} \operatorname{erfc}\left(\frac{\mathrm{E}[y^{(j)} \mid (d^{(j)} = +1)]}{\sqrt{2 \operatorname{var}[y^{(j)} \mid (d^{(j)} = +1)]}} \right) \qquad (3.3.3)$$

$$\operatorname{Prob}\big(y^{(j)} > 0 \mid (d^{(j)} = -1)\big) = \frac{1}{2} \operatorname{erfc}\left(\frac{-\mathrm{E}[y^{(j)} \mid (d^{(j)} = -1)]}{\sqrt{2 \operatorname{var}[y^{(j)} \mid (d^{(j)} = -1)]}} \right) \qquad (3.3.4)$$

where $\operatorname{var}[\cdot]$ denotes the variance operator and the complementary error function, $\operatorname{erfc}(\cdot)$, is defined as [Proakis and Salehi, 1994]

$$\operatorname{erfc}(\psi) \equiv \frac{2}{\sqrt{\pi}} \int_{\psi}^{\infty} e^{-\varpi^2} \, d\varpi. \qquad (3.3.5)$$

Suppose the jth user is sending a "+1". The mean value of $y^{(j)} \mid (d^{(j)} = +1)$ is given by

$$\mathrm{E}\big[y^{(j)} \mid (d^{(j)} = +1)\big] = 2\beta \mathrm{E}\big[(x_k^{(j)})^2\big]. \qquad (3.3.6)$$

Also, from the independence of the chaotic sequences, the variance of $y^{(j)} \mid (d^{(j)} = +1)$ is readily shown equal to

$$\begin{aligned}\mathrm{var}\big[y^{(j)} \mid (d^{(j)} = +1)\big] &= 2\beta\,\mathrm{var}\big[(x_k^{(j)})^2\big] + \sum_{k=1}^{2\beta} \sum_{m=1,m\neq k}^{2\beta} \mathrm{cov}\big[(x_k^{(j)})^2, (x_m^{(j)})^2\big] \\ &\quad + 2\beta \mathrm{E}\big[(x_k^{(j)})^2\big] \sum_{i=1,i\neq j}^{N} \mathrm{E}\big[(x_k^{(i)})^2\big] \\ &\quad + \sum_{i=1,i\neq j}^{N} \sum_{k=1}^{2\beta} \sum_{m=1,m\neq k}^{2\beta} \mathrm{E}\big[x_k^{(i)} x_m^{(i)}\big]\mathrm{E}\big[x_k^{(j)} x_m^{(j)}\big] \\ &\quad + \beta N_0 \mathrm{E}\big[(x_k^{(j)})^2\big], \end{aligned} \tag{3.3.7}$$

where $\mathrm{cov}[\psi, \varpi]$, defined as

$$\mathrm{cov}[\psi, \varpi] = \mathrm{E}[\psi\varpi] - \mathrm{E}[\psi]\mathrm{E}[\varpi], \tag{3.3.8}$$

denotes the covariance of ψ and ϖ. Notice that in the derivation of the mean and variance of $y^{(j)} \mid (d^{(j)} = +1)$, the expectation is taken over both the x's and the other d's except $d^{(j)}$. Now, using

$$\mathrm{E}\big[y^{(j)} \mid (d^{(j)} = -1)\big] = -\mathrm{E}\big[y^{(j)} \mid (d^{(j)} = +1)\big], \tag{3.3.9}$$

$$\mathrm{var}\big[y^{(j)} \mid (d^{(j)} = -1)\big] = \mathrm{var}\big[y^{(j)} \mid (d^{(j)} = +1)\big], \tag{3.3.10}$$

it is seen that the two conditional probabilities in (3.3.3) and (3.3.4) are equal, and the BER for the jth user is simplified to

$$\begin{aligned}\mathrm{BER}_{\mathrm{GA}}^{(j)} &= \mathrm{Prob}\big(y^{(j)} \leqslant 0 \,\big|\, (d^{(j)} = +1)\big) \\ &= \frac{1}{2}\,\mathrm{erfc}\left(\frac{\mathrm{E}[y^{(j)} \mid (d^{(j)} = +1)]}{\sqrt{2\,\mathrm{var}[y^{(j)} \mid (d^{(j)} = +1)]}}\right).\end{aligned} \tag{3.3.11}$$

Clearly, the accuracy of the above formulae is subject to the validity of the Gaussian distribution assumption and its moments capturing the exact probabilities. Notice that only low order moments of the chaotic spreading sequences have been considered, which are not enough to characterize them. Thus the formulae could give identical results for quite distinct chaotic spreading. This is the case for Chebyshev spreading with different orders, as will be demonstrated in Section 3.5. When the spreading factor is sufficiently large, it gives reasonably good estimates of the BERs, but when the spreading factor is small and the distribution of the conditional decision parameter deviates significantly from Gaussian, the above formulae give rather disappointing results.

Application example

Suppose the Chebyshev map is used to generate the chaotic signals [Geisel and Fairen, 1984]. In order to provide each user a different chaotic signal, we may either use one Chebyshev map with N different initial conditions or use N Chebyshev maps of different degrees. In either case, the second and fourth terms in (3.3.7) can be eliminated since the chaotic signals generated by Chebyshev maps satisfy the following conditions (see Appendix 3A):

$$\mathrm{cov}\big[\big(x_k^{(j)}\big)^2, \big(x_m^{(j)}\big)^2\big] = 0, \quad k \neq m, \tag{3.3.12}$$

$$\mathrm{E}\big[x_k^{(j)} x_m^{(j)}\big] = 0, \quad k \neq m. \tag{3.3.13}$$

Also $\mathrm{E}[(x_k^{(i)})^2] = P_s$ for all i, indicating that all users are transmitting with the same average power. The variance of $y^{(j)} \mid (d^{(j)} = +1)$ can then be shown equal to

$$\mathrm{var}\big[y^{(j)} \mid \big(d^{(j)} = +1\big)\big] = 2\beta\,\mathrm{var}\big[\big(x_k^{(j)}\big)^2\big] + 2(N-1)\beta P_s^2 + \beta N_0 P_s. \tag{3.3.14}$$

Writing

$$\Psi = \frac{\mathrm{var}[(x_k^{(1)})^2]}{P_s^2} = \frac{\mathrm{var}[(x_k^{(2)})^2]}{P_s^2} = \cdots = \frac{\mathrm{var}[(x_k^{(N)})^2]}{P_s^2}, \tag{3.3.15}$$

we have the BER for the jth user given by

$$\mathrm{BER}_{\mathrm{GA}}^{(j)} = \frac{1}{2}\,\mathrm{erfc}\left(\left[\frac{\Psi}{\beta} + \frac{(N-1)}{\beta} + \left(\frac{E_b}{N_0}\right)^{-1}\right]^{-1/2}\right), \tag{3.3.16}$$

where E_b denotes the average bit energy and equals $2\beta P_s$, and E_b/N_0 denotes the bit-energy-to-noise-power-spectral-density ratio.

3.4 EXACT BIT ERROR RATE ANALYSIS OF THE MA-ACSK SYSTEM

In this section we apply the approach of Lawrance and Ohama [2002], which gave exact BERs for single-user antipodal CSK systems, to derive exact analytical results for BERs in multiple-access antipodal CSK systems [Tam *et al.*, 2004a]. This approach enables full dynamics of the chaotic spreading to be utilized and not just low order moments of the spreading. Firstly, an exact Gaussian mean and variance result is available for the bit error rate $\mathrm{BER}^{(j)}$ when considered conditionally on the spreading and on the bits of the other users; this follows only from the Gaussian distribution of the noise. Secondly, the full dynamics of the spreading are used to uncondition the result and hence obtain the fully exact $\mathrm{BER}^{(j)}$; this is explicitly obtained as an N-dimensional integral over the independent invariant distributions of the chaotic spreading.

To simplify the notations, we denote the vector of the symbols sent by Users $1, 2, \ldots, j-1, j+1, \ldots, N$ by $\boldsymbol{d}^{(j)}$ and the initial conditions of the unmodulated chaotic signals by the vector $\boldsymbol{x}_0$, i.e.,

$$\boldsymbol{d}^{(j)} = \begin{bmatrix} d^{(1)} & d^{(2)} & \cdots & d^{(j-1)} & d^{(j+1)} & \cdots & d^{(N)} \end{bmatrix}, \tag{3.4.1}$$

$$\boldsymbol{x}_0 = \begin{bmatrix} x_1^{(1)} & x_1^{(2)} & \cdots & x_1^{(N)} \end{bmatrix}. \tag{3.4.2}$$

Given the vectors $\boldsymbol{d}^{(j)}$ and $\boldsymbol{x}_0$, the value of $y^{(j)}$ in (3.3.1), which is now conditioned on all spreading and transmissions of all other users, has a Gaussian distribution exactly. The mean of the conditional decision parameter, $y^{(j)} \mid (d^{(j)}, \boldsymbol{d}^{(j)}, \boldsymbol{x}_0)$, can be written as

$$\mathrm{E}\big[y^{(j)} \mid (d^{(j)}, \boldsymbol{d}^{(j)}, \boldsymbol{x}_0)\big] = d^{(j)} \sum_{k=1}^{2\beta} \big(x_k^{(j)}\big)^2 + \sum_{i=1, i\neq j}^{N} d^{(i)} \sum_{k=1}^{2\beta} x_k^{(i)} x_k^{(j)}. \tag{3.4.3}$$

Also, because the chaotic sequences are independent, the variance of the decision parameter is readily shown equal to

$$\mathrm{var}\big[y^{(j)} \mid (d^{(j)}, \boldsymbol{d}^{(j)}, \boldsymbol{x}_0)\big] = \mathrm{var}\left[\sum_{k=1}^{2\beta} \xi_k x_k^{(j)}\right] = \frac{N_0}{2} \sum_{k=1}^{2\beta} \big(x_k^{(j)}\big)^2. \tag{3.4.4}$$

Since $d^{(i)}$ and $x_k^{(i)}$ are independent for all i, the BER for the jth user, $\mathrm{BER}^{(j)}$, can be written as

$$\begin{aligned}
\mathrm{BER}_{\text{exact}}^{(j)} = &\int_{-\infty}^{\infty}\int_{-\infty}^{\infty}\cdots\int_{-\infty}^{\infty} \sum_{\boldsymbol{d}^{(j)}} \Big[\mathrm{Prob}\big(y^{(j)} \leqslant 0 \mid (d^{(j)} = +1, \boldsymbol{d}^{(j)}, \boldsymbol{x}_0)\big) \\
&\times \mathrm{Prob}\big(d^{(j)} = +1\big)\,\mathrm{Prob}\big(\boldsymbol{d}^{(j)}\big)\rho(\boldsymbol{x}_0) \\
&+ \mathrm{Prob}\big(y^{(j)} > 0 \mid (d^{(j)} = -1, \boldsymbol{d}^{(j)}, \boldsymbol{x}_0)\big) \\
&\times \mathrm{Prob}\big(d^{(j)} = -1\big)\,\mathrm{Prob}\big(\boldsymbol{d}^{(j)}\big)\rho(\boldsymbol{x}_0)\Big]\, dx_1^{(1)}\, dx_1^{(2)} \cdots dx_1^{(N)} \\
= &\int_{-\infty}^{\infty}\int_{-\infty}^{\infty}\cdots\int_{-\infty}^{\infty} \sum_{\boldsymbol{d}^{(j)}} \Big[\mathrm{Prob}\big(y^{(j)} \leqslant 0 \mid (d^{(j)} = +1, \boldsymbol{d}^{(j)}, \boldsymbol{x}_0)\big)
\end{aligned}$$

$$
\begin{aligned}
&\times \operatorname{Prob}\big(d^{(j)} = +1\big) \operatorname{Prob}\big(\boldsymbol{d}^{(j)}\big)\rho(\boldsymbol{x}_0) \\
&+ \operatorname{Prob}\big(y^{(j)} > 0 \,\big|\, \big(d^{(j)} = -1, -\boldsymbol{d}^{(j)}, \boldsymbol{x}_0\big)\big) \\
&\times \operatorname{Prob}\big(d^{(j)} = -1\big) \operatorname{Prob}\big(-\boldsymbol{d}^{(j)}\big)\rho(\boldsymbol{x}_0)\Big]\, dx_1^{(1)}\, dx_1^{(2)} \cdots dx_1^{(N)} \\
= &\int_{-\infty}^{\infty}\int_{-\infty}^{\infty}\cdots\int_{-\infty}^{\infty} \frac{1}{2^N} \sum_{\boldsymbol{d}^{(j)}} \Big[\operatorname{Prob}\big(y^{(j)} \leqslant 0 \,\big|\, \big(d^{(j)} = +1, \boldsymbol{d}^{(j)}, \boldsymbol{x}_0\big)\big) \\
&+ \operatorname{Prob}\big(y^{(j)} > 0 \,\big|\, \big(d^{(j)} = -1, -\boldsymbol{d}^{(j)}, \boldsymbol{x}_0\big)\big)\Big] \\
&\times \rho(\boldsymbol{x}_0)\, dx_1^{(1)}\, dx_1^{(2)} \cdots dx_1^{(N)}
\end{aligned}
\tag{3.4.5}
$$

where $\rho(\boldsymbol{x}_0)$ denotes the joint probability density function of $\boldsymbol{x}_0$. Using

$$
\mathrm{E}\big[y^{(j)} \,\big|\, \big(d^{(j)} = -1, -\boldsymbol{d}^{(j)}, \boldsymbol{x}_0\big)\big] = -\mathrm{E}\big[y^{(j)} \,\big|\, \big(d^{(j)} = +1, \boldsymbol{d}^{(j)}, \boldsymbol{x}_0\big)\big] \tag{3.4.6}
$$

and

$$
\operatorname{var}\big[y^{(j)} \,\big|\, \big(d^{(j)} = -1, -\boldsymbol{d}^{(j)}, \boldsymbol{x}_0\big)\big] = \operatorname{var}\big[y^{(j)} \,\big|\, \big(d^{(j)} = +1, \boldsymbol{d}^{(j)}, \boldsymbol{x}_0\big)\big], \tag{3.4.7}
$$

the $\mathrm{BER}_{\text{exact}}^{(j)}$ given in (3.4.5) can be simplified to

$$
\begin{aligned}
\mathrm{BER}_{\text{exact}}^{(j)} &= \frac{1}{2^{N-1}} \int_{-\infty}^{\infty}\int_{-\infty}^{\infty}\cdots\int_{-\infty}^{\infty} \sum_{\boldsymbol{d}^{(j)}} \operatorname{Prob}\big(y^{(j)} \leqslant 0 \,\big|\, \big(d^{(j)} = +1, \boldsymbol{d}^{(j)}, \boldsymbol{x}_0\big)\big) \\
&\times \rho(\boldsymbol{x}_0)\, dx_1^{(1)}\, dx_1^{(2)} \cdots dx_1^{(N)} \\
&= \frac{1}{2^{N-1}} \int_{-\infty}^{\infty}\int_{-\infty}^{\infty}\cdots\int_{-\infty}^{\infty} \\
&\times \sum_{\boldsymbol{d}^{(j)}} \frac{1}{2} \operatorname{erfc}\left(\frac{\mathrm{E}[y^{(j)} \mid (d^{(j)} = +1, \boldsymbol{d}^{(j)}, \boldsymbol{x}_0)]}{\sqrt{2 \operatorname{var}[y^{(j)} \mid (d^{(j)} = +1, \boldsymbol{d}^{(j)}, \boldsymbol{x}_0)]}}\right) \\
&\times \rho(\boldsymbol{x}_0)\, dx_1^{(1)}\, dx_1^{(2)} \cdots dx_1^{(N)}.
\end{aligned}
\tag{3.4.8}
$$

Since $x_1^{(i)}$'s are independent variables, $\rho(\boldsymbol{x}_0)$ is equal to

$$
\rho(\boldsymbol{x}_0) = \rho_1\big(x_1^{(1)}\big)\rho_2\big(x_1^{(2)}\big) \cdots \rho_N\big(x_1^{(N)}\big), \tag{3.4.9}
$$

where $\rho_i(x_1^{(i)})$ represents the natural invariant probability density function of $x_1^{(i)}$. Thus, the desired exact result equals

$$\mathrm{BER}_{\mathrm{exact}}^{(j)} = \frac{1}{2^N}\int_{-\infty}^{\infty}\int_{-\infty}^{\infty}\cdots\int_{-\infty}^{\infty}\sum_{\boldsymbol{d}^{(j)}}\mathrm{erfc}\left(\frac{\mathrm{E}[y^{(j)} \mid (d^{(j)}=+1,\boldsymbol{d}^{(j)},\boldsymbol{x}_0)]}{\sqrt{2\,\mathrm{var}[y^{(j)} \mid (d^{(j)}=+1,\boldsymbol{d}^{(j)},\boldsymbol{x}_0)]}}\right)$$
$$\times\,\rho_1\big(x_1^{(1)}\big)\rho_2\big(x_1^{(2)}\big)\cdots\rho_N\big(x_1^{(N)}\big)\,dx_1^{(1)}\,dx_1^{(2)}\cdots dx_1^{(N)}. \tag{3.4.10}$$

This result consists of 2^{N-1} N-dimensional integrals which are numerically explicit, although will clearly not be computationally practical for large N. The result could also be used to enable the lower limits of accuracy of the moment-based analytical results to be ascertained by comparison.

Application example
We consider again the use of the Chebyshev map for generating chaotic sequences for all users. To provide each user a different chaotic signal, we may either use one Chebyshev map with N different initial conditions or use N Chebyshev maps of different degrees.

Since the initial values $x_1^{(i)}$'s lie between -1 and $+1$, to simplify the mathematics, we substitute $x_1^{(i)} = \cos\phi^{(i)}$ into (3.4.10), where $0 < \phi^{(i)} < \pi$ and $\phi^{(m)} \neq \phi^{(n)}$ for $m \neq n$. We write

$$\boldsymbol{\phi} = \begin{bmatrix}\phi^{(1)} & \phi^{(2)} & \cdots & \phi^{(N)}\end{bmatrix} \tag{3.4.11}$$

for notational clarity. Now, the conditional mean and variance of $y^{(j)}$ can be rewritten as (see Appendix 3A), respectively,

$$\begin{aligned}
&\mathrm{E}\big[y^{(j)} \mid (d^{(j)}=+1,\boldsymbol{d}^{(j)},\boldsymbol{\phi})\big]\\
&\quad= d^{(j)}\sum_{k=1}^{2\beta}\Big(\cos\big((\chi^{(j)})^k\phi^{(j)}\big)\Big)^2\\
&\qquad+\sum_{i=1,i\neq j}^{N} d^{(i)}\sum_{k=1}^{2\beta}\cos\big((\chi^{(i)})^k\phi^{(i)}\big)\cos\big((\chi^{(j)})^k\phi^{(j)}\big)
\end{aligned} \tag{3.4.12}$$

and

$$\mathrm{var}\big[y^{(j)} \mid \big(d^{(j)}=+1,\boldsymbol{d}^{(j)},\boldsymbol{\phi}\big)\big] = \frac{N_0}{2}\sum_{k=1}^{2\beta}\Big(\cos\big((\chi^{(j)})^k\phi^{(j)}\big)\Big)^2, \tag{3.4.13}$$

where $\chi^{(i)}$ is an integer larger than two and it represents the degree of the Chebyshev map used by the ith user. In general, $\chi^{(i)}$ can be different for different users. For the case where only one Chebyshev map is used and each user is assigned with a different initial condition, we have the same $\chi^{(i)}$ value for all i. In either case, $\text{BER}^{(j)}$ is given by

$$\begin{aligned}
\text{BER}^{(j)}_{\text{exact}} &= \frac{1}{2^N}\int_{\pi}^{0}\int_{\pi}^{0}\cdots\int_{\pi}^{0}\sum_{\boldsymbol{d}^{(j)}}\operatorname{erfc}\left(\frac{\text{E}[y^{(j)} \mid (d^{(j)}=+1,\boldsymbol{d}^{(j)},\boldsymbol{\phi})]}{\sqrt{2\operatorname{var}[y^{(j)} \mid (d^{(j)}=+1,\boldsymbol{d}^{(j)},\boldsymbol{\phi})]}}\right)\\
&\quad\times\frac{1}{\pi\sin(\phi^{(1)})}\frac{1}{\pi\sin(\phi^{(2)})}\cdots\frac{1}{\pi\sin(\phi^{(N)})}\\
&\quad\times\left(-\sin(\phi^{(1)})\right)\left(-\sin(\phi^{(2)})\right)\cdots\left(-\sin(\phi^{(N)})\right)\\
&\quad\times d\phi^{(1)}\,d\phi^{(2)}\cdots d\phi^{(N)}\\
&= \frac{1}{(2\pi)^N}\int_{0}^{\pi}\int_{0}^{\pi}\cdots\int_{0}^{\pi}\sum_{\boldsymbol{d}^{(j)}}\operatorname{erfc}\left(\frac{\text{E}[y^{(j)} \mid (d^{(j)}=+1,\boldsymbol{d}^{(j)},\boldsymbol{\phi})]}{\sqrt{2\operatorname{var}[y^{(j)} \mid (d^{(j)}=+1,\boldsymbol{d}^{(j)},\boldsymbol{\phi})]}}\right)\\
&\quad\times d\phi^{(1)}\,d\phi^{(2)}\cdots d\phi^{(N)}.
\end{aligned}\tag{3.4.14}$$

The final result thus consists of 2^{N-1} N-dimensional integrals which we can evaluate by some computer software.

3.5 RESULTS AND DISCUSSIONS

In this section, we evaluate the two aforementioned methods of computing BERs. In particular, we study the following aspects of performance evaluation:

1. BER of single-user system.
2. BERs of multi-user system with chaotic sequences generated by one Chebyshev map with different initial conditions.
3. BER of multi-user system with chaotic sequences generated by Chebyshev maps of different degrees.
4. Effect of the number of users on the BER calculations.

In each of the first three cases, we compare BERs calculated from the Gaussian-approximation-based (GA) method and the exact analytical method, with the control data provided by brute-force (BF) numerical simulations. For the last case, we only show the results found from the GA method and BF simulations.

Figure 3.2 compares the BERs found from the GA method, the exact method and BF simulations for the single-user CSK system. Spreading factors varying from 1 to 5 are used ($2\beta = 1$ represents no spreading) and Chebyshev maps of

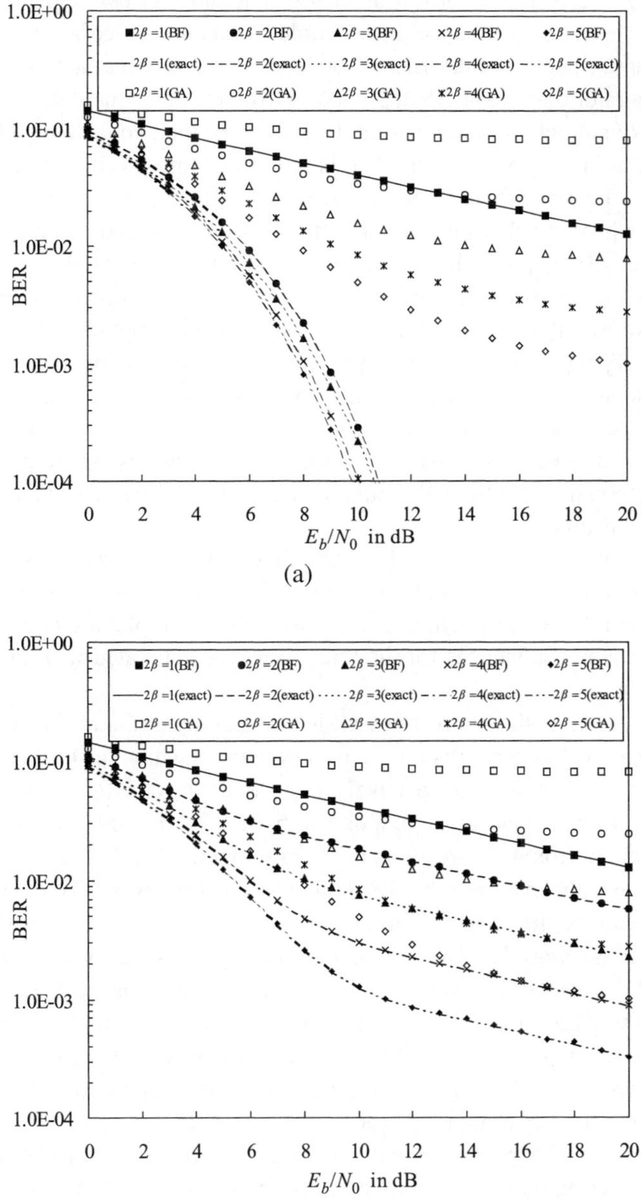

Figure 3.2. BERs from the Gaussian-approximation-based method (GA), exact analytical method, and brute-force simulation (BF) for a single-user system, with $2\beta = 1$ to 5, and (a) Chebyshev map of degree 2 as the chaos generator; (b) Chebyshev map of degree 3 as the chaos generator.

degrees 2 and 3 are employed as chaos generator (in two separate studies). It is found, because of reliance on low order moments which are the same for all Chebyshev orders, that the BERs obtained by the GA method are the same for both generators, whereas the BERs found from the exact method fully employing the map dynamics show significant difference for the two different chaos generators. Furthermore, the exact method is in perfect agreement with the BF simulations for all cases, whereas the GA method gives deviating results especially for the case where the chaos generator is the Chebyshev map of degree 2.

Figure 3.3 compares the BERs found from the GA method, the exact method and BF simulations for the multi-user ACSK system employing one chaos generator. Here, we show results for a 2-user system for simplicity. Again, spreading factors varying from 1 to 5 are used, and Chebyshev maps of degrees 2 and 3 are employed as chaos generator (in two separate studies). Since one chaos generator is used for the system, two different initial conditions are used to generate chaotic sequences for the two users. Similar observations can be made here. The exact method matches the BF simulations perfectly, whereas the GA methods give deviating results.

When the number of users increases, the approximation by the GA method tends to improve, as shown in Fig. 3.4 which compares the GA method, the exact method and the BF simulations for a 3-user system employing one chaos generator. The chaotic sequences for the three users are generated with different initial conditions.

Next, we study the multi-user ACSK system employing multiple chaos generators. Here, we show results for a 2-user system, in which the Chebyshev maps of degrees 2 and 3 are used, respectively, to generate the chaotic sequences for Users 1 and 2. Again, spreading factors of 1 to 5 are used. Figure 3.5 compares the results obtained from the GA method, the exact method and the BF simulations. In this case, we observe from the BF simulations that the BERs for the two users are different, due to the use of different chaos generators. The exact method tells this difference, whereas the GA method gives identical results for the two users.

Finally, to show the effect of the number of users on the BER calculations, we plot the BERs against the number of users N for the multi-user system employing one chaos generator. The chaotic sequences for the users are generated with different initial conditions. Figure 3.6 shows only the results from the GA method and the BF simulations for $E_b/N_0 = 10$ dB and $2\beta = 5$. Here, the chaos generator is the Chebyshev map of degree 2. It is found that the GA method gives better approximation as the number of users increases. Note that the exact BERs are not plotted in this case. As can be judged from (3.4.14), the computational intensity required by the exact method increases exponentially with the number of users N as well as the spreading factor 2β. Thus, when the number of users or the spreading factor is large, the computational requirement becomes too excessive

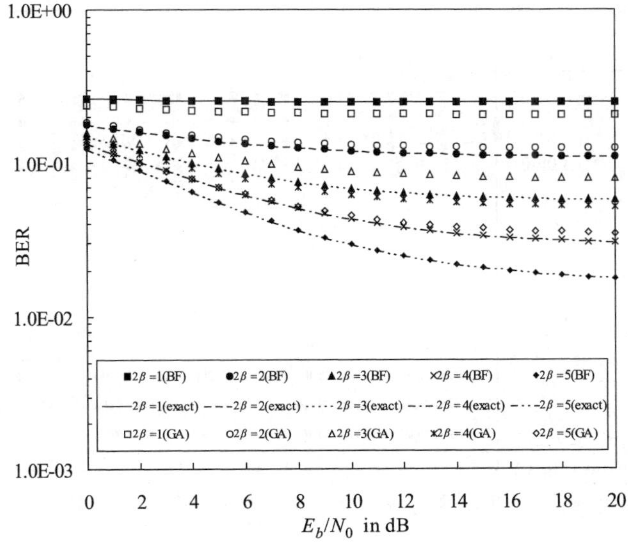

(a)

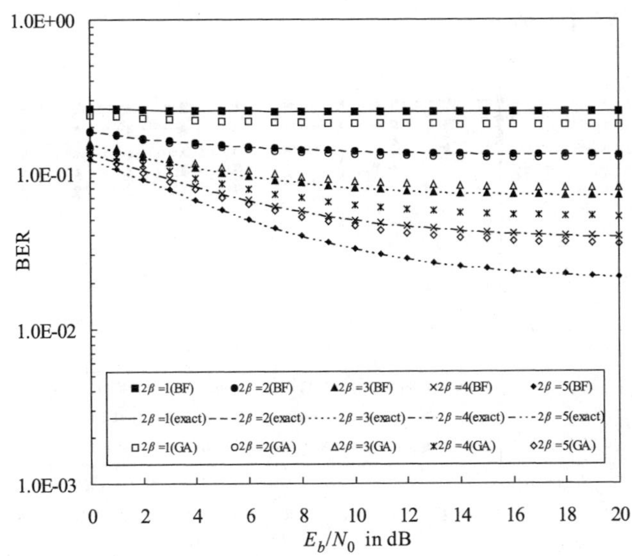

(b)

Figure 3.3. BERs from the Gaussian-approximation-based method (GA), exact analytical method, and brute-force simulation (BF) for a 2-user system using one chaos generator with different initial conditions, with $2\beta = 1$ to 5 and (a) Chebyshev map of degree 2 as the chaos generator; (b) Chebyshev map of degree 3 as the chaos generator.

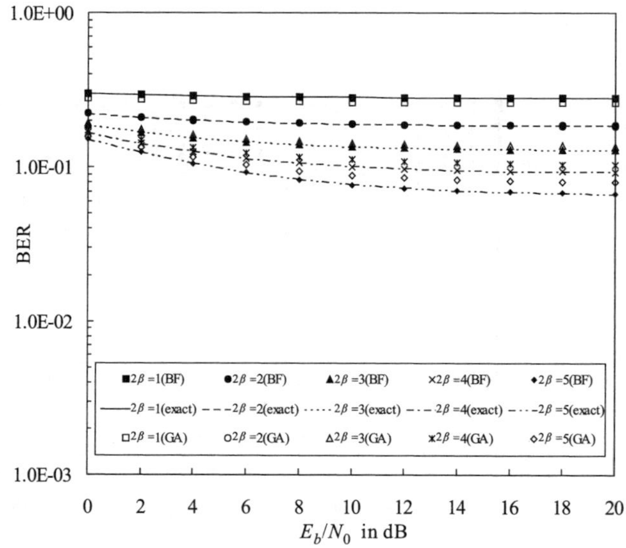

(a)

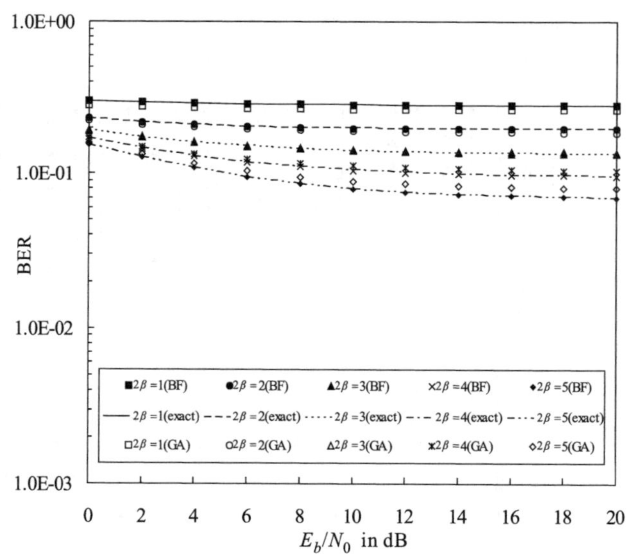

(b)

Figure 3.4. BERs from the Gaussian-approximation-based method (GA), exact analytical method, and brute-force simulation (BF) for a 3-user system using one chaos generator with different initial conditions, with $2\beta = 1$ to 5 and (a) Chebyshev map of degree 2 as the chaos generator; (b) Chebyshev map of degree 3 as the chaos generator.

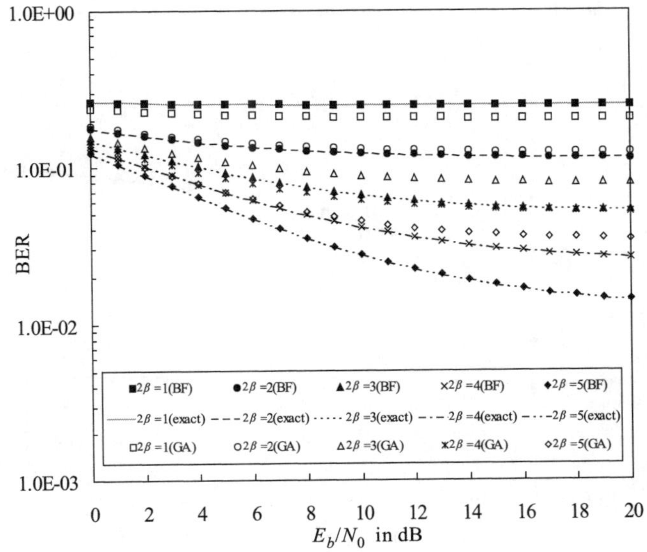

(a)

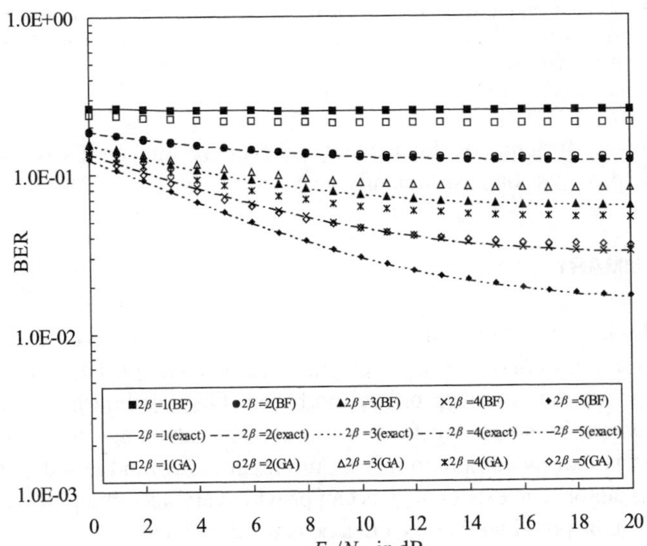

(b)

Figure 3.5. BERs from the Gaussian-approximation-based method (GA), exact analytical method, and brute-force simulation (BF) for a 2-user system using two chaos generators for different users, with $2\beta = 1$ to 5. (a) User 1 using Chebyshev map of degree 2; (b) User 2 using Chebyshev map of degree 3.

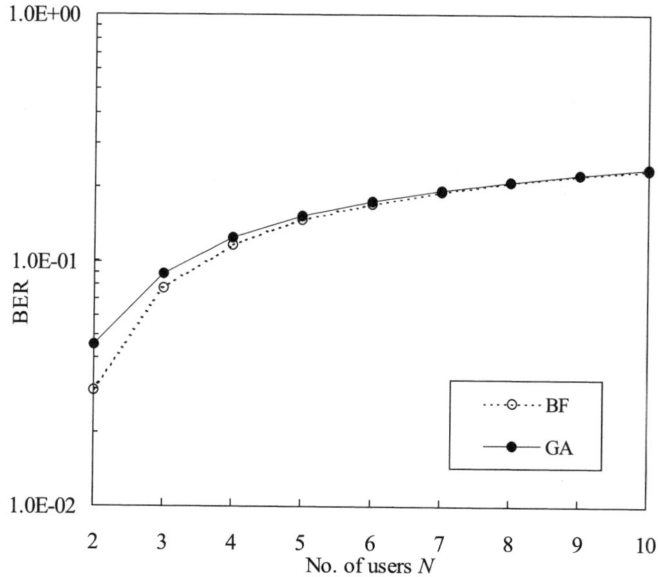

Figure 3.6. BERs versus the number of users N from the Gaussian-approximation-based method (GA) and brute-force simulation (BF) for multi-user system using one chaos generator (Chebyshev map of degree 2) with different initial conditions, at $E_b/N_0 = 10$ dB and $2\beta = 5$.

to be feasible; fortunately, this is when the exact results are not required because the GA approximation is sufficiently accurate.

3.6 SUMMARY

The multiple-access technique for use with coherent antipodal chaos shift keying under a noisy condition has been analyzed thoroughly. The results reveal that the traditional Gaussian-approximation-based (GA) method gives good approximation when the spreading factor is large. The advantage of the GA method, moreover, is that its computational requirement is low and is independent of the spreading factor. The exact analysis can provide very accurate prediction of BERs even for small spreading factors, but a possible drawback is that the computational intensity required increases exponentially with the number of users as well as the spreading factor.

In the study of the coherent antipodal chaos shift keying system with multiple users, it is also observed that the system performance degrades with the number of users. The main reason is that the finite correlation between different chaotic signals gives rise to co-channel interference between users, and such interference increases as more and more users are making use of the same channel for communications. To alleviate such inter-user interference, multi-user detection techniques can be applied, which is the subject of the next chapter.

APPENDIX 3A. DERIVATION OF THE STATISTICAL PROPERTIES FOR THE CHAOTIC SEQUENCES GENERATED BY THE CHEBYSHEV MAP

Consider the Chebyshev map of degree χ, defined as

$$g_\chi(x) = \cos\bigl(\chi \cos^{-1}(x)\bigr), \quad -1 \leqslant x \leqslant 1. \tag{3.A.1}$$

For $\chi = 2$, and 3 (cubic map), we have, respectively,

$$g_2(x) = 2x^2 - 1 \tag{3.A.2}$$

and

$$g_3(x) = 4x^3 - 3x. \tag{3.A.3}$$

The invariant probability density function of x, denoted by $\rho(x)$, is [Kohda and Tsuneda, 1994]

$$\rho(x) = \begin{cases} \dfrac{1}{\pi\sqrt{1-x^2}} & \text{if } |x| < 1, \\ 0 & \text{otherwise.} \end{cases} \tag{3.A.4}$$

Define

$$\begin{aligned} g_\chi^{(2)}(x) &= g_\chi\bigl(g_\chi(x)\bigr), \\ g_\chi^{(3)}(x) &= g_\chi\bigl(g_\chi^{(2)}(x)\bigr), \\ &\vdots \\ g_\chi^{(k)}(x) &= g_\chi\bigl(g_\chi^{(k-1)}(x)\bigr). \end{aligned} \tag{3.A.5}$$

Putting $x = \cos\phi$ in (3.A.5), we get, for the case $\chi = 2$,

$$\begin{aligned} g_2(\cos\phi) &= 2\cos^2\phi - 1 = \cos 2\phi, \\ g_2^{(2)}(\cos\phi) &= g_2\bigl(g_2(\cos\phi)\bigr) = g_2(\cos 2\phi) = \cos\bigl(2^2\phi\bigr), \\ g_2^{(3)}(\cos\phi) &= g_2\bigl(g_2^{(2)}(\cos\phi)\bigr) = g_2\bigl(\cos\bigl(2^2\phi\bigr)\bigr) = \cos\bigl(2^3\phi\bigr), \end{aligned}$$

$$\vdots$$

$$g_2^{(k)}(\cos\phi) = g_2\big(g_2^{(k-1)}(\cos\phi)\big) = g_2\big(\cos(2^{k-1}\phi)\big) = \cos\big(2^k\phi\big). \quad (3.A.6)$$

Likewise, we have, for $\chi = 3$,

$$\begin{aligned}
g_3(\cos\phi) &= 4\cos^3\phi - 3\cos\phi = \cos 3\phi, \\
g_3^{(2)}(\cos\phi) &= g_3\big(g_3(\cos\phi)\big) = g_3(\cos 3\phi) = \cos\big(3^2\phi\big), \\
g_3^{(3)}(\cos\phi) &= g_3\big(g_3^{(2)}(\cos\phi)\big) = g_3\big(\cos(3^2\phi)\big) = \cos\big(3^3\phi\big), \\
&\vdots \\
g_3^{(k)}(\cos\phi) &= g_3\big(g_3^{(k-1)}(\cos\phi)\big) = g_3\big(\cos(3^{k-1}\phi)\big) = \cos\big(3^k\phi\big), \quad (3.A.7)
\end{aligned}$$

and in general, we can show that for the Chebyshev map of degree χ,

$$g_\chi^{(k)}(\cos\phi) = \cos\big(\chi^k\phi\big). \quad (3.A.8)$$

Moreover, the invariant probability density function of $\cos(\phi)$ can be shown equal to

$$\rho\big(\cos(\phi)\big) = \begin{cases} \dfrac{1}{\pi\sin(\phi)} & \text{if } 0 \leqslant \phi \leqslant \pi, \\ 0 & \text{otherwise.} \end{cases} \quad (3.A.9)$$

Derivation of $\mathrm{E}[x_k]$, $\mathrm{E}[x_k^2]$, $\mathrm{var}[x_k]$ *and* $\mathrm{var}[x_k^2]$

$$\mathrm{E}[x_k] = \int_{-\infty}^{\infty} x\rho(x)\,dx = \int_{-1}^{1} x\,\frac{1}{\pi\sqrt{1-x^2}}\,dx = 0, \quad (3.A.10)$$

$$\mathrm{var}[x_k] = \mathrm{E}\big[x_k^2\big] = \int_{-\infty}^{\infty} x^2\rho(x)\,dx = \int_{-1}^{1} x^2\,\frac{1}{\pi\sqrt{1-x^2}}\,dx = \frac{1}{2}, \quad (3.A.11)$$

$$\mathrm{var}\big[x_k^2\big] = \int_{-\infty}^{\infty}\big(x_k^2 - \mathrm{E}[x_k^2]\big)^2\rho(x)\,dx = \int_{-1}^{1}\left(x^2 - \frac{1}{2}\right)^2\frac{1}{\pi\sqrt{1-x^2}}\,dx = \frac{1}{8}. \quad (3.A.12)$$

Derivation of $\mathrm{cov}[x_k^2, x_m^2]$

The autovariance of $\{x_k^2\}$ is defined as

$$\mathrm{cov}\big[x_k^2, x_m^2\big] = \mathrm{E}\big[x_k^2 x_m^2\big] - \mathrm{E}\big[x_k^2\big]\mathrm{E}\big[x_m^2\big]. \quad (3.A.13)$$

We consider the case where $k \neq m$. Without loss of generality, we assume $m = k + \kappa$ for some positive integer κ. Then,

$$\begin{aligned} \mathrm{E}\left[x_k^2 x_m^2\right] &= \int_{-\infty}^{\infty} x^2 \left(g_\chi^{(\kappa)}(x)\right)^2 \rho(x)\,dx \\ &= \int_{-1}^{1} x^2 \left(g_\chi^{(\kappa)}(x)\right)^2 \frac{1}{\pi\sqrt{1-x^2}}\,dx. \end{aligned} \tag{3.A.14}$$

Using (3.A.8) and (3.A.9), and $x = \cos\phi$, (3.A.14) can be written as

$$\begin{aligned} \mathrm{E}\left[x_k^2 x_m^2\right] &= \int_{\pi}^{0} \cos^2\phi \left(g_\chi^{(\kappa)}(\cos\phi)\right)^2 \frac{1}{\pi\sin\phi}(-\sin\phi)\,d\phi \\ &= \frac{1}{\pi}\int_0^{\pi} \cos^2\phi \left(g_\chi^{(\kappa)}(\cos\phi)\right)^2 d\phi \\ &= \frac{1}{\pi}\int_0^{\pi} \cos^2\phi \cos^2\left(\chi^\kappa \phi\right) d\phi \\ &= \frac{1}{\pi}\int_0^{\pi} \frac{1+\cos 2\phi}{2}\,\frac{1+\cos(2\chi^\kappa\phi)}{2}\,d\phi \\ &= \frac{1}{4}. \end{aligned} \tag{3.A.15}$$

Substituting (3.A.11) and (3.A.15) into (3.A.13), we get

$$\operatorname{cov}\left[x_k^2, x_m^2\right] = 0, \quad k \neq m. \tag{3.A.16}$$

Derivation of $\mathrm{E}[x_k x_m]$

The autocorrelation of $\{x_k\}$ is denoted by $\mathrm{E}[x_k x_m]$. Consider the case where $k \neq m$ and if we assume $m = k + \kappa$ for some positive integer κ, we have

$$\begin{aligned} \mathrm{E}[x_k x_m] &= \int_{-\infty}^{\infty} x g_\chi^{(\kappa)}(x)\rho(x)\,dx \\ &= \frac{1}{\pi}\int_0^{\pi} \cos\phi\, g_\chi^{(\kappa)}(\cos\phi)\,d\phi \\ &= \frac{1}{\pi}\int_0^{\pi} \cos\phi \cos\left(\chi^\kappa\phi\right) d\phi \\ &= 0. \end{aligned} \tag{3.A.17}$$

Derivation of $\mathrm{E}[x_k x_{k+\kappa} x_{k+\zeta}]$
We consider the case where $\kappa, \zeta \in I^+$ and $0 < \kappa < \zeta$. Therefore,

$$\begin{aligned}
\mathrm{E}[x_k x_{k+\kappa} x_{k+\zeta}] &= \int_{-\infty}^{\infty} x g_\chi^{(\kappa)}(x) g_\chi^{(\zeta)}(x) \rho(x)\, dx \\
&= \frac{1}{\pi} \int_0^{\pi} \cos\phi\, g^{(\kappa)}(\cos\phi) g^{(\zeta)}(\cos\phi)\, d\phi \\
&= \frac{1}{\pi} \int_0^{\pi} \cos\phi \cos(\chi^\kappa \phi) \cos(\chi^\zeta \phi)\, d\phi \\
&= 0.
\end{aligned} \tag{3.A.18}$$

Derivation of $\mathrm{E}[x_k x_{k+\kappa} x_{k+\zeta} x_{k+\vartheta}]$
We consider the case where $\kappa, \zeta, \vartheta \in I^+$ and $0 < \kappa < \zeta < \vartheta$. Thus,

$$\begin{aligned}
\mathrm{E}[x_k x_{k+\kappa} x_{k+\zeta} x_{k+\vartheta}] &= \int_{-\infty}^{\infty} x g_\chi^{(\kappa)}(x) g_\chi^{(\zeta)}(x) g_\chi^{(\vartheta)}(x) \rho(x)\, dx \\
&= \frac{1}{\pi} \int_0^{\pi} \cos\phi\, g_\chi^{(\kappa)}(\cos\phi) g_\chi^{(\zeta)}(\cos\phi) g_\chi^{(\vartheta)}(\cos\phi)\, d\phi \\
&= \frac{1}{\pi} \int_0^{\pi} \cos\phi \cos(\chi^\kappa \phi) \cos(\chi^\zeta \phi) \cos(\chi^\vartheta \phi)\, d\phi \\
&= 0.
\end{aligned} \tag{3.A.19}$$

Derivation of $\mathrm{E}[x_k^2 x_m]$ *for the Chebyshev map of degree* 3
When $k = m$,

$$\mathrm{E}\left[x_k^2 x_m\right] = \mathrm{E}\left[x_k^3\right] = \int_{-\infty}^{\infty} x^3 \rho(x)\, dx = \int_{-1}^{1} x^3 \rho(x)\, dx = 0, \tag{3.A.20}$$

because $\rho(x)$ is an even function whereas x^3 is odd. Next, we consider the case where $k > m$. Assuming $k = m + \kappa$ for some positive integer κ, we obtain

$$\mathrm{E}\left[x_k^2 x_m\right] = \int_{-\infty}^{\infty} \left(g_3^{(\kappa)}(x)\right)^2 x \rho(x)\, dx. \tag{3.A.21}$$

Within the integral in (3.A.21), both $\rho(x)$ and $(g_3^{(\kappa)}(x))^2$ are even while x is odd. Thus, it can be concluded that $\mathrm{E}[x_k^2 x_m] = 0$ for $k > m$. Finally, for $k < m$, assuming $m = k + \kappa$ for some positive integer κ, we have

$$\mathrm{E}\left[x_k^2 x_{k+\kappa}\right] = \int_{-\infty}^{\infty} x^2 g_3^{(\kappa)}(x) \rho(x)\, dx$$

$$= \frac{1}{\pi}\int_0^{\pi} \cos^2\phi\cos(3^{\kappa}\phi)\,d\phi$$
$$= 0. \tag{3.A.22}$$

Derivation of $\mathrm{E}[x_k^3 x_m]$ *for the Chebyshev map of degree* 3
When $k = m$,

$$\mathrm{E}[x_k^3 x_m] = \mathrm{E}[x_k^4] = \int_{-\infty}^{\infty} x^4\rho(x)\,dx = \int_{-1}^{1} x^4\rho(x)\,dx = \frac{3}{8}. \tag{3.A.23}$$

Next, we consider the case where $k > m$. Assuming $k = m + \kappa$ for some positive integer κ, making the substitution $x = \cos\phi$ and applying (3.A.7), we have

$$\mathrm{E}[x_k^3 x_m] = \int_{-\infty}^{\infty} \left(g_3^{(\kappa)}(x)\right)^3 x\rho(x)\,dx$$
$$= \frac{1}{\pi}\int_0^{\pi} \cos^3(3^{\kappa}\phi)\cos\phi\,d\phi$$
$$= 0. \tag{3.A.24}$$

For $k < m$, we assume $m = k + \kappa$ for some positive integer κ.

$$\mathrm{E}[x_k^3 x_m] = \int_{-\infty}^{\infty} x^3 g_3^{(\kappa)}(x)\rho(x)\,dx$$
$$= \frac{1}{\pi}\int_0^{\pi} \cos^3\phi\cos(3^{\kappa}\phi)\,d\phi$$
$$= \begin{cases} 0 & \text{for } \kappa \geqslant 2, \\ 1/8 & \text{for } \kappa = 1. \end{cases} \tag{3.A.25}$$

Thus, we conclude that

$$\mathrm{E}[x_k^3 x_m] = \begin{cases} 3/8 & \text{for } k = m, \\ 1/8 & \text{for } k + 1 = m, \\ 0 & \text{otherwise.} \end{cases} \tag{3.A.26}$$

Derivation of $\mathrm{E}[x_k^2 x_m x_u]$ *for the Chebyshev map of degree* 3
We first consider the case where $k < m < u$. Assuming $m = k + \kappa$ and $u = k + \zeta$ for some positive integers κ and ζ $(0 < \kappa < \zeta)$, we have

$$\mathrm{E}[x_k^2 x_m x_u] = \int_{-\infty}^{\infty} x^2 g_3^{(\kappa)}(x) g_3^{(\zeta)}(x)\rho(x)\,dx$$

$$
\begin{aligned}
&= \frac{1}{\pi}\int_0^{\pi} (\cos\phi)^2 \cos(3^{\kappa}\phi)\cos(3^{\zeta}\phi)\,d\phi \\
&= 0. \qquad (3.A.27)
\end{aligned}
$$

Next, consider the case where $k > m > u$. Assuming $k = u + \zeta$ and $m = u + \kappa$ for some positive integers ζ and κ, we have

$$
\begin{aligned}
\mathrm{E}[x_k^2 x_m x_u] &= \int_{-\infty}^{\infty} (g_3^{(\zeta)}(x))^2 g_3^{(\kappa)}(x) x \rho(x)\,dx \\
&= \frac{1}{\pi}\int_0^{\pi} \cos(3^{\zeta}\phi)^2 \cos(3^{\kappa}\phi)\cos\phi\,d\phi \\
&= 0. \qquad (3.A.28)
\end{aligned}
$$

Using a likewise procedure, it is readily concluded that $\mathrm{E}[x_k^2 x_m x_u] = 0$ for non-equal integers k, m and u.

4

MULTIUSER DETECTION TECHNIQUES FOR MULTIPLE-ACCESS COHERENT ANTIPODAL CHAOS-SHIFT-KEYING SYSTEMS

In Chapter 3, the performance of multiple-access antipodal chaos-shift-keying system has been evaluated thoroughly. Because of the non-zero cross-correlations between the chaotic spreading signals, the co-channel interference increases as the number of users increases. Since the simple correlator-type receiver is not designed to combat such interference, increasing co-channel interference degrades the bit error performance of all users. Such multiple-access interference (MAI) limits the performance of the system even under a noiseless environment. In this chapter, we investigate the use of multiuser detectors to mitigating the effect of MAI, and hence improving the bit error performance.

4.1 INTRODUCTION

Multiuser detection has been shown effective to combat MAI in conventional direct-sequence code-division-multiple-access (DS-CDMA) systems. The operating principle of a multiuser detector is to, based on the available spreading codes, cancel the interference between users and jointly decode the transmitted symbols. The optimum multiuser detector was first proposed based on the Viterbi algorithm [Verdú, 1986]. The decoding is performed so as to maximize the joint posterior probability according to the maximum-likelihood criterion. However, the optimum multiuser receiver is not employed in practice because the complexity increases exponentially with the number of users. As a result, a class of suboptimum detectors have been used.

Suboptimum multiuser detectors can be broadly classified into linear detectors (decorrelating detector and minimum mean-square-error detector) and subtractive interference cancellation detectors (successive interference cancellation detector, parallel interference cancellation detector and zero-forcing decision-feedback detector). For the linear detectors, a linear transformation is performed at the receiver to mitigate the MAI. For the decorrelating detector, the interference can be completely eliminated [Lupas and Verdú, 1989, 1990]. However, the noise component of the remaining signal increases, especially for a large loading. In contrast to the decorrelating detector, the minimum mean-square-error detector allows residual interference without causing a significant increase in noise level [Xie *et al.*, 1990]. A parallel interference cancellation (PIC) detector is a non-linear multiuser detector that can cancel the total interference for all users simultaneously [Yoon *et al.*, 1993]. The detector requires exact knowledge of the spreading sequences of all users as well as an estimation of the transmitted symbols. Based on the estimation of the transmitted symbols, the mutual interference is computed and subtracted from the incoming signal before decoding is performed.

The application of multiuser detection techniques to multiple-access chaos-based communication systems, though an important topic, has not been studied until recently. In the work of Argüello *et al.* [2002], two nonlinear subtractive interference cancellation detectors have been applied to a multiple-access chaotic-sequence spread-spectrum (MA-CSSS) system. Based on simulations, their studies illustrated that multiuser detection techniques could substantially improve the bit error rate (BER) of the MA-CSSS system.

In this chapter, we study the use of linear and nonlinear multiuser detectors in a multiple-access antipodal CSK system (MA-ACSK). We will also develop analytical techniques to evaluate the BERs of the system. This chapter is organized as follows. Section 4.2 is focused on linear multiuser detection techniques. In particular, Section 4.2.1 presents an overview of the MA-ACSK system and the operations of the linear detectors. The linear detectors include single-user detector, decorrelating detector (DD) and minimum mean-square-error (MMSE) detector [Madsen and Cho, 1999; Tam *et al.*, 2002b, 2004c, 2004d]. In Section 4.2.2, we develop techniques to evaluate the BERs of the linear multiuser detectors. Brute-force simulation results are presented and compared with the analytical results in Section 4.2.3. Section 4.3 studies three types of linear detectors applied in conjunction with parallel interference cancellation (PIC) detector in the MA-ACSK communication system. A multistage PIC detector is described in Section 4.3.1 and the technique for deriving the approximate BERs for the PIC detectors (i.e., conventional/PIC, DD/PIC and MMSE/PIC detectors) is given in Section 4.3.2. In Section 4.3.3, the analytical and simulated BER results of the PIC detectors are presented.

4.2 LINEAR MULTIUSER DETECTION TECHNIQUES

4.2.1 System Description

4.2.1.1 *Transmitter Structure* Consider the MA-ACSK system with N users shown in Fig. 4.1. Define $d_l^{(i)} \in \{-1, +1\}$ as the lth transmitted symbol for the

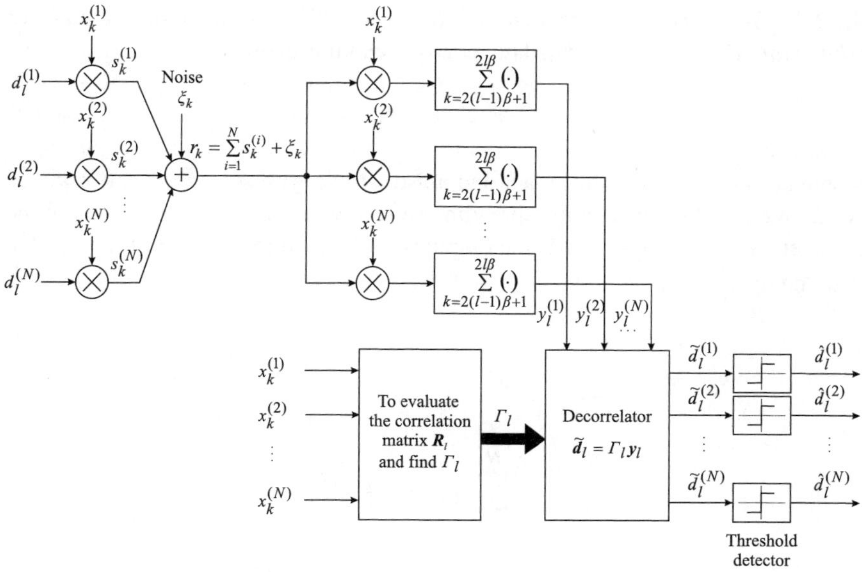

Figure 4.1. Multiple-access antipodal CSK system with a linear multiuser detector.

ith user which is assumed to be equiprobable for "+1" and "−1". Denote the chaotic sequence generated by the ith generator by $\{x_k^{(i)}\}$, which is used to spread the binary symbol sequence $\{d_l^{(i)}\}$. Assuming a spreading factor of 2β, the transmitted signal for the ith user at time $k = 2(l-1)\beta + 1, 2(l-1)\beta + 2, \ldots, 2l\beta$ can be expressed as

$$s_k^{(i)} = d_l^{(i)} x_k^{(i)}. \tag{4.2.1}$$

The overall transmitted signal of the system at time k is thus given by

$$s_k = \sum_{i=1}^{N} s_k^{(i)}. \tag{4.2.2}$$

Note that the chaotic sequences are independent of one another because they are derived from different generators. Also, the mean value of each chaotic sequence is zero in order to avoid transmitting any non-information-bearing dc components. In other words,

$$\mathrm{E}[x_k^{(i)}] = 0, \quad i = 1, 2, \ldots, N. \tag{4.2.3}$$

4.2.1.2 *Single-User and Multi-User Receivers* We assume a simple *additive white Gaussian noise channel.* The received signal is given by

$$r_k = s_k + \xi_k, \tag{4.2.4}$$

where ξ_k is an additive white Gaussian noise sample with zero mean and variance $N_0/2$. Assume that the chaotic spreading sequences are reproduced exactly at the receiver. As shown in Fig. 4.1, the output of the jth correlator ($j = 1, 2, \ldots, N$), denoted by $y_l^{(j)}$, equals

$$\begin{aligned} y_l^{(j)} &= \sum_{k=2(l-1)\beta+1}^{2l\beta} r_k x_k^{(j)} \\ &= \sum_{k=2(l-1)\beta+1}^{2l\beta} \left(\sum_{i=1}^{N} s_k^{(i)} + \xi_k \right) x_k^{(j)} \\ &= \sum_{i=1}^{N} d_l^{(i)} \sum_{k=2(l-1)\beta+1}^{2l\beta} x_k^{(i)} x_k^{(j)} + \sum_{k=2(l-1)\beta+1}^{2l\beta} \xi_k x_k^{(j)}. \end{aligned} \tag{4.2.5}$$

Define the correlation matrix $\boldsymbol{R}_l$ as

$$\boldsymbol{R}_l = \begin{bmatrix} R_{1,1,l} & R_{1,2,l} & \cdots & R_{1,N,l} \\ R_{2,1,l} & R_{2,2,l} & \cdots & R_{2,N,l} \\ \vdots & \vdots & \ddots & \vdots \\ R_{N,1,l} & R_{N,2,l} & \cdots & R_{N,N,l} \end{bmatrix}, \tag{4.2.6}$$

where

$$R_{j,i,l} = \sum_{k=2(l-1)\beta+1}^{2l\beta} x_k^{(i)} x_k^{(j)} \tag{4.2.7}$$

denotes the correlation (mutual interference) between the ith and the jth users during the lth symbol duration. Note that $\boldsymbol{R}_l$ is a symmetric matrix. The outputs of the correlators can now be expressed as

$$\boldsymbol{y}_l = \boldsymbol{R}_l \boldsymbol{d}_l + \boldsymbol{\eta}_l, \tag{4.2.8}$$

where

$$\boldsymbol{y}_l = \left[y_l^{(1)} \quad y_l^{(2)} \quad \cdots \quad y_l^{(N)} \right]^{\mathrm{T}}, \tag{4.2.9}$$

$$\boldsymbol{d}_l = \begin{bmatrix} d_l^{(1)} & d_l^{(2)} & \cdots & d_l^{(N)} \end{bmatrix}^{\mathrm{T}}, \tag{4.2.10}$$

$$\boldsymbol{\eta}_l = \begin{bmatrix} \eta_l^{(1)} & \eta_l^{(2)} & \cdots & \eta_l^{(N)} \end{bmatrix}^{\mathrm{T}}, \tag{4.2.11}$$

$$\eta_l^{(j)} = \sum_{k=2(l-1)\beta+1}^{2l\beta} \xi_k x_k^{(j)}, \tag{4.2.12}$$

and T denotes matrix transposition.

4.2.1.2.1 Conventional Single-User Detector For the conventional single-user detector, the decoded symbols, denoted by $\hat{\boldsymbol{d}}_l$, are computed from

$$\hat{\boldsymbol{d}}_l = \begin{bmatrix} \hat{d}_l^{(1)} & \hat{d}_l^{(2)} & \cdots & \hat{d}_l^{(N)} \end{bmatrix}^{\mathrm{T}} = \mathrm{sgn}[\boldsymbol{y}_l], \tag{4.2.13}$$

where sgn[·] represents the sign function. In other words, if $y_l^{(j)} > 0$, "+1" is detected for the jth user, otherwise, "−1" is decoded. Note that in a MA-ACSK communication system, the correlation matrix $\boldsymbol{R}_l$ changes from bit to bit. Therefore, $\boldsymbol{R}_l$ needs to be re-calculated for each bit duration.

4.2.1.2.2 Decorrelating Detector For a given $\boldsymbol{d}_l$, we assume that $\boldsymbol{y}_l$ is Gaussian with mean $\boldsymbol{R}_l\boldsymbol{d}_l$ and covariance matrix $\frac{N_0}{2}\boldsymbol{R}_l$ [Xie *et al.*, 1990]. The N-dimensional probability density function is then given by [Proakis, 1995]

$$\rho(\boldsymbol{y}_l \mid \boldsymbol{d}_l) = \frac{1}{\sqrt{(\pi N_0)^N \det[\boldsymbol{R}_l]}} \exp\left[-\frac{1}{N_0}(\boldsymbol{y}_l - \boldsymbol{R}_l\boldsymbol{d}_l)^{\mathrm{T}}\boldsymbol{R}_l^{-1}(\boldsymbol{y}_l - \boldsymbol{R}_l\boldsymbol{d}_l)\right] \tag{4.2.14}$$

where det[·] represents the determinant operator. The optimum solution which maximizes the probability of a correct decision is equivalent to minimizing the likelihood function

$$\Omega(\boldsymbol{d}_l) = (\boldsymbol{y}_l - \boldsymbol{R}_l\boldsymbol{d}_l)^{\mathrm{T}}\boldsymbol{R}_l^{-1}(\boldsymbol{y}_l - \boldsymbol{R}_l\boldsymbol{d}_l). \tag{4.2.15}$$

From (4.2.15), we obtain the optimum solution

$$\tilde{\boldsymbol{d}}_l = \boldsymbol{R}_l^{-1}\boldsymbol{y}_l \tag{4.2.16}$$

where

$$\tilde{\boldsymbol{d}}_l = \begin{bmatrix} \tilde{d}_l^{(1)} & \tilde{d}_l^{(2)} & \cdots & \tilde{d}_l^{(N)} \end{bmatrix}^{\mathrm{T}} \tag{4.2.17}$$

denotes the estimated $\boldsymbol{d}_l$. The decoded symbols of a decorrelating detector can be evaluated using

$$\hat{\boldsymbol{d}}_l = \mathrm{sgn}[\tilde{\boldsymbol{d}}_l]. \tag{4.2.18}$$

4.2.1.2.3 Minimum Mean-Square-Error (MMSE) Detector In a communication system, both the noise and the interference components can degrade the bit error performance. The decorrelating detector described in the previous section can eliminate the mutual interference between users, but the price to pay is an increase in the number of noise components, which is proportional to the number of users. The overall system performance is limited by the resulting noise level. A minimum mean-square-error (MMSE) detector takes the effects of both co-channel interference and noise into consideration during the detection process. The main idea is to find a mapping, represented by $\boldsymbol{H}_l$, that minimizes the mean-square-error, denoted by $J(\boldsymbol{d}_l)$, between the transmitted symbols $\boldsymbol{d}_l$ and the estimated $\boldsymbol{d}_l$ (represented by $\tilde{\boldsymbol{d}}_l$). The equation linking $\boldsymbol{H}_l$ and $J(\boldsymbol{d}_l)$ is given by

$$J(\boldsymbol{d}_l) = \mathrm{E}\big[(\boldsymbol{d}_l - \tilde{\boldsymbol{d}}_l)^{\mathrm{T}}(\boldsymbol{d}_l - \tilde{\boldsymbol{d}}_l)\big] \tag{4.2.19}$$

where

$$\tilde{\boldsymbol{d}}_l = \boldsymbol{H}_l \boldsymbol{y}_l. \tag{4.2.20}$$

The mapping that minimizes the mean-square-error is readily shown equal to [Xie *et al.*, 1990]

$$\boldsymbol{H}_l = \left(\boldsymbol{R}_l + \frac{N_0}{2}\boldsymbol{I}\right)^{-1} \tag{4.2.21}$$

where $\boldsymbol{I}$ is the identity matrix and $\boldsymbol{H}_l$ is a square matrix defined as

$$\boldsymbol{H}_l = \begin{bmatrix} H_{1,1,l} & H_{1,2,l} & \cdots & H_{1,N,l} \\ H_{2,1,l} & H_{2,2,l} & \cdots & H_{2,N,l} \\ \vdots & \vdots & \ddots & \vdots \\ H_{N,1,l} & H_{N,2,l} & \cdots & H_{N,N,l} \end{bmatrix}. \tag{4.2.22}$$

The symbols are then decoded according to

$$\hat{\boldsymbol{d}}_l = \mathrm{sgn}[\tilde{\boldsymbol{d}}_l]. \tag{4.2.23}$$

In general, the symbols decoded using the linear multiuser detector shown in Fig. 4.1 can be expressed as

$$\hat{\boldsymbol{d}}_l = \mathrm{sgn}[\Gamma_l \boldsymbol{y}_l], \tag{4.2.24}$$

where Γ_l is the linear mapping. For the conventional single-user, decorrelating and MMSE detectors, Γ_l is represented by $\boldsymbol{I}$, $\boldsymbol{R}_l^{-1}$ and $(\boldsymbol{R}_l + \frac{N_0}{2}\boldsymbol{I})^{-1}$, respectively. In a noiseless condition, the MMSE detector is equivalent to the decorrelating detector (DD). When the noise level is large, the DD and MMSE detectors have similar performance as the conventional single-user detector.

4.2.2 Performance Analysis

In this section, we develop techniques to evaluate the bit error rate (BER) of the aforementioned multiuser systems. It is assumed that all users use the same chaotic generator but each uses a different set of initial conditions. We further assume that all users have identical BERs because they receive similar interference. Without loss of generality, we consider the first user in a 2-user system and the jth user in an N-user system, and we derive the probability of error for the first symbol. For brevity, the subscripts of the variables $d_l^{(j)}$, $\tilde{d}_l^{(j)}$, $\hat{d}_l^{(j)}$, $y_l^{(j)}$, $\eta_l^{(j)}$, $\boldsymbol{R}_l$ and $\boldsymbol{H}_l$ are omitted.

4.2.2.1 *Analysis of the Decorrelating Detector*

4.2.2.1.1 Two-User System We consider the first user. It can be readily shown that the inverse of the correlation matrix $\boldsymbol{R}$ equals

$$\boldsymbol{R}^{-1} = \frac{1}{R_{1,1}R_{2,2} - R_{1,2}^2}\begin{bmatrix} R_{2,2} & -R_{1,2} \\ -R_{1,2} & R_{1,1} \end{bmatrix}. \tag{4.2.25}$$

The outputs of the correlators for the first and second users are given, respectively, by

$$y^{(1)} = \sum_{k=1}^{2\beta} r_k x_k^{(1)} = d^{(1)}R_{1,1} + d^{(2)}R_{1,2} + \eta^{(1)} \tag{4.2.26}$$

and

$$y^{(2)} = \sum_{k=1}^{2\beta} r_k x_k^{(2)} = d^{(1)}R_{2,1} + d^{(2)}R_{2,2} + \eta^{(2)}. \tag{4.2.27}$$

The estimated $\boldsymbol{d}$ is then obtained by pre-multiplying $\boldsymbol{y}$ by $\boldsymbol{R}^{-1}$, i.e.,

$$\tilde{\boldsymbol{d}} = \boldsymbol{R}^{-1}\boldsymbol{y} \tag{4.2.28}$$

and the input to the first threshold detector equals

$$\tilde{d}^{(1)} = \frac{1}{R_{1,1}R_{2,2} - R_{1,2}^2}\left[d^{(1)}\left(R_{1,1}R_{2,2} - R_{1,2}^2\right) + \left(R_{2,2}\eta^{(1)} - R_{1,2}\eta^{(2)}\right)\right]. \tag{4.2.29}$$

Define

$$Q_1 = R_{1,1}R_{2,2} - R_{1,2}^2, \tag{4.2.30}$$

$$Q_2 = R_{2,2}\eta^{(1)} - R_{1,2}\eta^{(2)}, \tag{4.2.31}$$

$$\begin{aligned} Q &= d^{(1)}\big(R_{1,1}R_{2,2} - R_{1,2}^2\big) + \big(R_{2,2}\eta^{(1)} - R_{1,2}\eta^{(2)}\big) \\ &= d^{(1)}Q_1 + Q_2. \end{aligned} \tag{4.2.32}$$

Using the Schwarz's inequality[1] and (4.2.7), it is readily shown that $Q_1 > 0$.

Suppose a "+1" is sent by the first user, i.e., $d^{(1)} = +1$. The mean and variance of $Q \mid (d^{(1)} = +1)$ can be calculated, respectively, by

$$\mathrm{E}\big[Q \mid \big(d^{(1)} = +1\big)\big] = \mathrm{E}[Q_1] = \mathrm{E}[R_{1,1}]\mathrm{E}[R_{2,2}] - \mathrm{E}\big[R_{1,2}^2\big] \tag{4.2.33}$$

and

$$\mathrm{var}\big[Q \mid \big(d^{(1)} = +1\big)\big] = \mathrm{var}[Q_1] + \mathrm{var}[Q_2] + 2\,\mathrm{cov}[Q_1, Q_2]. \tag{4.2.34}$$

It is also readily shown that the covariance term in (4.2.34) equals zero. Thus, (4.2.34) can be simplified to

$$\begin{aligned} &\mathrm{var}\big[Q \mid \big(d^{(1)} = +1\big)\big] \\ &\quad = \mathrm{var}[Q_1] + \mathrm{var}[Q_2] \\ &\quad = \mathrm{var}[R_{1,1}]\,\mathrm{var}[R_{2,2}] + \mathrm{E}^2[R_{1,1}]\,\mathrm{var}[R_{2,2}] + \mathrm{E}^2[R_{2,2}]\,\mathrm{var}[R_{1,1}] \\ &\qquad + \mathrm{var}\big[R_{1,2}^2\big] - 2\big(\mathrm{E}\big[R_{1,1}R_{2,2}R_{1,2}^2\big] - \mathrm{E}[R_{1,1}]\mathrm{E}[R_{2,2}]\mathrm{E}\big[R_{1,2}^2\big]\big) \\ &\qquad + \frac{N_0}{2}\big(\mathrm{E}[R_{1,1}]\mathrm{E}\big[R_{2,2}^2\big] - \mathrm{E}\big[R_{1,2}^2 R_{2,2}\big]\big). \end{aligned} \tag{4.2.35}$$

Since $Q_1 > 0$, we can decode the symbol based only on the sign of Q, i.e., when $Q \leqslant 0$, "−1" is decoded, otherwise, "+1" is recovered. Furthermore, $Q \mid (d^{(1)} = +1)$ is the sum of a large number of random variables. We may therefore assume that it is normally distributed. Given "+1" has been sent, an error occurs when $Q \leqslant 0$ and the probability is given by

$$\begin{aligned} \mathrm{Prob}\big(\tilde{d}^{(1)} \leqslant 0 \mid \big(d^{(1)} = +1\big)\big) &= \mathrm{Prob}\big(Q \leqslant 0 \mid \big(d^{(1)} = +1\big)\big) \\ &= \frac{1}{2}\,\mathrm{erfc}\left(\frac{\mathrm{E}[Q \mid (d^{(1)} = +1)]}{\sqrt{2\,\mathrm{var}[Q \mid (d^{(1)} = +1)]}}\right). \end{aligned} \tag{4.2.36}$$

---

[1] Schwarz's inequality states that for two sets of M numbers given by $\{a_m; m = 1, 2, \ldots, M\}$ and $\{b_m; m = 1, 2, \ldots, M\}$, $(\sum_{m=1}^{M} a_m^2)(\sum_{m=1}^{M} b_m^2) \geqslant (\sum_{m=1}^{M} a_m b_m)^2$. Strict inequality holds unless $a_m/a_n = b_m/b_n$ for all values of m and n [Briggs *et al.*, 1985].

Likewise, it can be shown that when a "-1" is sent, i.e., $d^{(1)} = -1$, the mean and variance of $Q \mid (d^{(1)} = -1)$ equal, respectively,

$$\mathrm{E}\big[Q \mid \big(d^{(1)} = -1\big)\big] = -\mathrm{E}\big[Q \mid \big(d^{(1)} = +1\big)\big] \tag{4.2.37}$$

and

$$\mathrm{var}\big[Q \mid \big(d^{(1)} = -1\big)\big] = \mathrm{var}\big[Q \mid \big(d^{(1)} = +1\big)\big]. \tag{4.2.38}$$

Assuming that $Q \mid (d^{(1)} = -1)$ is normal, the conditional error probability is evaluated from

$$\begin{aligned}\mathrm{Prob}\big(\tilde{d}^{(1)} > 0 \mid \big(d^{(1)} = -1\big)\big) &= \mathrm{Prob}\big(Q > 0 \mid \big(d^{(1)} = -1\big)\big) \\ &= \frac{1}{2}\,\mathrm{erfc}\left(\frac{-\mathrm{E}[Q \mid (d^{(1)} = -1)]}{\sqrt{2\,\mathrm{var}[Q \mid (d^{(1)} = -1)]}}\right).\end{aligned} \tag{4.2.39}$$

The overall bit error rate (BER) for the first user, denoted by $\mathrm{BER}^{(1)}$, can be evaluated by using (4.2.36) through (4.2.39), i.e.,

$$\begin{aligned}\mathrm{BER}^{(1)} &= \frac{1}{2}\mathrm{Prob}\big(\tilde{d}^{(1)} \leqslant 0 \mid \big(d^{(1)} = +1\big)\big) + \frac{1}{2}\mathrm{Prob}\big(\tilde{d}^{(1)} > 0 \mid \big(d^{(1)} = -1\big)\big) \\ &= \frac{1}{2}\mathrm{Prob}\big(Q \leqslant 0 \mid \big(d^{(1)} = +1\big)\big) + \frac{1}{2}\mathrm{Prob}\big(Q > 0 \mid \big(d^{(1)} = -1\big)\big) \\ &= \frac{1}{2}\,\mathrm{erfc}\left(\frac{\mathrm{E}[Q \mid (d^{(1)} = +1)]}{\sqrt{2\,\mathrm{var}[Q \mid (d^{(1)} = +1)]}}\right).\end{aligned} \tag{4.2.40}$$

The values of the means and variances ($\mathrm{E}[R_{j,j}]$, $\mathrm{E}[R_{j,j}^2]$, $\mathrm{E}[R_{1,2}^2]$, $\mathrm{E}[R_{1,2}^2 R_{2,2}]$, $\mathrm{E}[R_{1,1}\ R_{2,2} R_{1,2}^2]$, $\mathrm{var}[R_{1,2}^2]$ and $\mathrm{var}[R_{j,j}]$) can be obtained by numerical simulation. The values are then substituted into (4.2.33), (4.2.35) and (4.2.40) to evaluate the BER. We refer to this method as *mixed analysis–simulation* technique.

Example — Chaotic sequences generated by the cubic map
Suppose all users use the cubic map

$$x_{k+1} = 4x_k^3 - 3x_k \tag{4.2.41}$$

to generate the chaotic sequences and each user uses a different initial condition. The invariant probability density function of $\{x_k\}$, denoted by $\rho(x)$, is given by

[Kohda and Tsuneda, 1994]

$$\rho(x) = \begin{cases} \dfrac{1}{\pi\sqrt{1-x^2}} & \text{if } |x| < 1, \\ 0 & \text{otherwise,} \end{cases} \tag{4.2.42}$$

and the chaotic sequences $\{x_k^{(j)}; j = 1, 2, \ldots, N\}$ possess the following statistical properties (see Appendix 3A).

$$\mathrm{E}\big[x_k^{(j)}\big] = 0, \tag{4.2.43}$$

$$\mathrm{E}\big[\big(x_k^{(j)}\big)^2\big(x_m^{(j)}\big)^2\big] = \mathrm{E}\big[\big(x_k^{(j)}\big)^2\big]\mathrm{E}\big[\big(x_m^{(j)}\big)^2\big] \quad \text{for } k \neq m, \tag{4.2.44}$$

$$\mathrm{E}\big[x_k^{(j)} x_m^{(j)}\big] = 0 \quad \text{for } k \neq m, \tag{4.2.45}$$

$$\mathrm{E}\big[x_k^{(j)} x_m^{(j)} x_u^{(j)}\big] = 0 \quad \text{for } k < m < u, \tag{4.2.46}$$

$$\mathrm{E}\big[x_k^{(j)} x_m^{(j)} x_u^{(j)} x_v^{(j)}\big] = 0 \quad \text{for } k < m < u < v, \tag{4.2.47}$$

$$\mathrm{E}\big[\big(x_k^{(j)}\big)^2 x_m^{(j)}\big] = 0 \quad \text{for } k \neq m, \tag{4.2.48}$$

$$\mathrm{E}\big[\big(x_k^{(j)}\big)^3 x_m^{(j)}\big] = \begin{cases} 0 & \text{for } m \neq k+1 \text{ and } m \neq k, \\ 1/8 & \text{for } m = k+1, \end{cases} \tag{4.2.49}$$

$$\mathrm{E}\big[\big(x_k^{(j)}\big)^2 x_m^{(j)} x_u^{(j)}\big] = 0 \quad \text{for non-equal integers } k, m \text{ and } u. \tag{4.2.50}$$

Denote the average transmission power by P_s where

$$P_s = \mathrm{E}\big[\big(x_k^{(1)}\big)^2\big] = \mathrm{E}\big[\big(x_k^{(2)}\big)^2\big] = \cdots = \mathrm{E}\big[\big(x_k^{(N)}\big)^2\big]. \tag{4.2.51}$$

Also, define

$$\Lambda = \frac{\mathrm{E}[(x_k^{(1)})^4]}{P_s^2} = \frac{\mathrm{E}[(x_k^{(2)})^4]}{P_s^2} = \cdots = \frac{\mathrm{E}[(x_k^{(N)})^4]}{P_s^2}, \tag{4.2.52}$$

which is a constant dependent only upon the type of chaotic sequence but not on the average power. Using (4.2.43) through (4.2.52), it is readily shown that the mean and variance of $Q \mid (d^{(1)} = +1)$ are given, respectively, by (see Appendix 4A)

$$\mathrm{E}\big[Q \mid \big(d^{(1)} = +1\big)\big] = 2\beta(2\beta - 1)P_s^2 \tag{4.2.53}$$

and

$$\begin{aligned}
&\mathrm{var}\big[Q \mid (d^{(1)} = +1)\big] \\
&= 2\beta(2\beta-1)\Big[\mathrm{E}^2\big[(x_k^{(1)})^4\big] + 4(\beta-1)\mathrm{E}\big[(x_k^{(1)})^4\big]P_s^2 + (-4\beta+5)P_s^4\Big] \\
&\quad + 2\beta(2\beta-1)\frac{N_0}{2}\Big[\mathrm{E}\big[(x_k^{(1)})^4\big]P_s + (2\beta-1)P_s^3\Big]. \qquad (4.2.54)
\end{aligned}$$

Substituting (4.2.53) and (4.2.54) into (4.2.40), the analytical BER for the first user equals

$$\begin{aligned}
\mathrm{BER}^{(1)} = \frac{1}{2}\,\mathrm{erfc}\Bigg(\Bigg[\frac{\Lambda^2}{\beta(2\beta-1)} + \frac{4(\beta-1)\Lambda}{\beta(2\beta-1)} + \frac{-4\beta+5}{\beta(2\beta-1)} \\
+ \left(\frac{\Lambda}{(2\beta-1)} + 1\right)\left(\frac{E_b}{N_0}\right)^{-1}\Bigg]^{-1/2}\Bigg), \qquad (4.2.55)
\end{aligned}$$

where E_b represents the average bit energy defined as

$$E_b = 2\beta P_s. \qquad (4.2.56)$$

4.2.2.1.2 N-User System We define the inverse of the correlation matrix $\boldsymbol{R}$ as

$$\boldsymbol{R}^{-1} = \boldsymbol{P} = \begin{bmatrix} P_{1,1} & P_{1,2} & \cdots & P_{1,N} \\ P_{2,1} & P_{2,2} & \cdots & P_{2,N} \\ \vdots & \vdots & \ddots & \vdots \\ P_{N,1} & P_{N,2} & \cdots & P_{N,N} \end{bmatrix}. \qquad (4.2.57)$$

For the jth user, using (4.2.8), (4.2.16) and (4.2.57), the estimated value of $d^{(j)}$, denoted by $\tilde{d}^{(j)}$, is found equal to

$$\tilde{d}^{(j)} = d^{(j)} + \sum_{i=1}^{N} P_{j,i}\eta^{(i)}. \qquad (4.2.58)$$

Based on the estimated $d^{(j)}$, the symbol is decoded according to the following rule:

$$\hat{d}^{(j)} = \begin{cases} +1, & \tilde{d}^{(j)} > 0, \\ -1, & \tilde{d}^{(j)} \leqslant 0. \end{cases} \qquad (4.2.59)$$

For a given $d^{(j)} \in \{-1, +1\}$, the mean value of $\tilde{d}^{(j)}$ is given by

$$\mathrm{E}\big[\tilde{d}^{(j)} \mid d^{(j)}\big] = d^{(j)}, \qquad (4.2.60)$$

and the variance equals [Lupas and Verdú, 1989; Yoon and Ness, 2002] (see Appendix 4B)

$$\begin{aligned}\operatorname{var}\big[\tilde{d}^{(j)} \mid d^{(j)}\big] &= \operatorname{var}\left[\sum_{i=1}^{N} P_{j,i}\eta^{(i)}\right] = \mathrm{E}\left[\left(\sum_{i=1}^{N} P_{j,i}\eta^{(i)}\right)^2\right] \\ &= \frac{N_0}{2}\mathrm{E}[P_{j,j}] = \frac{N_0}{2}\mathrm{E}\left[\frac{\operatorname{adj}[\boldsymbol{R}]_{j,j}}{\det[\boldsymbol{R}]}\right],\end{aligned} \tag{4.2.61}$$

where $\operatorname{adj}[\boldsymbol{R}]_{j,j}$ represents the (j, j)th element of the matrix $\operatorname{adj}[\boldsymbol{R}]$, and $\operatorname{adj}[\boldsymbol{R}]$ denotes the adjoint of $\boldsymbol{R}$, i.e., $\operatorname{adj}[\boldsymbol{R}] = \det[\boldsymbol{R}]\boldsymbol{R}^{-1}$. It should be noted that because $\boldsymbol{R}$ is independent of $d^{(j)}$, so is $\operatorname{var}[\tilde{d}^{(j)} \mid d^{(j)}]$. Assuming that for a given $d^{(j)}$, $\tilde{d}^{(j)}$ is normal. The BER of the jth user can thus be expressed as

$$\begin{aligned}\mathrm{BER}^{(j)} &= \mathrm{Prob}\big(\tilde{d}^{(j)} \leqslant 0 \,\big|\, (d^{(j)} = +1)\big)\,\mathrm{Prob}\big(d^{(j)} = +1\big) \\ &\quad + \mathrm{Prob}\big(\tilde{d}^{(j)} > 0 \,\big|\, (d^{(j)} = -1)\big)\,\mathrm{Prob}\big(d^{(j)} = -1\big) \\ &= \frac{1}{2}\mathrm{Prob}\big(\tilde{d}^{(j)} \leqslant 0 \,\big|\, (d^{(j)} = +1)\big) + \frac{1}{2}\mathrm{Prob}\big(\tilde{d}^{(j)} > 0 \,\big|\, (d^{(j)} = -1)\big) \\ &= \frac{1}{4}\operatorname{erfc}\left(\frac{\mathrm{E}[\tilde{d}^{(j)} \mid (d^{(j)} = +1)]}{\sqrt{2\operatorname{var}[\tilde{d}^{(j)} \mid (d^{(j)} = +1)]}}\right) \\ &\quad + \frac{1}{4}\operatorname{erfc}\left(\frac{-\mathrm{E}[\tilde{d}^{(j)} \mid (d^{(j)} = -1)]}{\sqrt{2\operatorname{var}[\tilde{d}^{(j)} \mid (d^{(j)} = -1)]}}\right).\end{aligned} \tag{4.2.62}$$

Since

$$\mathrm{E}\big[\tilde{d}^{(j)} \,\big|\, (d^{(j)} = +1)\big] = -\mathrm{E}\big[\tilde{d}^{(j)} \,\big|\, (d^{(j)} = -1)\big] = 1 \tag{4.2.63}$$

and

$$\operatorname{var}\big[\tilde{d}^{(j)} \,\big|\, (d^{(j)} = +1)\big] = \operatorname{var}\big[\tilde{d}^{(j)} \,\big|\, (d^{(j)} = -1)\big], \tag{4.2.64}$$

the BER equation can be rewritten as

$$\mathrm{BER}^{(j)} = \frac{1}{2}\operatorname{erfc}\left(\frac{\mathrm{E}[\tilde{d}^{(j)} \mid (d^{(j)} = +1)]}{\sqrt{2\operatorname{var}[\tilde{d}^{(j)} \mid (d^{(j)} = +1)]}}\right) = \frac{1}{2}\operatorname{erfc}\left(\frac{1}{\sqrt{N_0\mathrm{E}[P_{j,j}]}}\right). \tag{4.2.65}$$

The term $E[P_{j,j}]$ cannot be evaluated analytically and has to be found by numerical simulation. Then, (4.2.65) can be used to calculate the BER value (mixed analysis-simulation technique).

4.2.2.2 *Analysis of the MMSE detector*

4.2.2.2.1 Two-User System From (4.2.21), it can be shown that the map $\boldsymbol{H}$ minimizing the mean-squared error is equal to

$$\boldsymbol{H} = \frac{1}{4R_{1,1}R_{2,2} - 4R_{1,2}^2 + 2N_0R_{1,1} + 2N_0R_{2,2} + N_0^2} \times \begin{bmatrix} 2(2R_{2,2}+N_0) & -4R_{1,2} \\ -4R_{1,2} & 2(2R_{1,1}+N_0) \end{bmatrix}. \tag{4.2.66}$$

Define

$$Z_1 = 4R_{1,1}R_{2,2} - 4R_{1,2}^2 + 2N_0R_{1,1} + 2N_0R_{2,2} + N_0^2. \tag{4.2.67}$$

Using the Schwarz's inequality, it can be shown that $R_{1,1}R_{2,2} - R_{1,2}^2 > 0$ and hence $Z_1 > 0$. The estimated $\boldsymbol{d}$ is given by

$$\tilde{\boldsymbol{d}} = \boldsymbol{H}\boldsymbol{y}. \tag{4.2.68}$$

For the first user, the estimated d is computed from

$$\begin{aligned} \tilde{d}^{(1)} &= \frac{1}{Z_1}\big[2(2R_{2,2}+N_0)\big(d^{(1)}R_{1,1} + d^{(2)}R_{1,2} + \eta^{(1)}\big) \\ &\quad - 4R_{1,2}\big(d^{(1)}R_{2,1} + d^{(2)}R_{2,2} + \eta^{(2)}\big)\big] \\ &= \frac{1}{Z_1}\big[4d^{(1)}\big(R_{1,1}R_{2,2} - R_{1,2}^2\big) + 4\big(R_{2,2}\eta^{(1)} - R_{1,2}\eta^{(2)}\big) \\ &\quad + 2N_0\big(d^{(1)}R_{1,1} + d^{(2)}R_{1,2}\big) + 2N_0\eta^{(1)}\big] \\ &\equiv \frac{Z}{Z_1}, \end{aligned} \tag{4.2.69}$$

where

$$\begin{aligned} Z &= 4d^{(1)}\big(R_{1,1}R_{2,2} - R_{1,2}^2\big) + 4\big(R_{2,2}\eta^{(1)} - R_{1,2}\eta^{(2)}\big) \\ &\quad + 2N_0\big(d^{(1)}R_{1,1} + d^{(2)}R_{1,2}\big) + 2N_0\eta^{(1)}. \end{aligned} \tag{4.2.70}$$

The decoded symbol $\hat{d}^{(1)}$ is given by the sign of $\tilde{d}^{(1)}$ which in turn equals the sign of Z.

For a given $\boldsymbol{d} = [d^{(1)}\ d^{(2)}]^{\mathrm{T}}$, the mean value of $Z \mid \boldsymbol{d}$ equals

$$\begin{aligned}\mathrm{E}[Z \mid \boldsymbol{d}] &= \mathrm{E}\big[4d^{(1)}\big(R_{1,1}R_{2,2} - R_{1,2}^2\big) + 2N_0 d^{(1)} R_{1,1}\big]\\ &= 4d^{(1)}\mathrm{E}[R_{1,1}]\mathrm{E}[R_{2,2}] - 4d^{(1)}\mathrm{E}\big[R_{1,2}^2\big]\\ &\quad + 2d^{(1)} N_0 \mathrm{E}[R_{1,1}]. \end{aligned} \tag{4.2.71}$$

Define

$$Q_3 = 2N_0 R_{1,1}, \tag{4.2.72}$$

$$Q_4 = 2N_0 R_{1,2}, \tag{4.2.73}$$

$$Q_5 = 2N_0 \eta^{(1)}. \tag{4.2.74}$$

The variance of $Z \mid \boldsymbol{d}$ is given by

$$\begin{aligned}\mathrm{var}[Z \mid \boldsymbol{d}] &= \mathrm{var}\big[4d^{(1)}Q_1 + 4Q_2 + d^{(1)}Q_3 + d^{(2)}Q_4 + Q_5\big]\\ &= 16\,\mathrm{var}[Q] + \mathrm{var}[Q_3] + \mathrm{var}[Q_4] + \mathrm{var}[Q_5]\\ &\quad + 8\,\mathrm{cov}[Q_1, Q_3] + 8d^{(1)}d^{(2)}\,\mathrm{cov}[Q_1, Q_4] + 8d^{(1)}\,\mathrm{cov}[Q_1, Q_5]\\ &\quad + 8d^{(1)}\,\mathrm{cov}[Q_2, Q_3] + 8d^{(2)}\,\mathrm{cov}[Q_2, Q_4] + 8\,\mathrm{cov}[Q_2, Q_5]\\ &\quad + 2d^{(1)}d^{(2)}\,\mathrm{cov}[Q_3, Q_4] + 2d^{(1)}\,\mathrm{cov}[Q_3, Q_5]\\ &\quad + 2d^{(2)}\,\mathrm{cov}[Q_4, Q_5], \end{aligned} \tag{4.2.75}$$

where Q_1, Q_2 and Q are defined as in (4.2.30) through (4.2.32). It can also be shown (see Appendix 4C) that

$$\mathrm{cov}[Q_1, Q_5] = 0, \tag{4.2.76}$$

$$\mathrm{cov}[Q_2, Q_3] = 0, \tag{4.2.77}$$

$$\mathrm{cov}[Q_2, Q_4] = 0, \tag{4.2.78}$$

$$\mathrm{cov}[Q_3, Q_5] = 0, \tag{4.2.79}$$

$$\mathrm{cov}[Q_4, Q_5] = 0, \tag{4.2.80}$$

$$\mathrm{cov}[Q_3, Q_4] = 0. \tag{4.2.81}$$

Thus the variance of $Z \mid \boldsymbol{d}$ can be simplified to

$$
\begin{aligned}
\mathrm{var}[Z \mid \boldsymbol{d}] &= 16\,\mathrm{var}[Q] + 4N_0^2\,\mathrm{var}[R_{1,1}] + 4N_0^2\,\mathrm{var}[R_{1,2}] + 4N_0^2\,\mathrm{var}\big[\eta^{(1)}\big] \\
&\quad + 16N_0\,\mathrm{cov}\big[\big(R_{1,1}R_{2,2} - R_{1,2}^2\big), R_{1,1}\big] \\
&\quad + 16d^{(1)}d^{(2)}N_0\,\mathrm{cov}\big[\big(R_{1,1}R_{2,2} - R_{1,2}^2\big), R_{1,2}\big] \\
&\quad + 16N_0\,\mathrm{cov}\big[\big(R_{2,2}\eta^{(1)} - R_{1,2}\eta^{(2)}\big), \eta^{(1)}\big] \\
&= 16\,\mathrm{var}[Q] + 4N_0^2\,\mathrm{var}[R_{1,1}] + 4N_0^2\,\mathrm{var}[R_{1,2}] + 2N_0^3\mathrm{E}[R_{1,1}] \\
&\quad + 16N_0\big(\mathrm{var}[R_{1,1}]\mathrm{E}[R_{2,2}] - \mathrm{E}\big[R_{1,2}^2R_{1,1}\big] + \mathrm{E}\big[R_{1,2}^2\big]\mathrm{E}[R_{1,1}]\big) \\
&\quad + 16d^{(1)}d^{(2)}N_0\big(\mathrm{E}[R_{1,1}R_{2,2}R_{1,2}] - \mathrm{E}\big[R_{1,2}^3\big]\big) \\
&\quad + 8N_0^2\big(\mathrm{E}[R_{2,2}]\mathrm{E}[R_{1,1}] - \mathrm{E}\big[R_{1,2}^2\big]\big).
\end{aligned} \tag{4.2.82}
$$

Since $Z \mid \boldsymbol{d}$ is the sum of a large number of random variables, we can assume that it is normally distributed. The bit error rate for the first user is thus given by

$$
\begin{aligned}
\mathrm{BER}^{(1)} &= \frac{1}{2}\mathrm{Prob}\big(\tilde{d}^{(1)} \leqslant 0 \,\big|\, \big(d^{(1)} = +1\big)\big) + \frac{1}{2}\mathrm{Prob}\big(\tilde{d}^{(1)} > 0 \,\big|\, \big(d^{(1)} = -1\big)\big) \\
&= \frac{1}{2}\mathrm{Prob}\big(Z \leqslant 0 \,\big|\, \big(d^{(1)} = +1\big)\big) + \frac{1}{2}\mathrm{Prob}\big(Z > 0 \,\big|\, \big(d^{(1)} = -1\big)\big) \\
&= \frac{1}{4}\mathrm{Prob}\big(Z \leqslant 0 \,\big|\, \big(d^{(1)} = +1, d^{(2)} = +1\big)\big) \\
&\quad + \frac{1}{4}\mathrm{Prob}\big(Z \leqslant 0 \,\big|\, \big(d^{(1)} = +1, d^{(2)} = -1\big)\big) \\
&\quad + \frac{1}{4}\mathrm{Prob}\big(Z > 0 \,\big|\, \big(d^{(1)} = -1, d^{(2)} = +1\big)\big) \\
&\quad + \frac{1}{4}\mathrm{Prob}\big(Z > 0 \,\big|\, \big(d^{(1)} = -1, d^{(2)} = -1\big)\big) \\
&= \frac{1}{4}\left[\frac{1}{2}\,\mathrm{erfc}\left(\frac{\mathrm{E}[Z \mid (d^{(1)} = +1, d^{(2)} = +1)]}{\sqrt{2\,\mathrm{var}[Z \mid (d^{(1)} = +1, d^{(2)} = +1)]}}\right)\right. \\
&\quad + \frac{1}{2}\,\mathrm{erfc}\left(\frac{\mathrm{E}[Z \mid (d^{(1)} = +1, d^{(2)} = -1)]}{\sqrt{2\,\mathrm{var}[Z \mid (d^{(1)} = +1, d^{(2)} = -1)]}}\right) \\
&\quad + \frac{1}{2}\,\mathrm{erfc}\left(\frac{-\mathrm{E}[Z \mid (d^{(1)} = -1, d^{(2)} = +1)]}{\sqrt{2\,\mathrm{var}[Z \mid (d^{(1)} = -1, d^{(2)} = +1)]}}\right) \\
&\quad \left. + \frac{1}{2}\,\mathrm{erfc}\left(\frac{-\mathrm{E}[Z \mid (d^{(1)} = -1, d^{(2)} = -1)]}{\sqrt{2\,\mathrm{var}[Z \mid (d^{(1)} = -1, d^{(2)} = -1)]}}\right)\right].
\end{aligned} \tag{4.2.83}
$$

Similar to the case of the decorrelating detector, the values of the means and variances in (4.2.71) and (4.2.82) can be obtained by numerical simulations and the BER can be calculated by the mixed analysis-simulation technique.

Example — Chaotic sequences generated by the cubic map
Suppose all users use the cubic map defined in (4.2.41) to generate the chaotic sequences. Given a "+1" is sent by the first user, i.e., $d^{(1)} = +1$, and applying (4.2.43) through (4.2.52), the mean and variance of $Z \mid \boldsymbol{d}$ can be shown equal to

$$\mathrm{E}\big[Z \mid (d^{(1)} = +1, d^{(2)})\big] = 8\beta(2\beta - 1)P_s^2 + 4\beta N_0 P_s, \tag{4.2.84}$$

$$\begin{aligned}\mathrm{var}\big[Z \mid (d^{(1)} = +1, d^{(2)})\big] = {} & 32\beta(2\beta - 1)\Big[\mathrm{E}^2\big[(x_k^{(1)})^4\big] \\ & + 4(\beta - 1)\mathrm{E}\big[(x_k^{(1)})^4\big]P_s^2 + (-4\beta + 5)P_s^4\Big] \\ & + 16\beta(2\beta - 1)N_0\Big[3\mathrm{E}\big[(x_k^{(1)})^4\big]P_s + (2\beta - 3)P_s^3\Big] \\ & + 8\beta N_0^2\Big[\mathrm{E}\big[(x_k^{(1)})^4\big] + 2(2\beta - 1)P_s^2\Big] + 4\beta N_0^3 P_s.\end{aligned} \tag{4.2.85}$$

It can be observed that both the mean and variance of $Z \mid \boldsymbol{d}$ do not depend on $d^{(2)}$. In other words,

$$\mathrm{E}\big[Z \mid (d^{(1)} = +1, d^{(2)})\big] = \mathrm{E}\big[Z \mid (d^{(1)} = +1)\big], \tag{4.2.86}$$

$$\mathrm{var}\big[Z \mid (d^{(1)} = +1, d^{(2)})\big] = \mathrm{var}\big[Z \mid (d^{(1)} = +1)\big]. \tag{4.2.87}$$

It is also readily shown that when a "−1" is sent by the first user, the mean and variance are, respectively,

$$\mathrm{E}\big[Z \mid (d^{(1)} = -1, d^{(2)})\big] = \mathrm{E}\big[Z \mid (d^{(1)} = -1)\big] = -\mathrm{E}\big[Z \mid (d^{(1)} = +1)\big] \tag{4.2.88}$$

and

$$\mathrm{var}\big[Z \mid (d^{(1)} = -1, d^{(2)})\big] = \mathrm{var}\big[Z \mid (d^{(1)} = -1)\big] = \mathrm{var}\big[Z \mid (d^{(1)} = +1)\big]. \tag{4.2.89}$$

Substituting (4.2.84) through (4.2.89) into (4.2.83), the analytical BER for the first user is simplified to

$$
\begin{aligned}
\mathrm{BER}^{(1)} &= \mathrm{Prob}\big(Z \leqslant 0 \,\big|\, (d^{(1)} = +1)\big) \\
&= \frac{1}{2}\,\mathrm{erfc}\Bigg(\Bigg[4\beta(2\beta-1) + 4\beta^2\left(\frac{E_b}{N_0}\right)^{-1}\Bigg] \\
&\quad \times \Bigg[16\beta(2\beta-1)\big(\Lambda^2 + 4(\beta-1)\Lambda + (-4\beta+5)\big) \\
&\quad + 16\beta^2(2\beta-1)\left(\frac{E_b}{N_0}\right)^{-1}\big(3\Lambda + (2\beta-3)\big) \\
&\quad + 16\beta^3\left(\frac{E_b}{N_0}\right)^{-2}\big(\Lambda + 2(2\beta-1)\big) + 16\beta^4\left(\frac{E_b}{N_0}\right)^{-3}\Bigg]^{-1/2}\Bigg).
\end{aligned}
\tag{4.2.90}
$$

4.2.2.2.2 N-User System Based on (4.2.8), and (4.2.20) to (4.2.22), the estimated value of d for the jth user can be shown equal to

$$
\begin{aligned}
\tilde{d}^{(j)} &= \sum_{i=1}^{N} H_{j,i}\, y^{(i)} = \sum_{i=1}^{N} H_{j,i}\left(\sum_{u=1}^{N} R_{i,u} d^{(u)} + \eta^{(i)}\right) \\
&= \sum_{u=1}^{N} d^{(u)} \sum_{i=1}^{N} H_{j,i} R_{i,u} + \sum_{i=1}^{N} H_{j,i}\eta^{(i)}.
\end{aligned}
\tag{4.2.91}
$$

From (4.2.21), we also have

$$
\sum_{i=1}^{N} H_{j,i}\left(R_{i,u} + \frac{N_0}{2}\delta_{iu}\right) = \delta_{ju},
\tag{4.2.92}
$$

where δ_{ju} denotes the Kronecker's delta defined as

$$
\delta_{ju} = \begin{cases} 1 & \text{when } j = u, \\ 0 & \text{when } j \neq u. \end{cases}
\tag{4.2.93}
$$

Putting (4.2.92) into (4.2.91), we obtain

$$
\tilde{d}^{(j)} = d^{(j)} - \frac{N_0}{2} \sum_{i=1}^{N} d^{(i)} H_{j,i} + \sum_{i=1}^{N} H_{j,i}\eta^{(i)}.
\tag{4.2.94}
$$

Given a fixed transmitted symbol vector $\boldsymbol{d}$, the mean and variance of $\tilde{d}^{(j)}$ are given, respectively, by (see Appendix 4C)

$$\mathrm{E}\big[\tilde{d}^{(j)} \mid \boldsymbol{d}\big] = \mathrm{E}\left[d^{(j)} - \frac{N_0}{2}\sum_{i=1}^{N} d^{(i)} H_{j,i}\right] = d^{(j)} - \frac{N_0}{2}\sum_{i=1}^{N} d^{(i)}\mathrm{E}[H_{j,i}] \quad (4.2.95)$$

and

$$\begin{aligned}\mathrm{var}\big[\tilde{d}^{(j)} \mid \boldsymbol{d}\big] &= \frac{N_0^2}{4}\,\mathrm{var}\left[\sum_{i=1}^{N} d^{(i)} H_{j,i}\right] + \mathrm{var}\left[\sum_{i=1}^{N} H_{j,i}\eta^{(i)}\right] \\ &\quad + 2\,\mathrm{cov}\left[-\frac{N_0}{2}\sum_{i=1}^{N} d^{(i)} H_{j,i}, \sum_{i=1}^{N} H_{j,i}\eta^{(i)}\right]. \end{aligned} \quad (4.2.96)$$

It is readily shown that the covariance term in (4.2.96) equals zero (see Appendix 4C). Thus, the variance of $\tilde{d}^{(j)} \mid \boldsymbol{d}$ is simplified to

$$\begin{aligned}\mathrm{var}\big[\tilde{d}^{(j)} \mid \boldsymbol{d}\big] &= \frac{N_0^2}{4}\,\mathrm{var}\left[\sum_{i=1}^{N} d^{(i)} H_{j,i}\right] + \mathrm{var}\left[\sum_{i=1}^{N} H_{j,i}\eta^{(i)}\right] \\ &= \frac{N_0^2}{4}\left(\sum_{i=1}^{N}\mathrm{var}[H_{j,i}] + \sum_{i=1}^{N}\sum_{u=1,u\neq i}^{N}\mathrm{cov}\big[d^{(i)}d^{(u)}H_{j,i}H_{j,u}\big]\right) \\ &\quad + \frac{N_0}{2}\mathrm{E}[H_{j,j}] - \frac{N_0^2}{4}\sum_{i=1}^{N}\mathrm{E}\big[H_{j,i}^2\big] \\ &= \frac{N_0}{2}\mathrm{E}[H_{j,j}] - \frac{N_0^2}{4}\sum_{i=1}^{N}\mathrm{E}^2[H_{j,i}] \\ &\quad + \frac{N_0^2}{4}\sum_{i=1}^{N}\sum_{u=1,u\neq i}^{N} d^{(i)}d^{(u)}\,\mathrm{cov}[H_{j,i}H_{j,u}]. \end{aligned} \quad (4.2.97)$$

Assuming that $\tilde{d}^{(j)} \mid \boldsymbol{d}$ is normal, the BER for the jth user is given by

$$\begin{aligned}\mathrm{BER}^{(j)} &= \sum_{\boldsymbol{d}^{(j)}}\Big[\mathrm{Prob}\big(\tilde{d}^{(j)} \leqslant 0 \,\big|\, (d^{(j)} = +1, \boldsymbol{d}^{(j)})\big)\,\mathrm{Prob}\big(d^{(j)} = +1\big)\,\mathrm{Prob}\big(\boldsymbol{d}^{(j)}\big) \\ &\quad + \mathrm{Prob}\big(\tilde{d}^{(j)} > 0 \,\big|\, (d^{(j)} = -1, \boldsymbol{d}^{(j)})\big)\,\mathrm{Prob}\big(d^{(j)} = -1\big)\,\mathrm{Prob}\big(\boldsymbol{d}^{(j)}\big)\Big]\end{aligned}$$

$$= \frac{1}{2^N} \sum_{\boldsymbol{d}^{(j)}} \Big[\text{Prob}\big(\tilde{d}^{(j)} \leqslant 0 \mid (d^{(j)} = +1, \boldsymbol{d}^{(j)})\big)$$

$$+ \text{Prob}\big(\tilde{d}^{(j)} > 0 \mid (d^{(j)} = -1, \boldsymbol{d}^{(j)})\big) \Big], \tag{4.2.98}$$

where the vector $\boldsymbol{d}^{(j)}$ denotes

$$\boldsymbol{d}^{(j)} = \big[d^{(1)}\, d^{(2)} \ldots d^{(j-1)}\, d^{(j+1)} \ldots d^{(N)} \big]. \tag{4.2.99}$$

Similar to the case of the decorrelating detector, the mean and covariance of the terms in (4.2.97) are obtained by simulation. Then they are substituted into (4.2.98) to obtain the BER of the users.

4.2.3 Results and Discussions

In all simulations, we assume that the cubic map defined in (4.2.41) is used by all users and each user uses a different initial condition. With the cubic map, it has been calculated in Appendix 3 that $\mathrm{E}[x_k^2] = 0.5$ and $\mathrm{E}[x_k^4] = 0.375$. The value of Λ in (4.2.52) thus equals

$$\Lambda = \frac{0.375}{0.5^2} = 1.5. \tag{4.2.100}$$

Three types of results are reported for the case of a two-user system. They are

- brute-force (BF) simulation results obtained by simulating the actual system;
- mixed analysis-simulation (MAS) results obtained by simulating the statistical values (mean, variance, covariance, etc.) and substituting the values into the analytical BER equations; and
- analytical solutions found from calculating the statistical values (mean, variance, covariance, etc.) and substituting the values into the analytical BER equations.

When there are more than two users, only the first two types of results are available. For comparison, the BER for a single-user system (interference-free case) is also given as a reference.

4.2.3.1 *Decorrelating Detector (DD)*

4.2.3.1.1 Two Users Figure 4.2 plots the BER values for spreading factors of 100 and 1000 when a decorrelating detector is used. The analytical BER is obtained by substituting (4.2.100) into (4.2.55) with appropriate average bit-energy-to-noise-power-spectral-density ratio (E_b/N_0). Here, we clearly see that the BF simulation

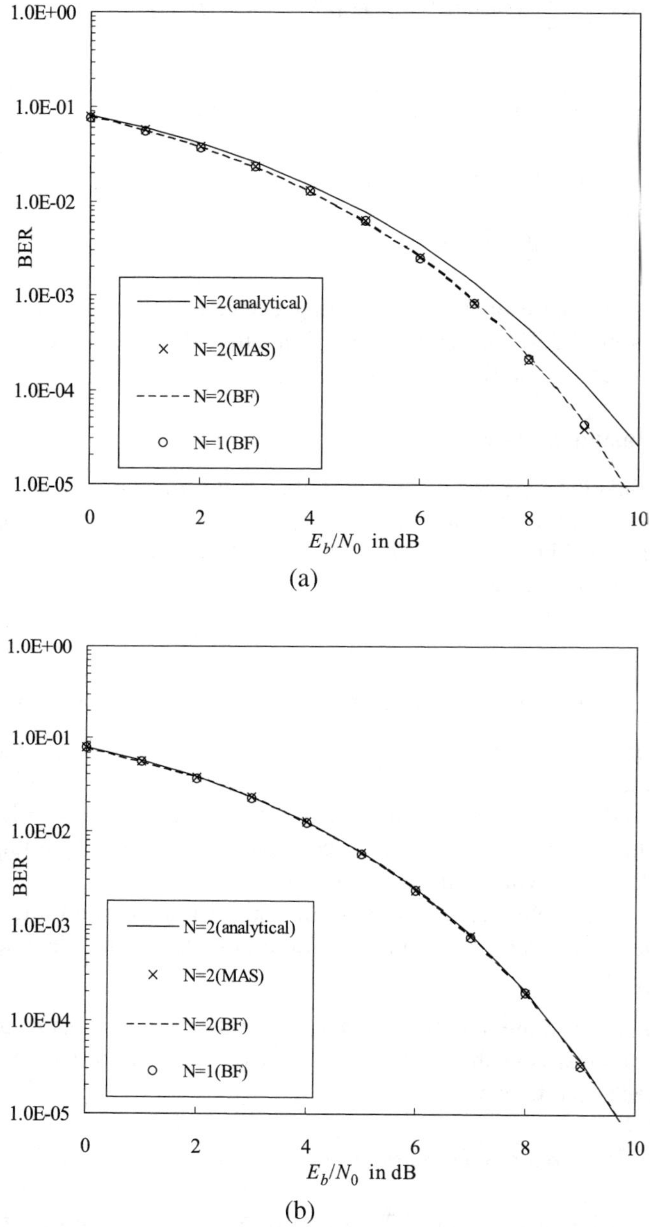

Figure 4.2. Analytical, mixed analysis-simulation (MAS) and brute-force simulation (BF) BERs versus E_b/N_0 for a MA-ACSK communication system using a decorrelating detector. (a) $2\beta = 100$; (b) $2\beta = 1000$.

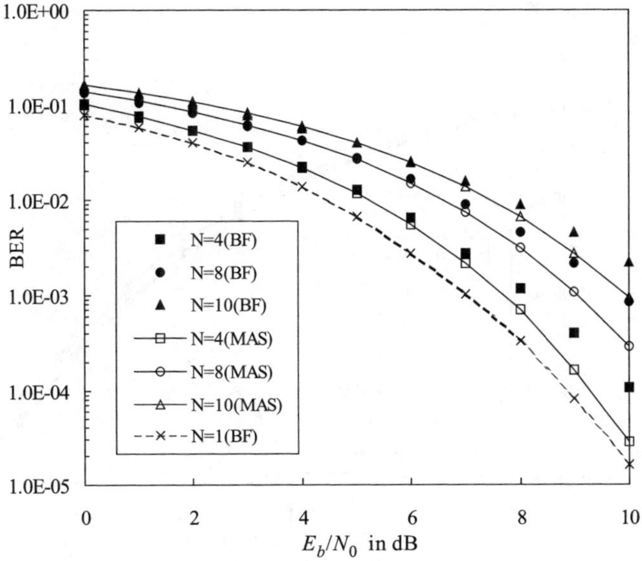

(a)

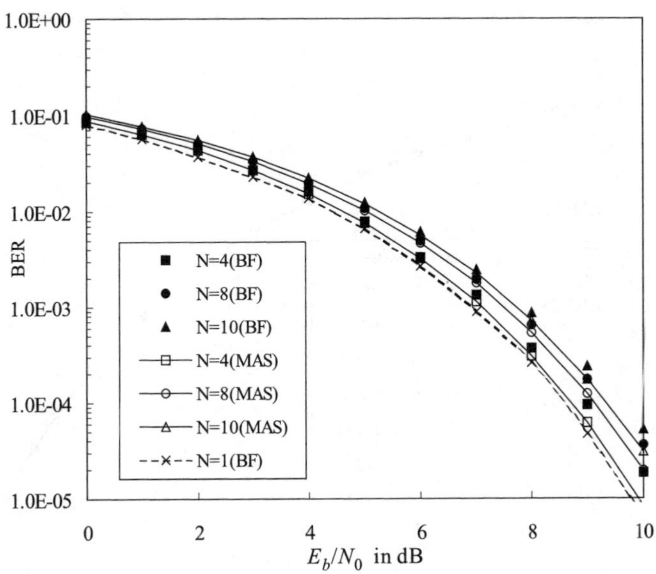

(b)

Figure 4.3. Mixed analysis-simulation (MAS) and brute-force simulation (BF) BERs versus E_b/N_0 for a MA-ACSK communication system using a decorrelating detector. (a) $2\beta = 20$; (b) $2\beta = 50$.

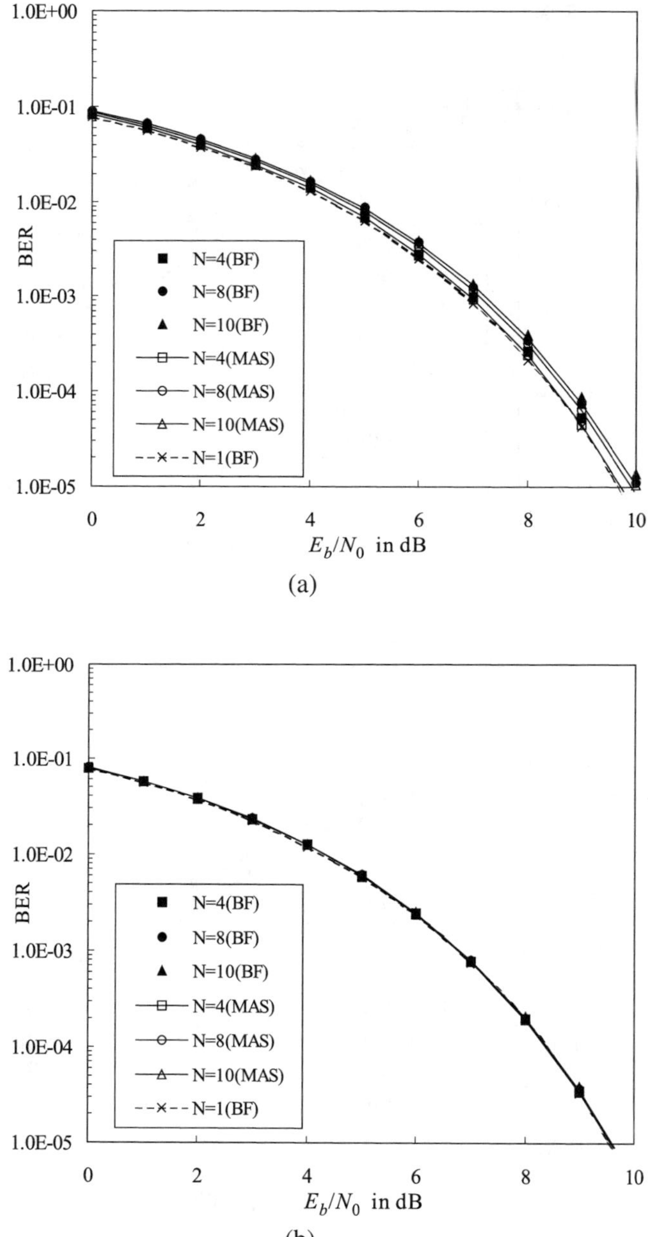

Figure 4.4. Mixed analysis-simulation (MAS) and brute-force simulation (BF) BERs versus E_b/N_0 for a MA-ACSK communication system using a decorrelating detector. (a) $2\beta = 100$; (b) $2\beta = 1000$.

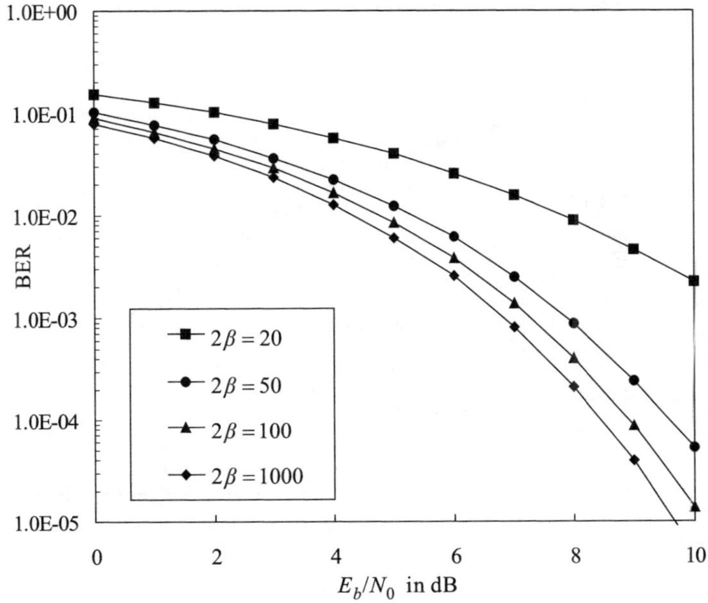

Figure 4.5. Brute-force simulation BER versus E_b/N_0 for a MA-ACSK system using a decorrelating detector. Number of users is 10.

results, the MAS results and the analytical solutions agree well with one another. When $2\beta = 1000$, all curves approach the single-user performance bound.

4.2.3.1.2 N Users Figures 4.3 and 4.4 plot the BF simulation results and the MAS results for 4-user, 8-user and 10-user systems. Spreading factors of 20, 50, 100 and 1000 are used. When the spreading factor increases, the BF simulation results and the MAS results get closer. It is because the assumption of normal distribution made in the MAS scheme becomes more realistic as the spreading factor increases. In Fig. 4.5, the simulated BERs are plotted for $2\beta = 20$, 50, 100 and 1000 when $N = 10$. It is observed that the BER reduces as the spreading factor increases. It is because when the spreading factor increases, the variation in the received bit energy (estimation problem quoted in [Kolumbán *et al.*, 1998]) reduces, causing less degradation to the system performance.

4.2.3.2 *MMSE Detector*

4.2.3.2.1 Two Users Figure 4.6 plots the BER values for spreading factors of 100 and 1000 when an MMSE detector is used. The analytical BER is found

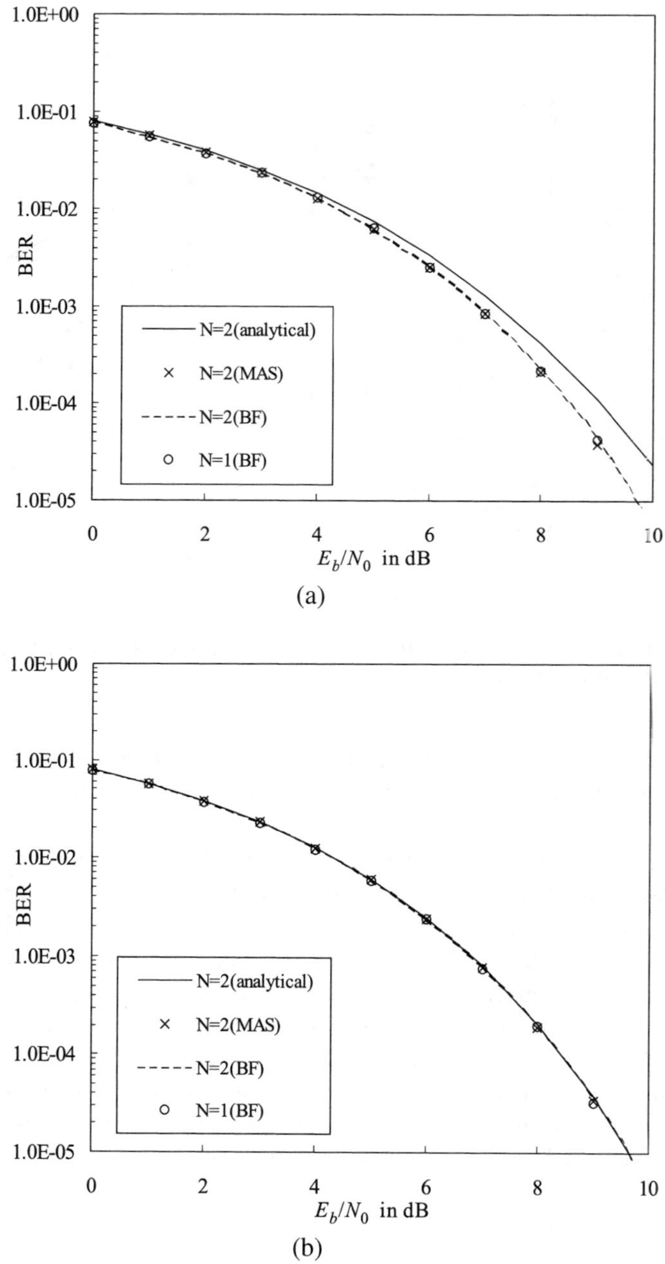

Figure 4.6. Analytical, mixed analysis-simulation (MAS) and brute-force simulation (BF) BERs versus E_b/N_0 for a MA-ACSK communication system using an MMSE detector. (a) $2\beta = 100$; (b) $2\beta = 1000$.

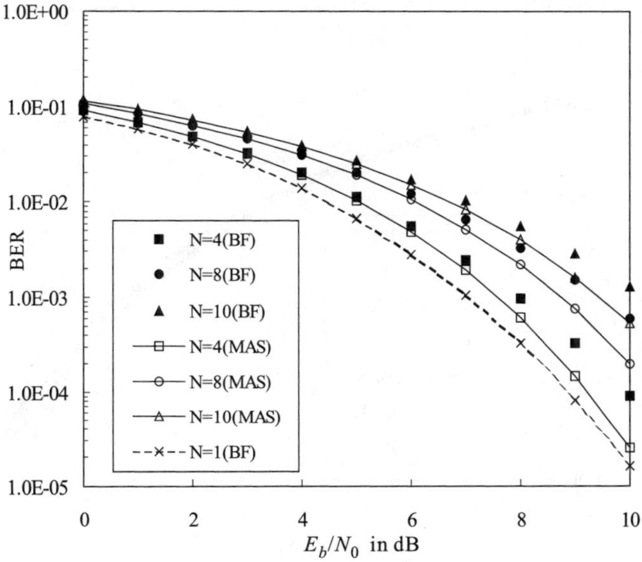

(a)

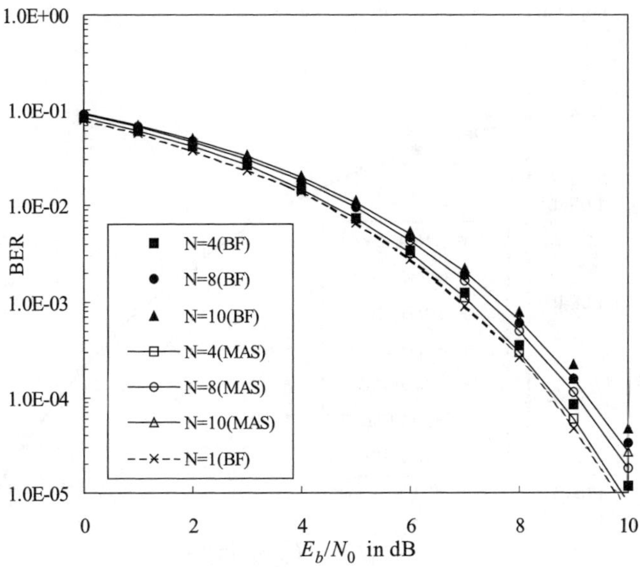

(b)

Figure 4.7. Mixed analysis-simulation (MAS) and brute-force simulation (BF) BERs versus E_b/N_0 for a MA-ACSK communication system using an MMSE detector. (a) $2\beta = 20$; (b) $2\beta = 50$.

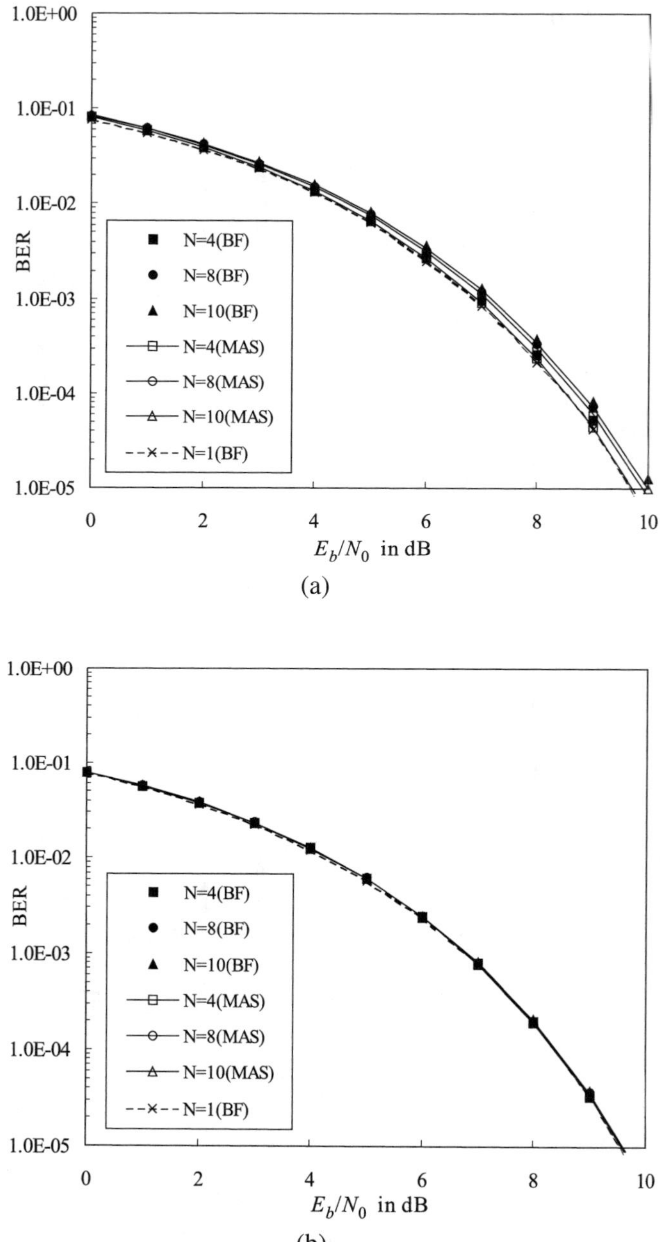

Figure 4.8. Mixed analysis-simulation (MAS) and brute-force simulation (BF) BERs versus E_b/N_0 for a MA-ACSK communication system using an MMSE detector. (a) $2\beta = 100$; (b) $2\beta = 1000$.

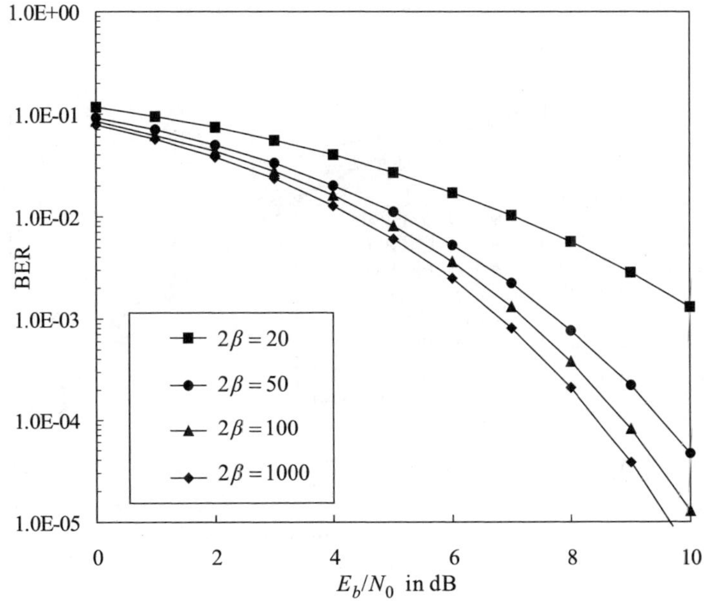

Figure 4.9. Brute-force simulation BER versus E_b/N_0 for a MA-ACSK communication system using an MMSE detector. Number of users is 10.

by substituting (4.2.100) into (4.2.90). It can be observed that the BF simulation results, the MAS results and the analytical solutions agree well with one another.

4.2.3.2.2 N Users In Figs. 4.7 and 4.8, the BER curves for the BF simulation results and the MAS results are shown for 4-user, 8-user and 10-user systems. Spreading factors of 20, 50, 100 and 1000 are used. The BER curves clearly indicate that the two sets of results are consistent. In Fig. 4.9, the simulated BERs are plotted for $2\beta = 20$, 50, 100 and 1000 when $N = 10$. As in the case of the decorrelating detector, the BER reduces as the spreading factor increases. This is also due to the reduction of the variation in the received bit energy as the spreading factor increases.

4.2.3.3 *Comparison of Conventional, Decorrelating and MMSE Detectors* Figure 4.10 plots the simulated BER versus E_b/N_0 for different number of users when the conventional single-user, decorrelating and MMSE detectors are used. From Fig. 4.10(a), it can be observed that the decorrelating and MMSE detectors achieve similar performance. For a low E_b/N_0, say below 5 dB, the decorrelating and MMSE detectors are slightly better than the conventional single-user detector. With a high E_b/N_0, say larger than 5 dB, the BER improvement of the multiuser

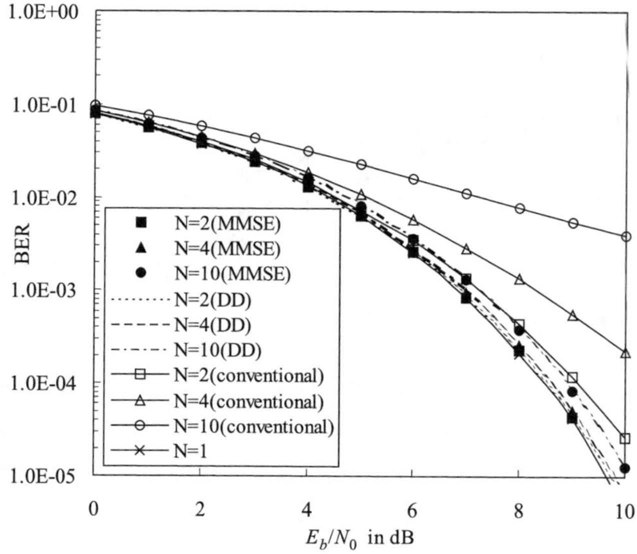

(a)

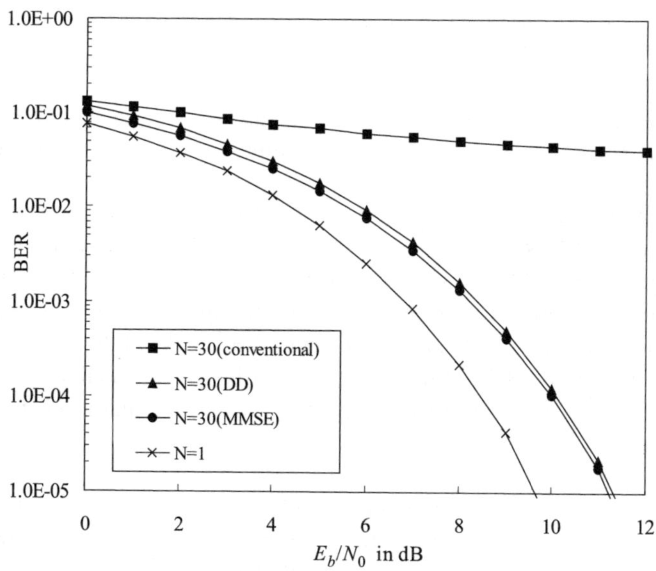

(b)

Figure 4.10. Brute-force simulation BER versus E_b/N_0 for a MA-ACSK communication system using conventional single-user (conventional), decorrelating (DD) and MMSE detectors when $2\beta = 100$. (a) $N = 1, 2, 4$ and 10; (b) $N = 30$.

detectors over the single-user detector is substantial. In particular, for E_b/N_0 values larger than 6 dB, the BERs for the decorrelating and MMSE detectors with 10 users is lower than those for the conventional single-user detector with only 2 users. For $N = 30$, the results are shown in Fig. 4.10(b). Here, we observe that the MMSE detector slightly outperforms the DD. When the E_b/N_0 increases, however, the difference becomes smaller.

4.3 NONLINEAR MULTIUSER DETECTION TECHNIQUES

4.3.1 Parallel Interference Cancellation (PIC) Detectors

A parallel interference cancellation (PIC) detector with multiple stages is shown in Fig. 4.11. At the 0th stage of the PIC detector, the transmitted symbols are first estimated using a linear detector such as a conventional single-user detector, decorrelating detector or MMSE detector. At each of the subsequent stages, the co-channel interference will be estimated and removed from the decision statistics.

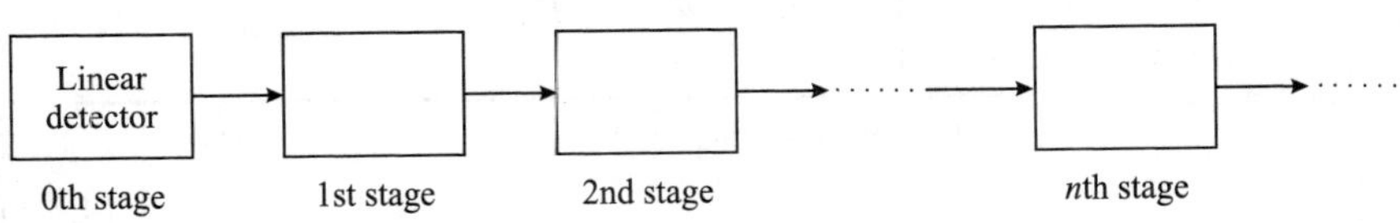

Figure 4.11. A multistage parallel interference cancellation (PIC) detector.

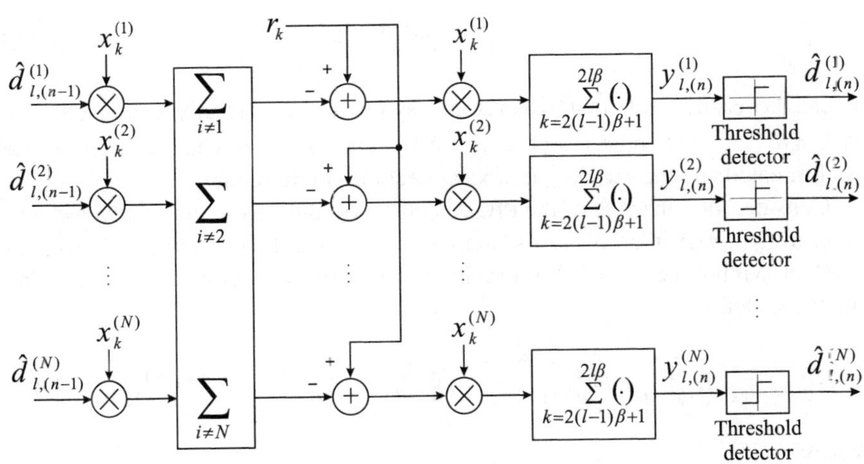

Figure 4.12. The nth stage of a PIC detector.

4.3.2 Performance Analysis of the PIC Detectors

The structure of the nth ($n \geqslant 1$) stage of the PIC detector is shown in Fig. 4.12. The transmitted symbols estimated by the previous stage are first spread by the corresponding chaotic sequences so as to approximate the transmitted signals for all users. Then, the co-channel interference is reconstructed and subtracted from the received signal for each of the users. At the nth stage, the output of the jth correlator is given by

$$\begin{aligned} y_{l,(n)}^{(j)} &= \sum_{k=2(l-1)\beta+1}^{2l\beta} \left(r_k - \sum_{i=1,i\neq j}^{N} \hat{d}_{l,(n-1)}^{(i)} x_k^{(i)} \right) x_k^{(j)} \\ &= \underbrace{d_l^{(j)} \sum_{k=2(l-1)\beta+1}^{2l\beta} \left(x_k^{(j)}\right)^2}_{\text{required signal}} \\ &\quad + \underbrace{\sum_{k=2(l-1)\beta+1}^{2l\beta} \sum_{i=1,i\neq j}^{N} \left(d_l^{(i)} - \hat{d}_{l,(n-1)}^{(i)}\right) x_k^{(i)} x_k^{(j)}}_{\text{inter-user interfering signal}} + \underbrace{\sum_{k=2(l-1)\beta+1}^{2l\beta} \xi_k x_k^{(j)}}_{\text{noise}}, \end{aligned} \tag{4.3.1}$$

and the symbol is decoded according to the sign of $y_{l,(n)}^{(j)}$, i.e.,

$$\hat{d}_{l,(n)}^{(j)} = \mathrm{sgn}\left[y_{l,(n)}^{(j)}\right]. \tag{4.3.2}$$

It can be easily seen that when some symbols are correctly estimated, i.e., $d_l^{(i)} = \hat{d}_{l,(n-1)}^{(i)}$ for some $i \in \{1, 2, \ldots, N\}$, some of the co-channel interference is eliminated and the estimation process becomes more reliable.

Consider the nth stage of the PIC detector. Without loss of generality, we consider the jth user in an N-user system and we derive the probability of error for the first symbol, i.e., $l = 1$. For brevity, we omit the subscript l in the following analysis. Define

$$\boldsymbol{D}_{(n-1)}^{(j)} = \left[D_{(n-1)}^{(1)} \quad D_{(n-1)}^{(2)} \quad \cdots \quad D_{(n-1)}^{(j-1)} \quad D_{(n-1)}^{(j+1)} \quad \cdots \quad D_{(n-1)}^{(N)} \right], \tag{4.3.3}$$

where

$$D_{(n-1)}^{(i)} = d^{(i)} - \hat{d}_{(n-1)}^{(i)} \tag{4.3.4}$$

denotes the difference between the transmitted symbol of the ith user and the estimated symbol at the $(n-1)$th stage of the PIC detector. The input to the jth threshold detector, i.e., the output of the j correlator in (4.3.1), is now rewritten as

$$y_{(n)}^{(j)} = d^{(j)} \sum_{k=1}^{2\beta} (x_k^{(j)})^2 + \sum_{i=1, i\neq j}^{N} D_{(n-1)}^{(i)} \sum_{k=1}^{2\beta} x_k^{(i)} x_k^{(j)} + \sum_{k=1}^{2\beta} \xi_k x_k^{(j)}. \tag{4.3.5}$$

Note that if the estimated symbol for the ith user is correct, i.e., $d^{(i)} = \hat{d}_{(n-1)}^{(i)}$, $D_{(n-1)}^{(i)}$ becomes zero and the interference from the ith user is eliminated. However, when $d^{(i)} \neq \hat{d}_{(n-1)}^{(i)}$, $D_{(n-1)}^{(i)}$ equals ± 2. Under such a condition, the interference may contribute positively or negatively to the required signal of the jth user, depending on the signs of $d^{(i)}$, $d^{(j)}$ and $\sum_{k=1}^{2\beta} x_k^{(i)} x_k^{(j)}$.

Assume that the transmitted symbol is "+1" for the jth user. For a given $\boldsymbol{D}_{(n-1)}^{(j)}$, the mean and variance of $y_{(n)}^{(j)}$ can be shown equal to, respectively,

$$\mathrm{E}\left[y_{(n)}^{(j)} \mid \left(d^{(j)} = +1, \boldsymbol{D}_{(n-1)}^{(j)}\right)\right] = 2\beta \mathrm{E}\left[\left(x_k^{(j)}\right)^2\right] \tag{4.3.6}$$

and

$$\begin{aligned}
&\mathrm{var}\left[y_{(n)}^{(j)} \mid \left(d^{(j)} = +1, \boldsymbol{D}_{(n-1)}^{(j)}\right)\right] \\
&= 2\beta\, \mathrm{var}\left[\left(x_k^{(j)}\right)^2\right] + \sum_{k=1}^{2\beta} \sum_{m=1, m\neq k}^{2\beta} \mathrm{cov}\left[\left(x_k^{(j)}\right)^2, \left(x_m^{(j)}\right)^2\right] \\
&\quad + 2\beta \mathrm{E}\left[\left(x_k^{(j)}\right)^2\right] \sum_{i=1, i\neq j}^{N} \left(D_{(n-1)}^{(i)}\right)^2 \mathrm{E}\left[\left(x_k^{(i)}\right)^2\right] \\
&\quad + \sum_{i=1, i\neq j}^{N} \sum_{k=1}^{2\beta} \sum_{m=1, m\neq k}^{2\beta} \mathrm{E}\left[x_k^{(i)} x_m^{(i)}\right] \mathrm{E}\left[x_k^{(j)} x_m^{(j)}\right] + \beta N_0 \mathrm{E}\left[\left(x_k^{(j)}\right)^2\right].
\end{aligned} \tag{4.3.7}$$

In the derivations of (4.3.6) and (4.3.7), we have assumed that the chaotic sequences are independent of one another because they are derived from different generators. Also, the mean value of each chaotic sequence is zero in order to avoid transmitting any non-information-bearing dc components.

Using the aforementioned procedures, we can derive $\mathrm{E}[y_{(n)}^{(j)} \mid (d^{(j)} = -1, \boldsymbol{D}_{(n-1)}^{(j)})]$ and $\mathrm{var}[y_{(n)}^{(j)} \mid (d^{(j)} = -1, \boldsymbol{D}_{(n-1)}^{(j)})]$. Assuming that both $y_{(n)}^{(j)} \mid (d^{(j)} =$

$+1, \boldsymbol{D}_{(n-1)}^{(j)})$ and $y_{(n)}^{(j)} \mid (d^{(j)} = -1, \boldsymbol{D}_{(n-1)}^{(j)})$ are normal when 2β is large, it can be readily shown that the conditional error probabilities of both cases are the same, i.e.,

$$\begin{aligned}
&\mathrm{Prob}\big(y_{(n)}^{(j)} \leqslant 0 \,\big|\, (d^{(j)} = +1, \boldsymbol{D}_{(n-1)}^{(j)})\big) \\
&\quad = \mathrm{Prob}\big(y_{(n)}^{(j)} > 0 \,\big|\, (d^{(j)} = -1, \boldsymbol{D}_{(n-1)}^{(j)})\big) \\
&\quad = \frac{1}{2}\,\mathrm{erfc}\left(\frac{\mathrm{E}[y_{(n)}^{(j)} \mid (d^{(j)} = +1, \boldsymbol{D}_{(n-1)}^{(j)})]}{\sqrt{2\,\mathrm{var}[y_{(n)}^{(j)} \mid (d^{(j)} = +1, \boldsymbol{D}_{(n-1)}^{(j)})]}}\right).
\end{aligned} \tag{4.3.8}$$

Therefore, the bit error rate for the j user at the nth stage of the PIC detector can be computed from

$$\begin{aligned}
\mathrm{BER}_{(n)}^{(j)} &= \sum_{\boldsymbol{D}_{(n-1)}^{(j)}} \Big[\mathrm{Prob}\big(y_{(n)}^{(j)} \leqslant 0 \,\big|\, (d^{(j)} = +1, \boldsymbol{D}_{(n-1)}^{(j)})\big)\,\mathrm{Prob}\big(d^{(j)} = +1\big) \\
&\qquad + \mathrm{Prob}\big(y_{(n)}^{(j)} > 0 \,\big|\, (d^{(j)} = -1, \boldsymbol{D}_{(n-1)}^{(j)})\big)\,\mathrm{Prob}\big(d^{(j)} = -1\big)\Big] \\
&\qquad \times \mathrm{Prob}\big(\boldsymbol{D}_{(n-1)}^{(j)}\big) \\
&= \sum_{\boldsymbol{D}_{(n-1)}^{(j)}} \frac{1}{2}\Big[\mathrm{Prob}\big(y_{(n)}^{(j)} \leqslant 0 \,\big|\, (d^{(j)} = +1, \boldsymbol{D}_{(n-1)}^{(j)})\big) \\
&\qquad + \mathrm{Prob}\big(y_{(n)}^{(j)} > 0 \,\big|\, (d^{(j)} = -1, \boldsymbol{D}_{(n-1)}^{(j)})\big)\Big]\,\mathrm{Prob}\big(\boldsymbol{D}_{(n-1)}^{(j)}\big) \\
&= \sum_{\boldsymbol{D}_{(n-1)}^{(j)}} \mathrm{Prob}\big(y_{(n)}^{(j)} \leqslant 0 \,\big|\, (d^{(j)} = +1, \boldsymbol{D}_{(n-1)}^{(j)})\big)\,\mathrm{Prob}\big(\boldsymbol{D}_{(n-1)}^{(j)}\big),
\end{aligned} \tag{4.3.9}$$

and $\mathrm{Prob}(\boldsymbol{D}_{(n-1)}^{(j)})$ is calculated using

$$\mathrm{Prob}\big(\boldsymbol{D}_{(n-1)}^{(j)}\big) = \sum_{\boldsymbol{d} \in \{-1,+1\}^N} \mathrm{Prob}\big(\boldsymbol{D}_{(n-1)}^{(j)} \,\big|\, \boldsymbol{d}\big)\,\mathrm{Prob}(\boldsymbol{d}) \tag{4.3.10}$$

with

$$\boldsymbol{d} = \big[d^{(1)}\ d^{(2)}\ \cdots\ d^{(N)}\big]^{\mathrm{T}} \tag{4.3.11}$$

denoting the transmitted symbol vector. Note that in (4.3.9) and (4.3.10), there are a lot of terms to compute because a large number of possible combinations have to be considered. Instead of dealing with such a large number of terms, we make use of a simpler approach to calculate the bit error rate.

Suppose under the given condition $\boldsymbol{D}_{(n-1)}^{(j)}$, at the $(n-1)$th stage, there are $Ne_{(n-1)}^{(j)} \in \{0, 1, \ldots, N-1\}$ symbols (excluding the one for the jth user) that have been incorrectly estimated, i.e., $d^{(i)} \neq \hat{d}_{(n-1)}^{(i)}$, $i \neq j$. For such cases, $(D_{(n-1)}^{(i)})^2 = 4$ (because $D_{(n-1)}^{(i)} = \pm 2$) and hence (4.3.7) can be rewritten as

$$
\begin{aligned}
&\operatorname{var}\big[y_{(n)}^{(j)} \mid \big(d^{(j)} = +1, \boldsymbol{D}_{(n-1)}^{(j)}\big)\big] \\
&= \operatorname{var}\big[y_{(n)}^{(j)} \mid \big(d^{(j)} = +1, Ne_{(n-1)}^{(j)}\big)\big] \\
&= 2\beta \operatorname{var}\big[\big(x_k^{(j)}\big)^2\big] + \sum_{k=1}^{2\beta} \sum_{m=1, m\neq k}^{2\beta} \operatorname{cov}\big[\big(x_k^{(j)}\big)^2, \big(x_m^{(j)}\big)^2\big] \\
&\quad + 8\beta Ne_{(n-1)}^{(j)} \mathrm{E}^2\big[\big(x_k^{(j)}\big)^2\big] \\
&\quad + \sum_{i=1, i\neq j}^{N} \sum_{k=1}^{2\beta} \sum_{m=1, m\neq k}^{2\beta} \mathrm{E}\big[x_k^{(i)} x_m^{(i)}\big] \mathrm{E}\big[x_k^{(j)} x_m^{(j)}\big] \\
&\quad + \beta N_0 \mathrm{E}\big[\big(x_k^{(j)}\big)^2\big].
\end{aligned}
\tag{4.3.12}
$$

With the introduction of $Ne_{(n-1)}^{(j)}$, the bit error probability in (4.3.9) can be expressed as

$$
\begin{aligned}
\mathrm{BER}_{(n)}^{(j)} &= \sum_{u=0}^{N-1} \operatorname{Prob}\big(y_{(n)}^{(j)} \leqslant 0 \mid \big(d^{(j)} = +1, Ne_{(n-1)}^{(j)} = u\big)\big) \operatorname{Prob}\big(Ne_{(n-1)}^{(j)} = u\big) \\
&= \sum_{u=0}^{N-1} \frac{1}{2} \operatorname{erfc}\left(\frac{\mathrm{E}[y_{(n)}^{(j)} \mid (d^{(j)} = +1, Ne_{(n-1)}^{(j)} = u)]}{\sqrt{2 \operatorname{var}[y_{(n)}^{(j)} \mid (d^{(j)} = +1, Ne_{(n-1)}^{(j)} = u)]}} \right) \\
&\quad \times \operatorname{Prob}\big(Ne_{(n-1)}^{(j)} = u\big),
\end{aligned}
\tag{4.3.13}
$$

where

$$
\begin{aligned}
\mathrm{E}\big[y_{(n)}^{(j)} \mid \big(d^{(j)} = +1, Ne_{(n-1)}^{(j)} = u\big)\big] &= \mathrm{E}\big[y_{(n)}^{(j)} \mid \big(d^{(j)} = +1, \boldsymbol{D}_{(n-1)}^{(j)}\big)\big] \\
&= 2\beta \mathrm{E}\big[\big(x_k^{(j)}\big)^2\big].
\end{aligned}
\tag{4.3.14}
$$

By computing the probability of occurrence of $Ne_{(n-1)}^{(j)} \in \{0, 1, \ldots, N-1\}$ and using (4.3.12), (4.3.13) and (4.3.14), the bit error rate at the nth stage of the PIC detector, $\mathrm{BER}_{(n)}^{(j)}$, can be found. In addition, $\operatorname{Prob}(Ne_{(n-1)}^{(j)} = u)$ is only significant when u is very small. Therefore, not more than a few values of u need to be considered.

4.3.3 Results and Discussions

Assume that all users use the cubic map defined as in (4.2.41) to generate the chaotic sequences and each user uses a different initial condition. Based on the statistical properties of the map (see Appendix 3A for details), the mean and variance derived in (4.3.14) and (4.3.12) can be simplified to, respectively,

$$\mathrm{E}\big[y_{(n)}^{(j)} \mid \big(d^{(j)} = +1, N_{e(n-1)}^{(j)} = u\big)\big] = 2\beta P_s \tag{4.3.15}$$

and

$$\mathrm{var}\big[y_{(n)}^{(j)} \mid \big(d^{(j)} = +1, N_{e(n-1)}^{(j)} = u\big)\big] = 2\beta\,\mathrm{var}\big[\big(x_k^{(j)}\big)^2\big] + 8\beta u P_s^2 + \beta N_0 P_s, \tag{4.3.16}$$

where P_s is the average transmission power defined as in (4.2.51). For brevity, we define

$$\Psi^{(j)} = \frac{\mathrm{var}[(x_k^{(j)})^2]}{P_s^2} = \frac{\mathrm{var}[(x_k^{(j)})^2]}{\mathrm{E}^2[(x_k^{(j)})^2]}. \tag{4.3.17}$$

In consequence, the bit error probability for the j user at the nth stage of the PIC detector can now be computed using

$$\mathrm{BER}_{(n)}^{(j)} = \frac{1}{2}\sum_{u=0}^{N-1} \mathrm{erfc}\left(\left[\frac{\Psi^{(j)}}{\beta} + \frac{4u}{\beta} + \left(\frac{E_b}{N_0}\right)^{-1}\right]^{-1/2}\right) \mathrm{Prob}\big(N_{e(n-1)}^{(j)} = u\big), \tag{4.3.18}$$

where E_b denotes the average bit energy defined as in (4.2.56).

Three linear detectors, namely conventional single-user detector, decorrelating detector (DD) and minimum mean-square-error (MMSE) detector, have been applied as the 1st stage (Stage 0) of the PIC detector. These combined detectors are termed as, respectively, conventional/PIC, DD/PIC and MMSE/PIC detectors. We assume that the cubic map defined as in (4.2.41) is used by all users, and each uses a different initial condition. Also, the spreading factor (2β) is 100 and the number of users (N) is 10. We further assume that $\mathrm{Prob}(N_{e(n-1)}^{(j)} = u)$ is small when u is large. Therefore, in our analysis, we assume that $\mathrm{Prob}(N_{e(n-1)}^{(j)} = u)$ is negligible under the following circumstances and the corresponding terms are neglected in the computation of (4.3.18):

- $u \geqslant 3$ for the conventional/PIC detector at the 1st stage ($n = 1$).
- $u \geqslant 2$ for the conventional/PIC detector at the 2nd stage and beyond.

- $u \geqslant 2$ for the DD/PIC detector at the 1st stage and beyond.
- $u \geqslant 2$ for the MMSE/PIC detector at the 1st stage and beyond.

Details on the calculations of $\text{Prob}(N_{e_{(n-1)}^{(j)}} = u)$ can be found in Appendix 4D.

Figures 4.13, 4.14 and 4.15, respectively, plot the simulated BERs for the conventional/PIC, DD/PIC and MMSE/PIC detectors, at various stages. The BER for a single-user system, which is equivalent to an interference-free system, is also given as a reference. For all the detectors under study, the BERs at Stage 1 are lower than those at Stage 0. For the conventional/PIC detector, the BER at Stage 2 is further reduced compared to that at Stage 1. The BER results for the conventional/PIC detector with 3 stages, the DD/PIC and the MMSE/PIC detectors with 2 stages, are close to that of the single-user system. When the number of stages is further increased, no improvement is observed. Overall, the results indicate that the parallel interference cancellation technique can further enhance the performance of linear multi-user detectors. With low BERs at Stage 0 and at subsequent stages, it is well justified that terms corresponding to large value of u can be ignored in the computation of (4.3.18).

Figures from 4.16 to 4.18 plot the approximate BERs (using (4.3.18)) and the brute-force (BF) simulated BERs when the PIC detectors are applied. It can be observed that the approximate and the simulated results agree with each other for all the PIC detectors under study. Finally, we compared the simulated BERs for

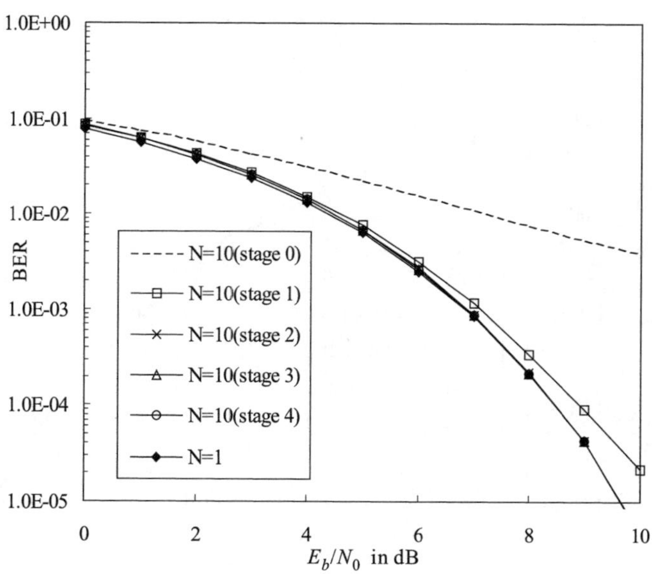

Figure 4.13. Simulated BER versus E_b/N_0 for the conventional/PIC detector at various stages. $2\beta = 100$ and $N = 10$.

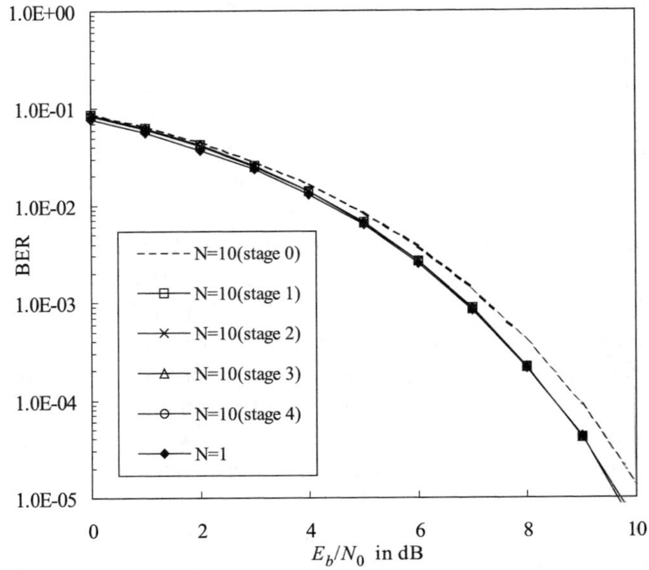

Figure 4.14. Simulated BER versus E_b/N_0 for the DD/PIC detector at various stages. $2\beta = 100$ and $N = 10$.

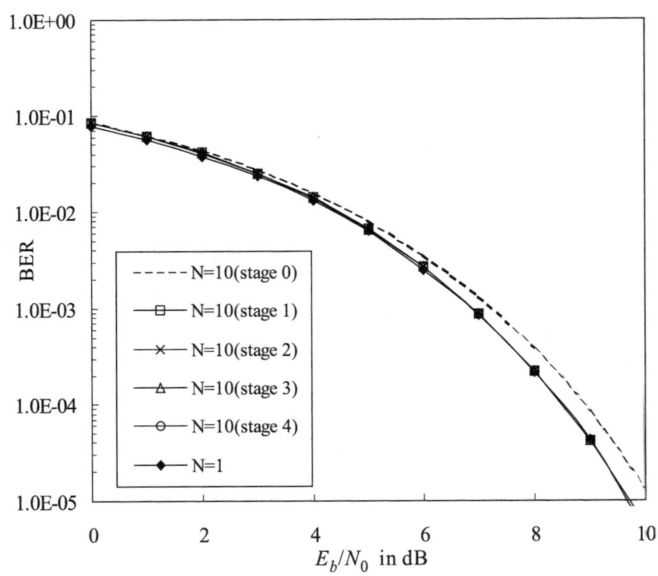

Figure 4.15. Simulated BER versus E_b/N_0 for the MMSE/PIC detector at various stages. $2\beta = 100$ and $N = 10$.

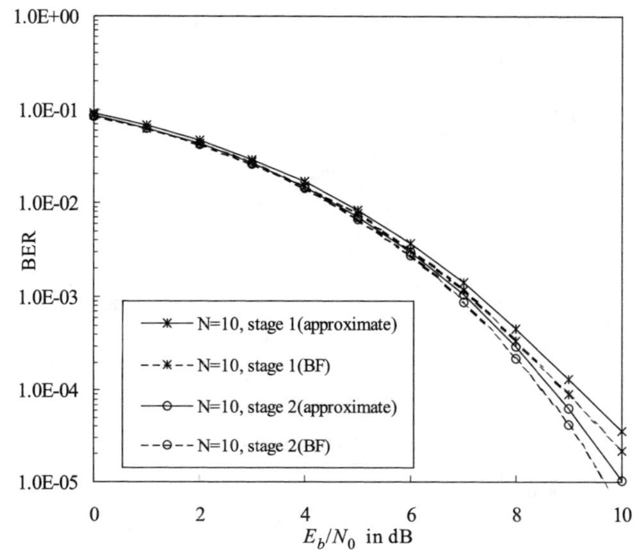

Figure 4.16. Approximate and brute-force (BF) simulated BERs versus E_b/N_0 for the conventional/PIC detector. $2\beta = 100$ and $N = 10$.

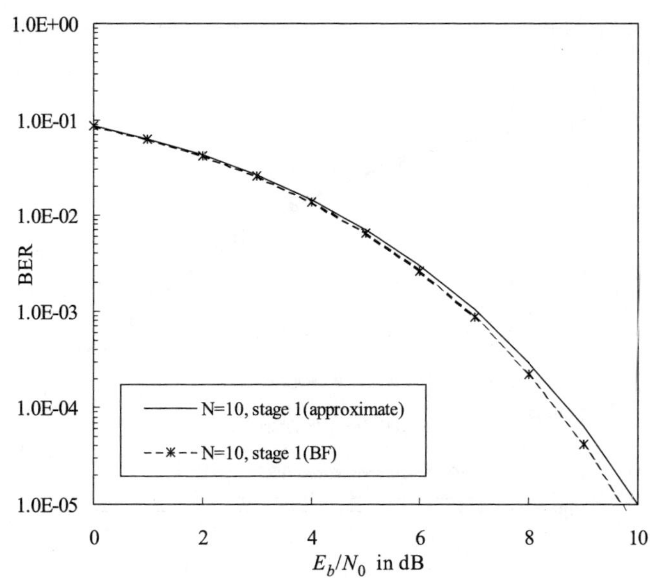

Figure 4.17. Approximate and brute-force (BF) simulated BERs versus E_b/N_0 for the DD/PIC detector. $2\beta = 100$ and $N = 10$.

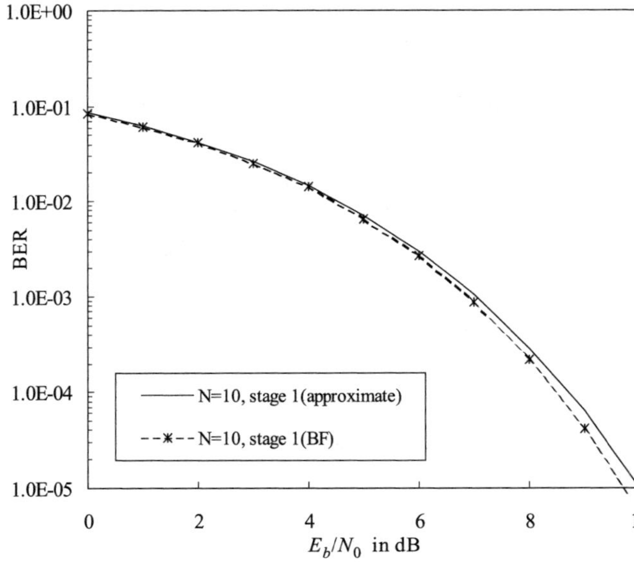

Figure 4.18. Approximate and brute-force (BF) simulated BERs versus E_b/N_0 for the MMSE/PIC detector. $2\beta = 100$ and $N = 10$.

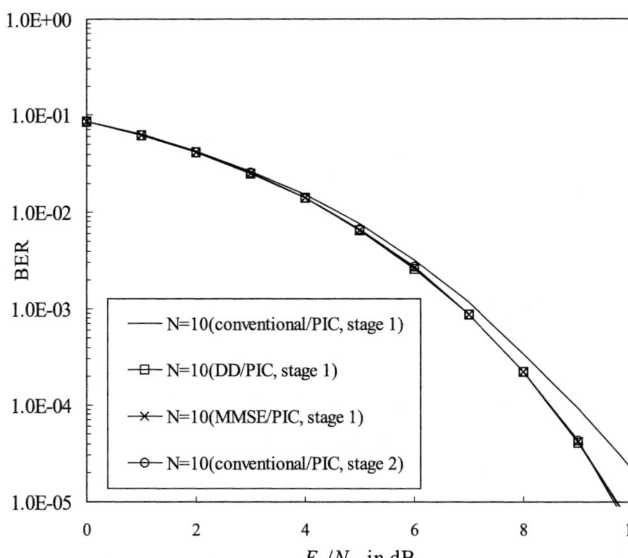

Figure 4.19. Comparison of the conventional/PIC, DD/PIC and MMSE/PIC detectors. $2\beta = 100$ and $N = 10$.

the conventional/PIC, DD/PIC and MMSE/PIC detectors. The results shown in Fig. 4.19 indicate that the BERs for the DD/PIC and MMSE/PIC detectors are almost the same. At Stage 1 of the PIC detector, the performances of the DD/PIC and MMSE/PIC detectors are better than that of the conventional/PIC detector. With an additional stage (Stage 2), the conventional/PIC can achieve the same performance as the DD/PIC and MMSE/PIC detectors.

4.4 SUMMARY

In this chapter, two linear multiuser detectors (decorrelating detector and minimum mean-square-error detector) and three nonlinear multiuser detectors (conventional/PIC, DD/PIC and MMSE/PIC detectors), have been applied to a multiple-access antipodal CSK communication system. Techniques for calculating the bit error rates (BERs) of the linear multiuser detectors and the approximate BERs of the nonlinear multiuser detectors have been presented. Results show that all multiuser detectors can substantially improve the BER performance of the multiple-access communication system. On the other hand, the complexity of the multiuser detectors increases as the number of users or the number of stages increases. As a result, there is a trade off between the system complexity and the performance.

In the previous chapter and in this chapter, we focus our study in coherent chaos-based communication systems, in which the chaotic carriers are assumed to be available at the receiving end. Although robust chaos synchronization is still not realizable, the results can provide information on the performance bounds for chaos-based communication systems under an additive white Gaussian noise environment. In the next few chapters, we will investigate noncoherent communication systems based on chaos. They are more practical because the receiver is not required to reproduce a chaotic signal in exact synchronization with that at the transmitting side. In most cases, the receiver only needs to determine the transmitted symbols based only from the incoming signals.

APPENDIX 4A. DERIVATION OF MEANS AND VARIANCES RELEVANT TO THE ANALYSES OF THE DECORRELATING DETECTOR AND MMSE DETECTOR

In this appendix, we derive the means and variances of some terms which are required in the analyses of the DD and MMSE detectors. All symbols are defined as in Section 4.2.

1. Mean value of the autocorrelation $R_{j,j}$ of the chaotic signal $\{x_k^{(j)}\}$:

$$E[R_{j,j}] = E\left[\sum_{k=1}^{2\beta}(x_k^{(j)})^2\right] = \sum_{k=1}^{2\beta} E\left[(x_k^{(j)})^2\right] = 2\beta E\left[(x_k^{(j)})^2\right]. \tag{4.A.1}$$

2. Mean value of $R_{j,j}^2$:

$$\begin{aligned} E\left[R_{j,j}^2\right] &= E\left[\left(\sum_{k=1}^{2\beta}(x_k^{(j)})^2\right)^2\right] \\ &= 2\beta E\left[(x_k^{(j)})^4\right] + \sum_{k=1}^{2\beta}\sum_{m=1,m\neq k}^{2\beta} E\left[(x_k^{(j)})^2(x_m^{(j)})^2\right]. \end{aligned} \tag{4.A.2}$$

If all users use the cubic map defined as in (4.2.41), the mean term is simplified to

$$E\left[R_{j,j}^2\right] = 2\beta E\left[(x_k^{(j)})^4\right] + 2\beta(2\beta-1)E^2\left[(x_k^{(j)})^2\right]. \tag{4.A.3}$$

3. Variance value of $R_{j,j}$:

$$\begin{aligned} \mathrm{var}[R_{j,j}] &= E\left[R_{j,j}^2\right] - E^2[R_{j,j}] \\ &= 2\beta E\left[(x_k^{(j)})^4\right] + \sum_{k=1}^{2\beta}\sum_{m=1,m\neq k}^{2\beta} E\left[(x_k^{(j)})^2(x_m^{(j)})^2\right] - 4\beta^2E^2\left[(x_k^{(j)})^2\right]. \end{aligned} \tag{4.A.4}$$

If all users use the cubic map, the variance term becomes

$$\mathrm{var}[R_{j,j}] = 2\beta E\left[(x_k^{(j)})^4\right] - 2\beta E^2\left[(x_k^{(j)})^2\right]. \tag{4.A.5}$$

4. Mean value of the cross-correlation $R_{j,i}$ between the chaotic signals $\{x_k^{(j)}\}$ and $\{x_k^{(i)}\}$ where $i \neq j$:

$$E[R_{j,i}] = E\left[\sum_{k=1}^{2\beta} x_k^{(j)}x_k^{(i)}\right] = \sum_{k=1}^{2\beta} E\left[(x_k^{(j)})\right]E\left[(x_k^{(i)})\right] = 0. \tag{4.A.6}$$

5. Mean value of $R^2_{j,i}$:

$$\mathrm{E}\left[R^2_{j,i}\right] = \mathrm{E}\left[\left(\sum_{k=1}^{2\beta} x_k^{(j)} x_k^{(i)}\right)^2\right]$$
$$= 2\beta \mathrm{E}\left[\left(x_k^{(j)}\right)^2\right]\mathrm{E}\left[\left(x_k^{(i)}\right)^2\right] + \sum_{k=1}^{2\beta} \sum_{m=1, m\neq k}^{2\beta} \mathrm{E}\left[x_k^{(j)} x_m^{(j)}\right]\mathrm{E}\left[x_k^{(i)} x_m^{(i)}\right]. \tag{4.A.7}$$

If all users use the cubic map, the mean term is simplified to

$$\mathrm{E}\left[R^2_{j,i}\right] = 2\beta \mathrm{E}^2\left[\left(x_k^{(j)}\right)^2\right]. \tag{4.A.8}$$

6. Variance value of $R_{j,i}$:

$$\mathrm{var}[R_{j,i}] = \mathrm{E}\left[R^2_{j,i}\right]. \tag{4.A.9}$$

7. Mean value of $R^3_{j,i}$:

$$\mathrm{E}\left[R^3_{j,i}\right] = \mathrm{E}\left[\left(\sum_{k=1}^{2\beta} x_k^{(j)} x_k^{(i)}\right)^3\right]$$
$$= \mathrm{E}\left[\sum_{k=1}^{2\beta} x_k^{(j)} x_k^{(i)} \sum_{m=1}^{2\beta} x_m^{(j)} x_m^{(i)} \sum_{u=1}^{2\beta} x_u^{(j)} x_u^{(i)}\right]$$
$$= 3\sum_{k=1}^{2\beta} \sum_{m=1, m\neq k}^{2\beta} \mathrm{E}\left[\left(x_k^{(j)}\right)^2 x_m^{(j)}\right]\mathrm{E}\left[\left(x_k^{(i)}\right)^2 x_m^{(i)}\right]$$
$$+ \sum_{k=1}^{2\beta} \sum_{m=1, m\neq k}^{2\beta} \sum_{\substack{u=1, u\neq k\\ u\neq m}}^{2\beta} \mathrm{E}\left[x_k^{(j)} x_m^{(j)} x_u^{(j)}\right]\mathrm{E}\left[x_k^{(i)} x_m^{(i)} x_u^{(i)}\right]. \tag{4.A.10}$$

If all users use the cubic map, the mean term is equal to zero, i.e.,

$$\mathrm{E}\left[R^3_{j,i}\right] = 0. \tag{4.A.11}$$

8. Mean value of $R_{j,j}R_{j,i}$:

$$\mathrm{E}[R_{j,j}R_{j,i}] = \mathrm{E}\left[\sum_{k=1}^{2\beta}(x_k^{(j)})^2 \sum_{m=1}^{2\beta} x_m^{(j)} x_m^{(i)}\right] = 0. \tag{4.A.12}$$

9. Mean value of $R_{j,j}R_{i,i}$:

$$\mathrm{E}[R_{j,j}R_{i,i}] = \mathrm{E}[R_{j,j}]\mathrm{E}[R_{i,i}] = 4\beta^2 \mathrm{E}\big[(x_k^{(j)})^2\big]\mathrm{E}\big[(x_k^{(i)})^2\big]. \tag{4.A.13}$$

If all users use the cubic map, the mean term is simplified to

$$\mathrm{E}[R_{j,j}R_{i,i}] = 4\beta^2 \mathrm{E}^2\big[(x_k^{(j)})^2\big]. \tag{4.A.14}$$

10. Mean value of $R_{j,j}R_{i,i}^2$:

$$\begin{aligned}\mathrm{E}\big[R_{j,j}R_{i,i}^2\big] &= \mathrm{E}[R_{j,j}]\mathrm{E}\big[R_{i,i}^2\big] \\ &= 2\beta \mathrm{E}\big[(x_k^{(j)})^2\big]\left(2\beta \mathrm{E}[(x_k^{(i)})^4] + \sum_{k=1}^{2\beta}\sum_{m=1,m\neq k}^{2\beta} \mathrm{E}\big[(x_k^{(i)})^2 (x_m^{(i)})^2\big]\right).\end{aligned} \tag{4.A.15}$$

If all users use the cubic map, the mean term becomes

$$\mathrm{E}\big[R_{j,j}R_{i,i}^2\big] = 4\beta^2 \mathrm{E}\big[(x_k^{(j)})^4\big]\mathrm{E}\big[(x_k^{(j)})^2\big] + 4\beta^2(2\beta-1)\mathrm{E}^3\big[(x_k^{(j)})^2\big]. \tag{4.A.16}$$

11. Mean value of $R_{j,j}R_{j,i}^2$:

$$\begin{aligned}\mathrm{E}\big[R_{j,j}R_{j,i}^2\big] &= \mathrm{E}\left[\sum_{k=1}^{2\beta}(x_k^{(j)})^2\left(\sum_{m=1}^{2\beta} x_m^{(j)} x_m^{(i)}\right)^2\right] \\ &= \mathrm{E}\left[\sum_{k=1}^{2\beta}(x_k^{(j)})^2 \sum_{m=1}^{2\beta} x_m^{(j)} x_m^{(i)} \sum_{u=1}^{2\beta} x_u^{(j)} x_u^{(i)}\right] \\ &= 2\beta \mathrm{E}\big[(x_k^{(j)})^4\big]\mathrm{E}\big[(x_k^{(i)})^2\big] \\ &\quad + \sum_{k=1}^{2\beta}\sum_{m=1,m\neq k}^{2\beta} \mathrm{E}\big[(x_k^{(j)})^2 (x_m^{(j)})^2\big]\mathrm{E}\big[(x_m^{(i)})^2\big]\end{aligned}$$

$$+2\sum_{k=1}^{2\beta}\sum_{m=1,m\neq k}^{2\beta}\mathrm{E}\big[\big(x_k^{(j)}\big)^3 x_m^{(j)}\big]\mathrm{E}\big[x_k^{(i)}x_m^{(i)}\big]$$

$$+\sum_{k=1}^{2\beta}\sum_{m=1,m\neq k}^{2\beta}\sum_{\substack{u=1,u\neq k\\ u\neq m}}^{2\beta}\mathrm{E}\big[\big(x_k^{(j)}\big)^2 x_m^{(j)}x_u^{(j)}\big]\mathrm{E}\big[x_m^{(i)}x_u^{(i)}\big]. \quad (4.A.17)$$

If all users use the cubic map, the mean term equals

$$\mathrm{E}\big[R_{j,j}R_{j,i}^2\big]=2\beta\mathrm{E}\big[\big(x_k^{(j)}\big)^4\big]\mathrm{E}\big[\big(x_k^{(j)}\big)^2\big]+2\beta(2\beta-1)\mathrm{E}^3\big[\big(x_k^{(j)}\big)^2\big]. \quad (4.A.18)$$

12. Mean value of $R_{j,i}^4$:

$$\mathrm{E}\big[R_{j,i}^4\big]=\mathrm{E}\left[\sum_{k=1}^{2\beta}x_k^{(j)}x_k^{(i)}\sum_{m=1}^{2\beta}x_m^{(j)}x_m^{(i)}\sum_{u=1}^{2\beta}x_u^{(j)}x_u^{(i)}\sum_{v=1}^{2\beta}x_v^{(j)}x_v^{(i)}\right]$$

$$=2\beta\mathrm{E}\big[\big(x_k^{(j)}\big)^4\big]\mathrm{E}\big[\big(x_k^{(i)}\big)^4\big]+4\sum_{k=1}^{2\beta}\sum_{v=1,v\neq k}^{2\beta}\mathrm{E}\big[\big(x_k^{(j)}\big)^3x_v^{(j)}\big]\mathrm{E}\big[\big(x_k^{(i)}\big)^3x_v^{(i)}\big]$$

$$+3\sum_{k=1}^{2\beta}\sum_{u=1,u\neq k}^{2\beta}\mathrm{E}\big[\big(x_k^{(j)}\big)^2\big(x_u^{(j)}\big)^2\big]\mathrm{E}\big[\big(x_k^{(i)}\big)^2\big(x_u^{(i)}\big)^2\big]$$

$$+6\sum_{k=1}^{2\beta}\sum_{u=1,u\neq k}^{2\beta}\sum_{\substack{v=1,v\neq k\\ v\neq u}}^{2\beta}\mathrm{E}\big[\big(x_k^{(j)}\big)^2x_u^{(j)}x_v^{(j)}\big]\mathrm{E}\big[\big(x_k^{(i)}\big)^2x_u^{(i)}x_v^{(i)}\big]$$

$$+\sum_{k=1}^{2\beta}\sum_{m=1,m\neq k}^{2\beta}$$

$$\times\sum_{\substack{u=1,u\neq k\\ u\neq m}}^{2\beta}\sum_{\substack{v=1,v\neq k\\ v\neq m,v\neq u}}^{2\beta}\mathrm{E}\big[x_k^{(j)}x_m^{(j)}x_u^{(j)}x_v^{(j)}\big]\mathrm{E}\big[x_k^{(i)}x_m^{(i)}x_u^{(i)}x_v^{(i)}\big]. \quad (4.A.19)$$

If all users use the cubic map, the mean term is simplified to

$$\mathrm{E}\big[R_{j,i}^4\big]=2\beta\mathrm{E}^2\big[\big(x_k^{(j)}\big)^4\big]+4(2\beta-1)\mathrm{E}^2\big[\big(x_k^{(j)}\big)^3x_{k+1}^{(j)}\big]$$
$$+6\beta(2\beta-1)\mathrm{E}^4\big[\big(x_k^{(j)}\big)^2\big]. \quad (4.A.20)$$

13. Mean value of $R_{j,j}R_{i,i}R_{j,i}$:

$$\begin{aligned}
\mathrm{E}[R_{j,j}R_{i,i}R_{j,i}] &= \mathrm{E}\left[\sum_{k=1}^{2\beta}(x_k^{(j)})^2\sum_{m=1}^{2\beta}(x_m^{(i)})^2\sum_{u=1}^{2\beta}x_u^{(j)}x_u^{(i)}\right] \\
&= \sum_{k=1}^{2\beta}\sum_{u=1,u\neq k}^{2\beta}\mathrm{E}\big[(x_k^{(j)})^2x_u^{(j)}\big]\mathrm{E}\big[(x_k^{(i)})^2x_u^{(i)}\big] \\
&\quad +\sum_{k=1}^{2\beta}\sum_{m=1,m\neq k}^{2\beta}\sum_{\substack{u=1,u\neq k\\ u\neq m}}^{2\beta}\mathrm{E}\big[(x_k^{(j)})^2x_u^{(j)}\big]\mathrm{E}\big[(x_m^{(i)})^2x_u^{(i)}\big].
\end{aligned} \tag{4.A.21}$$

If all users use the cubic map, the mean term equals zero, i.e.,

$$\mathrm{E}[R_{j,j}R_{i,i}R_{j,i}] = 0. \tag{4.A.22}$$

14. Mean value of $R_{j,j}R_{i,i}R_{j,i}^2$:

$$\begin{aligned}
&\mathrm{E}\big[R_{j,j}R_{i,i}R_{j,i}^2\big] \\
&= \mathrm{E}\left[\sum_{k=1}^{2\beta}(x_k^{(j)})^2\sum_{m=1}^{2\beta}(x_m^{(i)})^2\sum_{u=1}^{2\beta}x_u^{(j)}x_u^{(i)}\sum_{v=1}^{2\beta}x_v^{(j)}x_v^{(i)}\right] \\
&= \sum_{k=1}^{2\beta}\mathrm{E}\big[(x_k^{(j)})^4\big]\mathrm{E}\big[(x_k^{(i)})^4\big] + 2\sum_{k=1}^{2\beta}\sum_{v=1,v\neq k}^{2\beta}\mathrm{E}\big[(x_k^{(j)})^3x_v^{(j)}\big]\mathrm{E}\big[(x_k^{(i)})^3x_v^{(i)}\big] \\
&\quad +\sum_{k=1}^{2\beta}\sum_{m=1,m\neq k}^{2\beta}\mathrm{E}\big[(x_k^{(j)})^4\big]\mathrm{E}\big[(x_m^{(i)})^2(x_k^{(i)})^2\big] \\
&\quad +\sum_{k=1}^{2\beta}\sum_{m=1,m\neq k}^{2\beta}\mathrm{E}\big[(x_k^{(j)})^2(x_m^{(j)})^2\big]\mathrm{E}\big[(x_m^{(i)})^4\big] \\
&\quad +\sum_{k=1}^{2\beta}\sum_{u=1,u\neq k}^{2\beta}\mathrm{E}\big[(x_k^{(j)})^2(x_u^{(j)})^2\big]\mathrm{E}\big[(x_k^{(i)})^2(x_u^{(i)})^2\big] \\
&\quad +2\sum_{k=1}^{2\beta}\sum_{m=1,m\neq k}^{2\beta}\mathrm{E}\big[(x_k^{(j)})^3x_m^{(j)}\big]\mathrm{E}\big[(x_m^{(i)})^3x_k^{(i)}\big]
\end{aligned}$$

$$+\sum_{k=1}^{2\beta}\sum_{u=1,u\neq k}^{2\beta}\sum_{\substack{v=1,v\neq k\\ v\neq u}}^{2\beta}\mathrm{E}\big[\big(x_k^{(j)}\big)^2 x_u^{(j)} x_v^{(j)}\big]\mathrm{E}\big[\big(x_k^{(i)}\big)^2 x_u^{(i)} x_v^{(i)}\big]$$

$$+2\sum_{k=1}^{2\beta}\sum_{m=1,m\neq k}^{2\beta}\sum_{\substack{v=1,v\neq k\\ v\neq m}}^{2\beta}\mathrm{E}\big[\big(x_k^{(j)}\big)^3 x_v^{(j)}\big]\mathrm{E}\big[\big(x_m^{(i)}\big)^2 x_k^{(i)} x_v^{(i)}\big]$$

$$+2\sum_{k=1}^{2\beta}\sum_{m=1,m\neq k}^{2\beta}\sum_{\substack{v=1,v\neq k\\ v\neq m}}^{2\beta}\mathrm{E}\big[\big(x_k^{(j)}\big)^2 x_m^{(j)} x_v^{(j)}\big]\mathrm{E}\big[\big(x_m^{(i)}\big)^3 x_v^{(i)}\big]$$

$$+\sum_{k=1}^{2\beta}\sum_{m=1,m\neq k}^{2\beta}\sum_{\substack{u=1,u\neq k\\ u\neq m}}^{2\beta}\mathrm{E}\big[\big(x_k^{(j)}\big)^2\big(x_u^{(j)}\big)^2\big]\,\mathrm{E}\big[\big(x_m^{(i)}\big)^2\big(x_u^{(i)}\big)^2\big]$$

$$+\sum_{k=1}^{2\beta}\sum_{m=1,m\neq k}^{2\beta}\sum_{\substack{u=1,u\neq k\\ u\neq m}}^{2\beta}\sum_{\substack{v=1,v\neq k\\ v\neq m,v\neq u}}^{2\beta}\mathrm{E}\big[\big(x_k^{(j)}\big)^2 x_u^{(j)} x_v^{(j)}\big]\mathrm{E}\big[\big(x_m^{(i)}\big)^2 x_u^{(i)} x_v^{(i)}\big].$$

(4.A.23)

If all users use the cubic map, the mean term is as follows:

$$\begin{aligned}\mathrm{E}\big[R_{j,j}R_{i,i}R_{j,i}^2\big] &= 2\beta\mathrm{E}^2\big[\big(x_k^{(j)}\big)^4\big]+2(2\beta-1)\mathrm{E}^2\big[\big(x_k^{(j)}\big)^3 x_{k+1}^{(j)}\big]\\ &\quad+4\beta(2\beta-1)\mathrm{E}\big[\big(x_k^{(j)}\big)^4\big]\mathrm{E}^2\big[\big(x_k^{(j)}\big)^2\big]\\ &\quad+2\beta(2\beta-1)^2\mathrm{E}^4\big[\big(x_k^{(j)}\big)^2\big].\end{aligned} \quad (4.A.24)$$

15. Mean value of $\eta^{(j)}$:

$$\mathrm{E}\big[\eta^{(j)}\big]=\mathrm{E}\left[\sum_{k=1}^{2\beta}\xi_k x_k^{(j)}\right]=\sum_{k=1}^{2\beta}\mathrm{E}\big[\xi_k x_k^{(j)}\big]=\sum_{k=1}^{2\beta}\mathrm{E}[\xi_k]\mathrm{E}\big[x_k^{(j)}\big]=0. \quad (4.A.25)$$

16. Mean value of $(\eta^{(j)})^2$:

$$\mathrm{E}\big[\big(\eta^{(j)}\big)^2\big]=\mathrm{E}\left[\left(\sum_{k=1}^{2\beta}\xi_k x_k^{(j)}\right)^2\right]=\sum_{k=1}^{2\beta}\mathrm{E}\big[\xi_k^2\big]\mathrm{E}\big[\big(x_k^{(j)}\big)^2\big]=\beta N_0\mathrm{E}\big[\big(x_k^{(j)}\big)^2\big].$$

(4.A.26)

17. Variance value of $\eta^{(j)}$:

$$\mathrm{var}\big[\eta^{(j)}\big] = \mathrm{E}\big[\big(\eta^{(j)}\big)^2\big] = \beta N_0 \mathrm{E}\big[\big(x_k^{(j)}\big)^2\big]. \tag{4.A.27}$$

18. Mean value of $R_{i,i}\eta^{(j)}$:

$$\mathrm{E}\big[R_{i,i}\eta^{(j)}\big] = \mathrm{E}\left[\sum_{k=1}^{2\beta}\big(x_k^{(i)}\big)^2 \sum_{m=1}^{2\beta} \xi_m x_m^{(j)}\right] = \sum_{k=1}^{2\beta}\sum_{m=1}^{2\beta} \mathrm{E}\big[\big(x_k^{(i)}\big)^2\big]\mathrm{E}\big[x_m^{(j)}\big]\mathrm{E}[\xi_m] = 0. \tag{4.A.28}$$

19. Mean value of $R_{j,i}\eta^{(i)}$:

$$\mathrm{E}\big[R_{j,i}\eta^{(i)}\big] = \mathrm{E}\left[\sum_{k=1}^{2\beta} x_k^{(j)} x_k^{(i)} \sum_{m=1}^{2\beta} \xi_m x_m^{(i)}\right] = \sum_{k=1}^{2\beta}\sum_{m=1}^{2\beta} \mathrm{E}\big[x_k^{(j)} x_k^{(i)} x_m^{(i)}\big]\mathrm{E}[\xi_m] = 0. \tag{4.A.29}$$

20. Mean value of $R_{j,i}\eta^{(j)}\eta^{(i)}$:

$$\begin{aligned}
&\mathrm{E}\big[R_{j,i}\eta^{(j)}\eta^{(i)}\big] \\
&= \mathrm{E}\left[\sum_{k=1}^{2\beta} x_k^{(j)} x_k^{(i)} \sum_{m=1}^{2\beta} \xi_m x_m^{(j)} \sum_{u=1}^{2\beta} \xi_u x_u^{(i)}\right] \\
&= \frac{N_0}{2}\sum_{k=1}^{2\beta}\sum_{m=1}^{2\beta} \mathrm{E}\big[x_k^{(j)} x_m^{(j)}\big]\mathrm{E}\big[x_k^{(i)} x_m^{(i)}\big] \\
&= \frac{N_0}{2}\sum_{k=1}^{2\beta} \mathrm{E}\big[\big(x_k^{(j)}\big)^2\big]\mathrm{E}\big[\big(x_k^{(i)}\big)^2\big] + \frac{N_0}{2}\sum_{k=1}^{2\beta}\sum_{m=1,m\neq k}^{2\beta} \mathrm{E}\big[x_k^{(j)} x_m^{(j)}\big]\mathrm{E}\big[x_k^{(i)} x_m^{(i)}\big].
\end{aligned} \tag{4.A.30}$$

If all users use the cubic map, the mean term is simplified as follows.

$$\mathrm{E}\big[R_{j,i}\eta^{(j)}\eta^{(i)}\big] = \beta N_0 \mathrm{E}^2\big[\big(x_k^{(j)}\big)^2\big]. \tag{4.A.31}$$

APPENDIX 4B. DERIVATION OF MEANS, VARIANCES AND COVARIANCES RELEVANT TO THE ANALYSIS OF THE DECORRELATING DETECTOR

In this appendix, we derive the means, variances and covariances of some terms which are required in analyzing an N-user system using decorrelating detectors. All symbols are defined as in Section 4.2.

1. Variance of $R_{j,j}R_{i,i}$:

$$\begin{aligned}\mathrm{var}[R_{j,j}R_{i,i}] &= \mathrm{var}[R_{j,j}]\,\mathrm{var}[R_{i,i}] + \mathrm{E}^2[R_{j,j}]\,\mathrm{var}[R_{i,i}] + \mathrm{E}^2[R_{i,i}]\,\mathrm{var}[R_{j,j}] \\ &= \mathrm{var}[R_{j,j}]\big(\mathrm{E}\big[R_{i,i}^2\big] + \mathrm{E}^2[R_{i,i}]\big) \\ &= \mathrm{E}^2\big[R_{j,j}^2\big] - \mathrm{E}^4[R_{j,j}]. \end{aligned} \quad (4.B.1)$$

If all users use the cubic map defined as in (4.2.41), the variance of $R_{j,j}R_{i,i}$ becomes

$$\begin{aligned}\mathrm{var}[R_{j,j}R_{i,i}] = {} & 4\beta^2\mathrm{E}^2\big[\big(x_k^{(j)}\big)^4\big] + 8\beta^2(2\beta-1)\mathrm{E}\big[\big(x_k^{(j)}\big)^4\big]\mathrm{E}^2\big[\big(x_k^{(j)}\big)^2\big] \\ & + 4\beta^2(-4\beta+1)\mathrm{E}^4\big[\big(x_k^{(j)}\big)^2\big]. \end{aligned} \quad (4.B.2)$$

2. Variance of $R_{j,i}^2$:

$$\mathrm{var}\big[R_{j,i}^2\big] = \mathrm{E}\big[R_{j,i}^4\big] - \mathrm{E}^2\big[R_{j,i}^2\big]. \quad (4.B.3)$$

If all users use the cubic map, the variance of $R_{j,i}^2$ is evaluated below:

$$\begin{aligned}\mathrm{var}\big[R_{j,i}^2\big] = {} & 2\beta\mathrm{E}^2\big[\big(x_k^{(j)}\big)^4\big] + 4(2\beta-1)\mathrm{E}^2\big[\big(x_k^{(j)}\big)^3 x_{k+1}^{(j)}\big] \\ & + 2\beta(4\beta-3)\mathrm{E}^4\big[\big(x_k^{(j)}\big)^2\big]. \end{aligned} \quad (4.B.4)$$

3. Covariance of $R_{j,j}R_{i,i}$ and $R_{j,i}^2$:

$$\mathrm{cov}\big[R_{j,j}R_{i,i},\,R_{j,i}^2\big] = \mathrm{E}\big[R_{j,j}R_{i,i},\,R_{j,i}^2\big] - \mathrm{E}[R_{j,j}R_{i,i}]\mathrm{E}\big[R_{j,i}^2\big]. \quad (4.B.5)$$

If all users use the cubic map, the covariance of $R_{j,j}R_{i,i}$ and $R_{j,i}^2$ equals

$$\begin{aligned}\mathrm{cov}\big[R_{j,j}R_{i,i},\,R_{j,i}^2\big] = {} & 2\beta\mathrm{E}^2\big[\big(x_k^{(j)}\big)^4\big] + 2(2\beta-1)\mathrm{E}^2\big[\big(x_k^{(j)}\big)^3 x_{k+1}^{(j)}\big] \\ & + 4\beta(2\beta-1)\mathrm{E}\big[\big(x_k^{(j)}\big)^4\big]\mathrm{E}^2\big[\big(x_k^{(j)}\big)^2\big] \\ & + 2\beta(-4\beta+1)\mathrm{E}^4\big[\big(x_k^{(j)}\big)^2\big]. \end{aligned} \quad (4.B.6)$$

4. Mean value of $(R_{i,i}\eta^{(j)} - R_{j,i}\eta^{(i)})$:

$$\mathrm{E}\big[R_{i,i}\eta^{(j)} - R_{j,i}\eta^{(i)}\big] = \mathrm{E}\big[R_{i,i}\eta^{(j)}\big] - \mathrm{E}\big[R_{j,i}\eta^{(i)}\big] = 0. \quad (4.B.7)$$

5. Variance of $(R_{i,i}\eta^{(j)} - R_{j,i}\eta^{(i)})$:

$$\begin{aligned}\mathrm{var}\big[R_{i,i}\eta^{(j)} - R_{j,i}\eta^{(i)}\big] &= \mathrm{E}\big[\big(R_{i,i}\eta^{(j)} - R_{j,i}\eta^{(i)}\big)^2\big] \\ &= \mathrm{E}\big[R_{i,i}^2\big(\eta^{(j)}\big)^2 + R_{j,i}^2\big(\eta^{(i)}\big)^2 - 2R_{i,i}R_{j,i}\eta^{(j)}\eta^{(i)}\big] \\ &= \frac{N_0}{2}\big(\mathrm{E}\big[R_{j,j}R_{i,i}^2\big] - \mathrm{E}\big[R_{j,i}^2R_{i,i}\big]\big). \end{aligned} \tag{4.B.8}$$

If all users use the cubic map, the variance of $(R_{i,i}\eta^{(j)} - R_{j,i}\eta^{(i)})$ is simplified to

$$\begin{aligned}&\mathrm{var}\big[R_{i,i}\eta^{(j)} - R_{j,i}\eta^{(i)}\big] \\ &= \frac{N_0}{2}\Big(2\beta(2\beta-1)\mathrm{E}\big[\big(x_k^{(j)}\big)^4\big]\mathrm{E}\big[\big(x_k^{(j)}\big)^2\big] + 2\beta(2\beta-1)^2\mathrm{E}^3\big[\big(x_k^{(j)}\big)^2\big]\Big).\end{aligned} \tag{4.B.9}$$

6. Covariance of $(R_{j,j}R_{i,i} - R_{j,i}^2)$ and $(R_{i,i}\eta^{(j)} - R_{j,i}\eta^{(i)})$:

$$\begin{aligned}&\mathrm{cov}\big[\big(R_{j,j}R_{i,i} - R_{j,i}^2\big), \big(R_{i,i}\eta^{(j)} - R_{j,i}\eta^{(i)}\big)\big] \\ &= \mathrm{E}\big[\big(R_{j,j}R_{i,i} - R_{j,i}^2\big)\big(R_{i,i}\eta^{(j)} - R_{j,i}\eta^{(i)}\big)\big] \\ &\quad - \mathrm{E}\big[\big(R_{j,j}R_{i,i} - R_{j,i}^2\big)\big]\mathrm{E}\big[\big(R_{i,i}\eta^{(j)} - R_{j,i}\eta^{(i)}\big)\big] \\ &= \mathrm{E}\big[R_{j,j}R_{i,i}^2\eta^{(j)} - R_{j,j}R_{i,i}R_{j,i}\eta^{(i)} - R_{j,i}^2R_{i,i}\eta^{(j)} + R_{j,i}^3\eta^{(i)}\big] \\ &= 0. \end{aligned} \tag{4.B.10}$$

7. Mean value of $(\sum_{i=1}^{N} P_{j,i}\eta^{(i)})^2$:

$$\begin{aligned}&\mathrm{E}\left[\left(\sum_{i=1}^{N} P_{j,i}\eta^{(i)}\right)^2\right] \\ &= \mathrm{E}\left[\sum_{i=1}^{N} P_{j,i}\eta^{(i)} \sum_{u=1}^{N} P_{j,u}\eta^{(u)}\right] \\ &= \mathrm{E}\left[\sum_{i=1}^{N} P_{j,i}^2\big(\eta^{(i)}\big)^2 + \sum_{i=1}^{N}\sum_{u=1,u\neq i}^{N} P_{j,i}\eta^{(i)}P_{j,u}\eta^{(u)}\right] \\ &= \mathrm{E}[\xi_k^2]\mathrm{E}\left[\sum_{i=1}^{N} P_{j,i}^2\sum_{k=1}^{2\beta}\big(x_k^{(i)}\big)^2 + \sum_{i=1}^{N}\sum_{u=1,u\neq i}^{N} P_{j,i}P_{j,u}\sum_{k=1}^{2\beta} x_k^{(i)}x_k^{(u)}\right]\end{aligned}$$

$$= \frac{N_0}{2}\mathrm{E}\left[P_{j,j}\left(P_{j,j}R_{j,j} + \sum_{u=1,u\neq j}^{N} P_{j,u}R_{u,j}\right)\right]$$

$$+ \frac{N_0}{2}\mathrm{E}\left[\sum_{i=1,i\neq j}^{N} P_{j,i}\left(P_{j,i}R_{i,i} + \sum_{u=1,u\neq i}^{N} P_{j,u}R_{u,i}\right)\right]$$

$$= \frac{N_0}{2}\mathrm{E}[P_{j,j}] \tag{4.B.11}$$

where the last equality comes from the fact that (see (4.2.57))

$$\sum_{u=1}^{N} P_{j,u}R_{u,i} = \delta_{ji} \tag{4.B.12}$$

and δ_{ji} denotes the Kronecker's delta as defined in (4.2.93).

APPENDIX 4C. DERIVATION OF COVARIANCES AND MEAN RELEVANT TO THE ANALYSIS OF THE MMSE DETECTOR

In this appendix, we derive the covariances and mean of some terms which are required in the analysis of the MMSE detector. All symbols are defined as in Section 4.2.

1. Using the fact that the mean value of the noise sample is zero, i.e., $\mathrm{E}[\xi_k] = 0$, it is readily shown that the following covariances are zero:

$$\mathrm{cov}\left[\left(R_{j,j}R_{i,i} - R_{j,i}^2\right), \eta^{(j)}\right] = 0, \tag{4.C.1}$$

$$\mathrm{cov}\left[\left(R_{i,i}\eta^{(j)} - R_{j,i}\eta^{(i)}\right), R_{j,j}\right] = 0, \tag{4.C.2}$$

$$\mathrm{cov}\left[\left(R_{i,i}\eta^{(j)} - R_{j,i}\eta^{(i)}\right), R_{j,i}\right] = 0, \tag{4.C.3}$$

$$\mathrm{cov}\left[R_{j,j}, \eta^{(j)}\right] = 0, \tag{4.C.4}$$

$$\mathrm{cov}\left[R_{j,i}, \eta^{(j)}\right] = 0. \tag{4.C.5}$$

2. Covariance of $(R_{j,j}R_{i,i} - R_{j,i}^2)$ and $R_{j,j}$:

$$\mathrm{cov}\left[\left(R_{j,j}R_{i,i} - R_{j,i}^2\right), R_{j,j}\right]$$
$$= \mathrm{E}\left[\left(R_{j,j}R_{i,i} - R_{j,i}^2\right)R_{j,j}\right] - \mathrm{E}\left[R_{j,j}R_{i,i} - R_{j,i}^2\right]\mathrm{E}[R_{j,j}]$$
$$= \mathrm{E}\left[R_{j,j}^2R_{i,i}\right] - \mathrm{E}\left[R_{j,i}^2R_{j,j}\right] - \mathrm{E}[R_{j,j}R_{i,i}]\mathrm{E}[R_{j,j}] + \mathrm{E}\left[R_{j,i}^2\right]\mathrm{E}[R_{j,j}]$$

$$= \mathrm{var}[R_{j,j}]\mathrm{E}[R_{i,i}] - \mathrm{E}\big[R_{j,i}^2 R_{j,j}\big] + \mathrm{E}\big[R_{j,i}^2\big]\mathrm{E}[R_{j,j}]. \tag{4.C.6}$$

If all users use the cubic map defined as in (4.2.41), the covariance term is simplified to

$$\begin{aligned} &\mathrm{cov}\big[\big(R_{j,j}R_{i,i} - R_{j,i}^2\big), R_{j,j}\big] \\ &= 2\beta(2\beta-1)\mathrm{E}\big[\big(x_k^{(j)}\big)^4\big]\mathrm{E}\big[\big(x_k^{(j)}\big)^2\big] - 2\beta(2\beta-1)\mathrm{E}^3\big[\big(x_k^{(j)}\big)^2\big]. \end{aligned} \tag{4.C.7}$$

3. Covariance of $(R_{j,j}R_{i,i} - R_{j,i}^2)$ and $R_{j,i}$:

$$\begin{aligned} &\mathrm{cov}\big[\big(R_{j,j}R_{i,i} - R_{j,i}^2\big), R_{j,i}\big] \\ &= \mathrm{E}\big[\big(R_{j,j}R_{i,i} - R_{j,i}^2\big)R_{j,i}\big] - \mathrm{E}\big[R_{j,j}R_{i,i} - R_{j,i}^2\big]\mathrm{E}[R_{j,i}] \\ &= \mathrm{E}[R_{j,j}R_{i,i}R_{j,i}] - \mathrm{E}\big[R_{j,i}^3\big]. \end{aligned} \tag{4.C.8}$$

If all users use the cubic map, the above equation equals zero, i.e.,

$$\mathrm{cov}\big[\big(R_{j,j}R_{i,i} - R_{j,i}^2\big), R_{j,i}\big] = 0. \tag{4.C.9}$$

4. Covariance of $(R_{i,i}\eta^{(j)} - R_{j,i}\eta^{(i)})$ and $\eta^{(j)}$:

$$\begin{aligned} &\mathrm{cov}\big[\big(R_{i,i}\eta^{(j)} - R_{j,i}\eta^{(i)}\big), \eta^{(j)}\big] \\ &= \mathrm{E}\big[\big(R_{i,i}\eta^{(j)} - R_{j,i}\eta^{(i)}\big)\eta^{(j)}\big] - \mathrm{E}\big[R_{i,i}\eta^{(j)} - R_{j,i}\eta^{(i)}\big]\mathrm{E}\big[\eta^{(j)}\big] \\ &= \mathrm{E}[R_{i,i}]\mathrm{E}\big[\big(\eta^{(j)}\big)^2\big] - \mathrm{E}\big[R_{j,i}\eta^{(i)}\eta^{(j)}\big]. \end{aligned} \tag{4.C.10}$$

If the cubic map defined as in (4.2.41) is used by all users, the covariance term is simplified to

$$\mathrm{cov}\big[\big(R_{i,i}\eta^{(j)} - R_{j,i}\eta^{(i)}\big), \eta^{(j)}\big] = \beta(2\beta-1)\mathrm{E}^2\big[\big(x_k^{(j)}\big)^2\big]N_0. \tag{4.C.11}$$

5. Covariance of $R_{j,j}$ and $R_{j,i}$:

$$\mathrm{cov}[R_{j,j}, R_{j,i}] = \mathrm{E}[R_{j,j}R_{j,i}] - \mathrm{E}[R_{j,j}]\mathrm{E}[R_{j,i}] = 0. \tag{4.C.12}$$

6. Covariance of $\sum_{i=1}^N d^{(i)}H_{j,i}$ and $\sum_{i=1}^N H_{j,i}\eta^{(i)}$:

$$\mathrm{cov}\left[\sum_{i=1}^N d^{(i)}H_{j,i}, \sum_{i=1}^N H_{j,i}\eta^{(i)}\right]$$

$$= \mathrm{E}\left[\sum_{i=1}^{N} d^{(i)} H_{j,i} \sum_{i=1}^{N} H_{j,i}\eta^{(i)}\right] - \mathrm{E}\left[\sum_{i=1}^{N} d^{(i)} H_{j,i}\right]\mathrm{E}\left[\sum_{i=1}^{N} H_{j,i}\eta^{(i)}\right]$$

$$= \mathrm{E}\left[\sum_{i=1}^{N} d^{(i)} H_{j,i} \sum_{i=1}^{N} H_{j,i} \sum_{k=1}^{2\beta} \xi_k x_k^{(i)}\right]$$

$$- \mathrm{E}\left[\sum_{i=1}^{N} d^{(i)} H_{j,i}\right]\mathrm{E}\left[\sum_{i=1}^{N} H_{j,i} \sum_{k=1}^{2\beta} \xi_k x_k^{(i)}\right]$$

$$= \sum_{i=1}^{N}\sum_{k=1}^{2\beta} \mathrm{E}\left[d^{(i)} H_{j,i} H_{j,i} x_k^{(i)}\right]\mathrm{E}[\xi_k]$$

$$- \mathrm{E}\left[\sum_{i=1}^{N} d^{(i)} H_{j,i}\right]\sum_{i=1}^{N}\sum_{k=1}^{2\beta} \mathrm{E}\left[H_{j,i} x_k^{(i)}\right]\mathrm{E}[\xi_k]$$

$$= 0. \tag{4.C.13}$$

7. Mean value of $(\sum_{i=1}^{N} H_{j,i}\eta^{(i)})^2$:

$$\mathrm{E}\left[\left(\sum_{i=1}^{N} H_{j,i}\eta^{(i)}\right)^2\right]$$

$$= \mathrm{E}\left[\sum_{i=1}^{N} H_{j,i}\eta^{(i)} \sum_{u=1}^{N} H_{j,u}\eta^{(u)}\right]$$

$$= \mathrm{E}[\xi_k^2]\mathrm{E}\left[\sum_{i=1}^{N} H_{j,i}^2 \sum_{k=1}^{2\beta} (x_k^{(i)})^2 + \sum_{i=1}^{N}\sum_{u=1,u\neq i}^{N} H_{j,i} H_{j,u} \sum_{k=1}^{2\beta} x_k^{(i)} x_k^{(u)}\right]$$

$$= \frac{N_0}{2}\mathrm{E}\left[H_{j,j}\left(H_{j,j} R_{j,j} + \sum_{u=1,u\neq j}^{N} H_{j,u} R_{u,j}\right)\right]$$

$$+ \frac{N_0}{2}\mathrm{E}\left[\sum_{i=1,i\neq j}^{N} H_{j,i}\left(H_{j,i} R_{i,i} + \sum_{u=1,u\neq i}^{N} H_{j,u} R_{u,i}\right)\right]. \tag{4.C.14}$$

Since

$$\sum_{u=1}^{N} H_{j,u}\left(R_{u,i} + \frac{N_0}{2}\delta_{ui}\right) = \delta_{ji}, \tag{4.C.15}$$

where δ_{ji} denotes the Kronecker's delta as defined in (4.2.93), (4.C.14) is simplified to

$$\mathrm{E}\left[\left(\sum_{i=1}^{N} H_{j,i}\eta^{(i)}\right)^2\right] = \frac{N_0}{2}\mathrm{E}[H_{j,j}] - \frac{N_0{}^2}{4}\sum_{i=1}^{N}\mathrm{E}\left[H_{j,i}^2\right]. \tag{4.C.16}$$

APPENDIX 4D. CALCULATION OF THE PROBABILITY OF OCCURRENCE OF $Ne_{(n-1)}^{(j)}$ FOR THE PIC DETECTOR

All symbols are defined as in Section 4.3. Recall in (4.3.11) that the transmitted symbol vector $\boldsymbol{d}$ is defined as

$$\boldsymbol{d} = \begin{bmatrix} d^{(1)} & d^{(2)} & \cdots & d^{(N)} \end{bmatrix}^{\mathrm{T}}. \tag{4.D.1}$$

Here, we denote $\hat{\boldsymbol{d}}_{(n-1)}$ as the estimated symbol vector at the $(n-1)$ stage of the PIC detector, i.e.,

$$\hat{\boldsymbol{d}}_{(n-1)} = \begin{bmatrix} \hat{d}_{(n-1)}^{(1)} & \hat{d}_{(n-1)}^{(2)} & \cdots & \hat{d}_{(n-1)}^{(N)} \end{bmatrix}^{\mathrm{T}}. \tag{4.D.2}$$

Also, when some of the elements are removed from $\boldsymbol{d}$, we denote the resultant vector by $\boldsymbol{d}^{\boldsymbol{v}}$ where $\boldsymbol{v}$ is the vector containing the indices of the elements that have been removed. For example, assuming $i < j$,

$$\boldsymbol{d}^{(i,j)} = \begin{bmatrix} d^{(1)} & d^{(2)} & \cdots & d^{(i-1)} & d^{(i+1)} & \cdots & d^{(j-1)} & d^{(j+1)} & \cdots & d^{(N)} \end{bmatrix}^{\mathrm{T}}. \tag{4.D.3}$$

Similarly, we denote the resultant vector by $\hat{\boldsymbol{d}}_{(n-1)}^{\boldsymbol{v}}$ where $\boldsymbol{v}$ contains the indices of the elements that have been removed from $\hat{\boldsymbol{d}}_{(n-1)}$.

Probability that $Ne_{(n-1)}^{(j)} = 1$

The probability that $Ne_{(n-1)}^{(j)} = 1$ can be shown equal to [Madsen and Cho, 1999]

$$\begin{aligned} \mathrm{Prob}\left(Ne_{(n-1)}^{(j)} = 1\right) &= \sum_{i=1, i\neq j}^{N} \mathrm{Prob}\left(d^{(i)} \neq \hat{d}_{(n-1)}^{(i)}, \boldsymbol{d}^{(i,j)} = \hat{\boldsymbol{d}}_{(n-1)}^{(i,j)}\right) \\ &\leqslant \sum_{i=1, i\neq j}^{N} \mathrm{Prob}\left(d^{(i)} \neq \hat{d}_{(n-1)}^{(i)}\right) \\ &= (N-1)\mathrm{BER}_{(n-1)}^{(j)}. \end{aligned} \tag{4.D.4}$$

We approximate $\text{Prob}(N_{e\,(n-1)}^{(j)} = 1)$ by its upper bound $(N-1)\text{BER}_{(n-1)}^{(j)}$. The approximation is good when the probability of error of each user is small, e.g., less than 0.05. Using the analytical BERs of the linear detectors, namely the conventional single-user detector, decorrelating detector and MMSE detector (see Section 4.2.2), as the BER values of the 0th stage (denoted by $\text{BER}_{(0)}^{(j)}$), and by applying (4.D.4) and (4.3.18) repeatedly, the probability that $N_{e\,(n-1)}^{(j)} = 1$ at the $(n-1)$th stage can be found.

Probability that $N_{e\,(n-1)}^{(j)} = 2$

Using a similar approach, the probability that $N_{e\,(n-1)}^{(j)} = 2$ is given by

$$\begin{aligned}
&\text{Prob}\big(N_{e\,(n-1)}^{(j)} = 2\big)\\
&= \sum_{i=1,i\neq j}^{N} \sum_{\substack{v=1,v\neq j\\ v<i}}^{N} \text{Prob}\big(d^{(i)} \neq \hat{d}_{(n-1)}^{(i)}, d^{(v)} \neq \hat{d}_{(n-1)}^{(v)}, \boldsymbol{d}^{(i,j,v)} = \hat{\boldsymbol{d}}_{(n-1)}^{(i,j,v)}\big)\\
&\leqslant \sum_{i=1,i\neq j}^{N} \sum_{\substack{v=1,v\neq j\\ v<i}}^{N} \text{Prob}\big(d^{(i)} \neq \hat{d}_{(n-1)}^{(i)}, d^{(v)} \neq \hat{d}_{(n-1)}^{(v)}\big).
\end{aligned} \tag{4.D.5}$$

In the case where a conventional single-user detector is used as the 0th stage of the PIC detector, i.e., $n = 1$, the probability that $N_{e\,(0)}^{(j)} = 2$ can be evaluated by

$$\begin{aligned}
&\text{Prob}\big(N_{e\,(0)}^{(j)} = 2\big)\\
&\leqslant \sum_{i=1,i\neq j}^{N} \sum_{\substack{v=1,v\neq j\\ v<i}}^{N} \text{Prob}\big(d^{(i)} \neq \hat{d}_{(0)}^{(i)}, d^{(v)} \neq \hat{d}_{(0)}^{(v)} \mid (d^{(i)}, d^{(v)})\big)\,\text{Prob}\big(d^{(i)}, d^{(v)}\big)\\
&= \sum_{i=1,i\neq j}^{N} \sum_{\substack{v=1,v\neq j\\ v<i}}^{N} \big[\text{Prob}\big(y_{(0)}^{(i)} \leqslant 0, y_{(0)}^{(v)} \leqslant 0 \mid (d^{(i)} = +1, d^{(v)} = +1)\big)\\
&\quad \times \text{Prob}\big(d^{(i)} = +1, d^{(v)} = +1\big)\\
&\quad + \text{Prob}\big(y_{(0)}^{(i)} \leqslant 0, y_{(0)}^{(v)} > 0 \mid (d^{(i)} = +1, d^{(v)} = -1)\big)\\
&\quad \times \text{Prob}\big(d^{(i)} = +1, d^{(v)} = -1\big)\\
&\quad + \text{Prob}\big(y_{(0)}^{(i)} > 0, y_{(0)}^{(v)} \leqslant 0 \mid (d^{(i)} = -1, d^{(v)} = +1)\big)
\end{aligned}$$

$$\times \operatorname{Prob}\big(d^{(i)} = -1, d^{(v)} = +1\big)$$
$$+ \operatorname{Prob}\big(y_{(0)}^{(i)} > 0, y_{(0)}^{(v)} > 0 \mid \big(d^{(i)} = -1, d^{(v)} = -1\big)\big)$$
$$\times \operatorname{Prob}\big(d^{(i)} = -1, d^{(v)} = -1\big)\big]$$
$$= \frac{1}{2}(N-1)(N-2)$$
$$\times \Bigg[\frac{1}{4}\int_{-\infty}^{0}\int_{-\infty}^{0} \rho\big(y_{(0)}^{(i)}, y_{(0)}^{(v)} \mid \big(d^{(i)} = +1, d^{(v)} = +1\big)\big)\, dy_{(0)}^{(i)}\, dy_{(0)}^{(v)}$$
$$+ \frac{1}{4}\int_{-\infty}^{0}\int_{0}^{\infty} \rho\big(y_{(0)}^{(i)}, y_{(0)}^{(v)} \mid \big(d^{(i)} = +1, d^{(v)} = -1\big)\big)\, dy_{(0)}^{(i)}\, dy_{(0)}^{(v)}$$
$$+ \frac{1}{4}\int_{0}^{\infty}\int_{-\infty}^{0} \rho\big(y_{(0)}^{(i)}, y_{(0)}^{(v)} \mid \big(d^{(i)} = -1, d^{(v)} = +1\big)\big)\, dy_{(0)}^{(i)}\, dy_{(0)}^{(v)}$$
$$+ \frac{1}{4}\int_{0}^{\infty}\int_{0}^{\infty} \rho\big(y_{(0)}^{(i)}, y_{(0)}^{(v)} \mid \big(d^{(i)} = -1, d^{(v)} = -1\big)\big)\, dy_{(0)}^{(i)}\, dy_{(0)}^{(v)}\Bigg] \tag{4.D.6}$$

where $y_{(0)}^{(i)}$ and $y_{(0)}^{(v)}$ represent the outputs of the ith and vth correlators, respectively, in the single-user detectors. In (4.D.6), $\rho(\psi, \varpi)$ denotes the two-dimensional (bivariate) normal probability density function (pdf) of ψ and ϖ and is given by

$$\rho(\psi, \varpi) = \frac{1}{2\pi\sqrt{\operatorname{var}[\psi]\operatorname{var}[\varpi](1-\iota^2)}}$$
$$\times \exp\left\{-\frac{1}{2(1-\iota^2)}\left[\frac{(\psi - \mathrm{E}[\psi])^2}{\operatorname{var}[\psi]}\right.\right.$$
$$\left.\left. - 2\iota\frac{(\psi - \mathrm{E}[\psi])(\varpi - \mathrm{E}[\varpi])}{\sqrt{\operatorname{var}[\psi]\operatorname{var}[\varpi]}} + \frac{(\varpi - \mathrm{E}[\varpi])^2}{\operatorname{var}[\varpi]}\right]\right\}, \tag{4.D.7}$$

where ι is the correlation coefficient between ψ and ϖ, and is defined as

$$\iota = \frac{\mathrm{E}[\psi\varpi] - \mathrm{E}[\psi]\mathrm{E}[\varpi]}{\sqrt{\operatorname{var}[\psi]\operatorname{var}[\varpi]}}. \tag{4.D.8}$$

It is readily shown that

$$\begin{aligned}
&\int_{-\infty}^{0}\int_{-\infty}^{0} \rho\big(y^{(i)}_{(0)}, y^{(v)}_{(0)} \mid (d^{(i)} = +1, d^{(v)} = +1)\big)\, dy^{(i)}_{(0)}\, dy^{(v)}_{(0)} \\
&= \int_{-\infty}^{0}\int_{0}^{\infty} \rho\big(y^{(i)}_{(0)}, y^{(v)}_{(0)} \mid (d^{(i)} = +1, d^{(v)} = -1)\big)\, dy^{(i)}_{(0)}\, dy^{(v)}_{(0)} \\
&= \int_{0}^{\infty}\int_{-\infty}^{0} \rho\big(y^{(i)}_{(0)}, y^{(v)}_{(0)} \mid (d^{(i)} = -1, d^{(v)} = +1)\big)\, dy^{(i)}_{(0)}\, dy^{(v)}_{(0)} \\
&= \int_{0}^{\infty}\int_{0}^{\infty} \rho\big(y^{(i)}_{(0)}, y^{(v)}_{(0)} \mid (d^{(i)} = -1, d^{(v)} = -1)\big)\, dy^{(i)}_{(0)}\, dy^{(v)}_{(0)}. \qquad (4.D.9)
\end{aligned}$$

Hence, $\text{Prob}(Ne^{(j)}_{(0)} = 2)$ can be simplified to

$$\begin{aligned}
&\text{Prob}\big(Ne^{(j)}_{(0)} = 2\big) \\
&\leqslant \frac{1}{2}(N-1)(N-2) \\
&\quad \times \int_{-\infty}^{0}\int_{-\infty}^{0} \rho\big(y^{(i)}_{(0)}, y^{(v)}_{(0)} \mid (d^{(i)} = +1, d^{(v)} = +1)\big)\, dy^{(i)}_{(0)}\, dy^{(v)}_{(0)}.
\end{aligned} \qquad (4.D.10)$$

Similar to the case for $Ne^{(j)}_{(n-1)} = 1$, we approximate $\text{Prob}(Ne^{(j)}_{(0)} = 2)$ by its upper bound given in (4.D.10).

Probability that $Ne^{(j)}_{(n-1)} \geqslant 3$
The probability that $Ne^{(j)}_{(n-1)} \geqslant 3$ is given by

$$\begin{aligned}
&\text{Prob}\big(Ne^{(j)}_{(n-1)} = u\big) \\
&= \sum_{j \notin \boldsymbol{u} \cup \tilde{\boldsymbol{u}}} \sum_{\boldsymbol{u} \cap \tilde{\boldsymbol{u}} = \oslash} \cdots \sum_{\boldsymbol{u} \cup \tilde{\boldsymbol{u}} \cup j = \{1,2,\ldots,N\}} \text{Prob}\big(\boldsymbol{d}^{(j,\boldsymbol{u})}_{(n-1)} \neq \hat{\boldsymbol{d}}^{(j,\boldsymbol{u})}_{(n-1)}, \boldsymbol{d}^{(j,\tilde{\boldsymbol{u}})}_{(n-1)} = \hat{\boldsymbol{d}}^{(j,\tilde{\boldsymbol{u}})}_{(n-1)}\big) \\
&\leqslant \sum_{j \notin \boldsymbol{u} \cup \tilde{\boldsymbol{u}}} \sum_{\boldsymbol{u} \cap \tilde{\boldsymbol{u}} = \oslash} \cdots \sum_{\boldsymbol{u} \cup \tilde{\boldsymbol{u}} \cup j = \{1,2,\ldots,N\}} \text{Prob}\big(\boldsymbol{d}^{(j,\boldsymbol{u})}_{(n-1)} \neq \hat{\boldsymbol{d}}^{(j,\boldsymbol{u})}_{(n-1)}\big), \qquad (4.D.11)
\end{aligned}$$

where the vectors $\boldsymbol{u}$ and $\tilde{\boldsymbol{u}}$ contain the indices of the users that have made, respectively, wrong decisions and correct decisions. Using similar procedures as in the case $Ne^{(j)}_{(n-1)} = 2$, if a conventional single-user detector is used as the 0th stage, the probability that $Ne^{(j)}_{(0)} \geqslant 3$ can be shown equal to

$$\text{Prob}\big(N e_{(0)}^{(j)} = u\big)$$
$$\leqslant \binom{N-1}{u} \sum_{\boldsymbol{d}' \in \{-1,+1\}^u} \left[\int_{y_{(0)}^{(u)}} \cdots \int_{y_{(0)}^{(2)}} \int_{y_{(0)}^{(1)}} \rho(\boldsymbol{y}_0 \mid \boldsymbol{d}') \right.$$
$$\left. \times dy_{(0)}^{(1)}\, dy_{(0)}^{(2)} \cdots dy_{(0)}^{(u)}\, \text{Prob}(\boldsymbol{d}') \right], \tag{4.D.12}$$

where $\binom{\psi}{\varpi}$ denotes the number of ways to choose ϖ items out of ψ. Also, the vectors $\boldsymbol{y}_0$ and $\boldsymbol{d}'$ are defined as

$$\boldsymbol{y}_0 = \begin{bmatrix} \hat{y}_{(0)}^{(1)} & \hat{y}_{(0)}^{(2)} & \cdots & \hat{y}_{(0)}^{(u)} \end{bmatrix}^{\mathrm{T}}, \tag{4.D.13}$$
$$\boldsymbol{d}' = \begin{bmatrix} d^{(1)} & d^{(2)} & \cdots & d^{(u)} \end{bmatrix}^{\mathrm{T}}. \tag{4.D.14}$$

The u-dimensional normal pdf of $\boldsymbol{y}_0$ for a given vector $\boldsymbol{d}'$ is denoted by $\rho(\boldsymbol{y}_0 \mid \boldsymbol{d}')$. It is represented by

$$\rho(\boldsymbol{\psi}) = \frac{1}{\sqrt{(2\pi)^u \det[\boldsymbol{U}]}} \exp\left[-\frac{1}{2} \big(\boldsymbol{\psi} - \mathrm{E}[\boldsymbol{\psi}]\big)^{\mathrm{T}} \boldsymbol{U}^{-1} \big(\boldsymbol{\psi} - \mathrm{E}[\boldsymbol{\psi}]\big) \right] \tag{4.D.15}$$

where

$$\boldsymbol{\psi} = \begin{bmatrix} \psi^{(1)} & \psi^{(2)} & \cdots & \psi^{(u)} \end{bmatrix}^{\mathrm{T}}, \tag{4.D.16}$$

$\boldsymbol{U}^{-1}$ is the $u \times u$ covariance matrix of the random variables $\psi^{(1)}, \psi^{(2)}, \ldots, \psi^{(u)}$, and $\det[\cdot]$ denotes the determinant operator.

Probability that $N e_{(n-1)}^{(j)} = 0$
Since

$$\sum_{m=0}^{N-1} \text{Prob}\big(N e_{(n-1)}^{(j)} = m\big) = 1, \tag{4.D.17}$$

the probability that $N e_{(n-1)}^{(j)} = 0$ can be evaluated by

$$\text{Prob}\big(N e_{(n-1)}^{(j)} = 0\big) = 1 - \sum_{m=1}^{N-1} \text{Prob}\big(N e_{(n-1)}^{(j)} = m\big). \tag{4.D.18}$$

5

MULTIPLE-ACCESS GENERALIZED CORRELATION DELAY-SHIFT-KEYING SYSTEMS

In the previous two chapters, we have studied the performance of a multiple-access antipodal CSK system in detail. The system has been investigated based on the assumption that the chaotic generators at the transmitters and receivers are perfectly synchronized. Although the feasibility of chaos synchronization under a noisy environment is yet to be proven, the results provide researchers and engineers with the theoretical performance limits that a multiple-access chaos communication system can ultimately achieve. Starting from this chapter, we consider and analyze noncoherent chaos-based digital communication systems with multiple-access capability. Such systems are comparatively more practical because chaos synchronization is not needed. We begin by investigating a generalized version of *correlation delay shift keying* (CDSK) scheme.

Correlation-delay-shift-keying scheme is similar to the differential chaos-shift-keying (DCSK) scheme in that a reference chaotic signal is embedded in the transmitted signal [Sushchik *et al.*, 2000]. Unlike in DCSK, however, the reference signal and the information-bearing signal are now added together with a certain time delay in CDSK. As a consequence, each transmitted signal sample includes one reference sample and one information-bearing sample, and the transmitted signal sample is never repeated. Since no individual reference signal is sent, the bandwidth efficiency is improved. Moreover, by eliminating the switch required to perform the switching between the reference chaotic signal and information-bearing signal in the DCSK system, CDSK allows a continuous operation of the transmitter. Also, the transmitted signal is more homogeneous and less prone to interception. However, because the sum of two chaotic signals is sent, greater uncertainty (interference) is produced when the received signal correlates with its delayed version at the receiving side. Therefore, the performance of CDSK is worse than that of DCSK.

In the generalized CDSK (GCDSK) scheme, the transmitted signal is composed of a reference chaotic signal and a number of delayed chaotic signals, some of which are modulated by the data being sent [Tam *et al.*, 2006]. Such a construction of the transmitted signal allows the transmission of more than one reference signals and more than one information-bearing signals simultaneously. The useful signal component, as well as the interference component, will be enhanced at the receiving side. We show that with appropriate choice of system parameters, the bit error performance of the GCDSK system improves over the CDSK scheme. The system also inherits the merits of the CDSK system such as being switchless and allowing continuous operation of the transmitter. Moreover, we introduce a multiple-access scheme for the GCDSK system. In a multiple-access GCDSK system, the delay blocks will be assigned with different sets of delay values for different users. The receiver structure remains the same except that the corresponding delay needs to be set to recover the appropriate data stream.

The organization of the chapter is as follows. In Section 5.1, the operation of a baseband implementation of the CDSK system is briefly reviewed, and in Section 5.2, the GCDSK scheme is described. Section 5.3 presents the multiple-access scheme for the GCDSK system. Assuming a Gaussian correlator output, an approximate bit error rate is derived analytically in terms of the spreading factor, the length of delay and the number of delay units. Also, using the Chebyshev map as the chaotic generator, we perform simulations for the single-user and multi-user GCDSK systems and present the results in Section 5.2 and Section 5.3, respectively. Besides comparing with the analytical results, the simulation results for the GCDSK scheme are also compared with those of the DCSK and CDSK schemes.

5.1 REVIEW OF CORRELATION-DELAY-SHIFT-KEYING (CDSK) SCHEME

Figure 5.1 shows the transmitter and receiver structures of a correlation-delay-shift-keying (CDSK) system [Sushchik *et al.*, 2000]. Denote the lth transmitted symbol by $d_l \in \{-1, +1\}$ and assume that "-1" and "$+1$" are transmitted with equal probabilities. First a chaotic signal, denoted by $\{x_k\}$, is generated in the transmitter. The transmitted signal, s_k, is the sum of the chaotic signal x_k and the delayed version of the signal, x_{k-L}, modulated by the symbol $d_l \in \{-1, +1\}$, where L denotes the delay, i.e.,

$$s_k = x_k + d_l x_{k-L}. \tag{5.1.1}$$

Assume that the transmitted signal is passed through an additive white Gaussian noise (AWGN) channel with a two-sided power spectral density $N_0/2$. The received signal, denoted by r_k, is given by

$$r_k = s_k + \xi_k, \tag{5.1.2}$$

where ξ_k denotes the AWGN signal with zero mean and variance $N_0/2$. At the receiving side, the incoming signal correlates with a delayed version of itself within

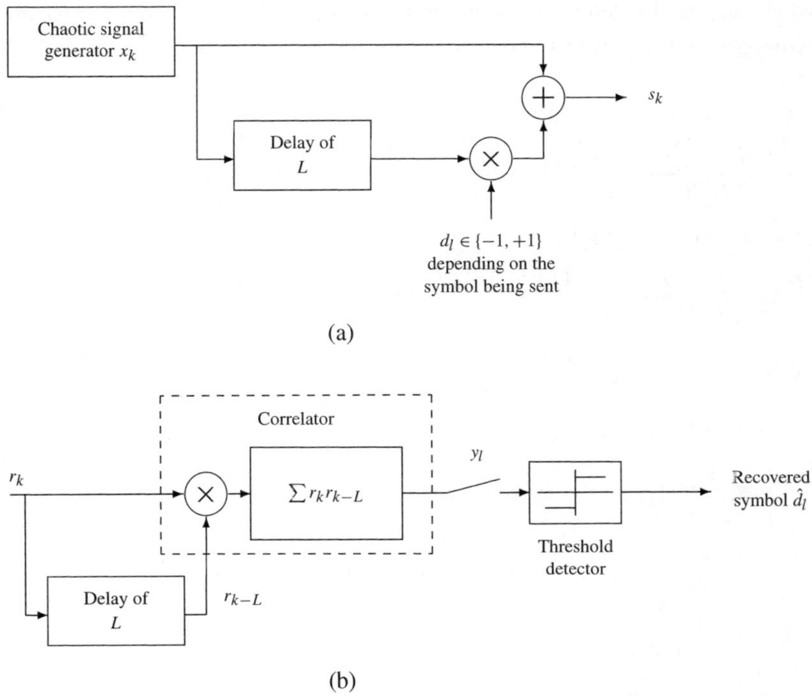

Figure 5.1. Correlation-delay-shift-keying system. (a) Transmitter; (b) receiver.

each symbol duration. Suppose the spreading factor is 2β, i.e., 2β samples are sent within one bit duration. The output of the correlator at the end of the lth symbol duration, denoted by y_l, equals

$$\begin{aligned} y_l &= \sum_{k=2(l-1)\beta+1}^{2l\beta} r_k r_{k-L} \\ &= \sum_{k=2(l-1)\beta+1}^{2(l-1)\beta+L-2\beta\lfloor\frac{L}{2\beta}\rfloor} (x_k + d_l x_{k-L} + \xi_k)\left(x_{k-L} + d_{l-\lfloor\frac{L}{2\beta}\rfloor-1} x_{k-2L} + \xi_{k-L}\right) \\ &\quad + \sum_{k=2(l-1)\beta+L-2\beta\lfloor\frac{L}{2\beta}\rfloor+1}^{2l\beta} (x_k + d_l x_{k-L} + \xi_k)\left(x_{k-L} + d_{l-\lfloor\frac{L}{2\beta}\rfloor} x_{k-2L} + \xi_{k-L}\right) \\ &= \lambda + \mu + \nu, \end{aligned} \tag{5.1.3}$$

where λ, μ and ν denote the *required signal*, the *intra-signal interference* and the *noise component*, respectively, and are given by

$$\lambda = d_l \sum_{k=2(l-1)\beta+1}^{2l\beta} x_{k-L}^2, \tag{5.1.4}$$

$$\begin{aligned}\mu = & \sum_{k=2(l-1)\beta+1}^{2(l-1)\beta+L-2\beta\lfloor\frac{L}{2\beta}\rfloor} \left(x_k x_{k-L} + d_{l-\lfloor\frac{L}{2\beta}\rfloor-1} x_k x_{k-2L} + d_l d_{l-\lfloor\frac{L}{2\beta}\rfloor-1} x_{k-L} x_{k-2L}\right) \\ & + \sum_{k=2(l-1)\beta+L-\lfloor\frac{L}{2\beta}\rfloor 2\beta+1}^{2l\beta} \left(x_k x_{k-L} + d_{l-\lfloor\frac{L}{2\beta}\rfloor} x_k x_{k-2L}\right. \\ & \left. + d_l d_{l-\lfloor\frac{L}{2\beta}\rfloor} x_{k-L} x_{k-2L}\right),\end{aligned} \tag{5.1.5}$$

$$\begin{aligned}\nu = & \sum_{k=2(l-1)\beta+1}^{2l\beta} (x_k \xi_{k-L} + d_l x_{k-L} \xi_{k-L} + x_{k-L} \xi_k + \xi_k \xi_{k-L}) \\ & + \sum_{k=2(l-1)\beta+1}^{2(l-1)\beta+L-2\beta\lfloor\frac{L}{2\beta}\rfloor} d_{l-\lfloor\frac{L}{2\beta}\rfloor-1} x_{k-2L} \xi_k \\ & + \sum_{k=2(l-1)\beta+L-2\beta\lfloor\frac{L}{2\beta}\rfloor+1}^{2l\beta} d_{l-\lfloor\frac{L}{2\beta}\rfloor} x_{k-2L} \xi_k.\end{aligned} \tag{5.1.6}$$

Note that in the above equations, the function $\lfloor\psi\rfloor$ computes the integral part of ψ. As in a DCSK system, the required signal in (5.1.4) is a time-varying component, depending upon the bit energy of the transmitted signal. The intra-signal interference, similar to the inter-user interference in a multiple-access system [Tam *et al.*, 2002a, 2003], originates from the correlation between the chaotic samples and may contribute positively or negatively to the required signal. The net effect is that more uncertainty on the correlator output is produced. Finally, the noise component comes from the noisy channel. Based on the correlator output, the symbol is decoded according to the following rule:

$$\hat{d}_l = \begin{cases} +1 & \text{if } y_l > 0, \\ -1 & \text{if } y_l \leqslant 0. \end{cases} \tag{5.1.7}$$

Because of the additional uncertainty due to the intra-signal interference, the performance of the CDSK system is always lower than that of the DCSK system. In particular, under a noiseless condition, the correlator for the CDSK receiver produces

$$y_l = \lambda + \mu. \tag{5.1.8}$$

Under the same condition, the correlator in the DCSK receiver gives only the required signal component, i.e., λ. By comparing these two expressions, we can conclude that DCSK outperforms CDSK in a noiseless environment.

5.2 GENERALIZED CORRELATION-DELAY-SHIFT-KEYING (GCDSK) SCHEME

5.2.1 Transmitter Structure of the GCDSK system

We introduce a generalized CDSK (GCDSK) communication system [Tam *et al.*, 2006], as shown in Fig. 5.2. The transmitter contains a chaotic signal generator and $(M-1)$ delay blocks. We assume that $M > 2$ because when $M = 2$, the GCDSK system degenerates to the CDSK system. We denote the minimum delay by L. The chaotic signals with delays $L, 3L, 5L, \ldots$ are modulated by the data sequence $\{d_l\}$, whereas the signals with delays $2L, 4L, 6L, \ldots$ are unmodulated. Finally, the transmitted signal is formed by adding the original chaotic signal and all the delayed signals. During the lth bit duration, i.e., for time $k = 2(l-1)\beta +$

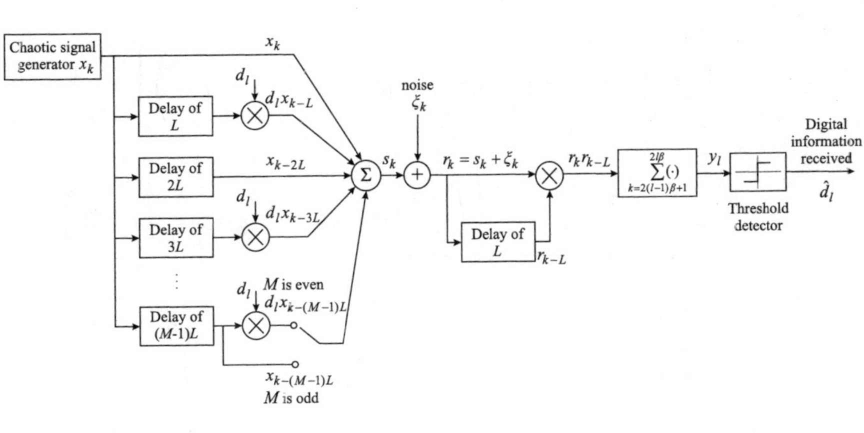

Figure 5.2. Block diagram of a generalized correlation-delay-shift-keying communication system.

$1, 2(l-1)\beta+2, \ldots, 2l\beta$, the transmitted signal is given by

$$s_k = \begin{cases} \sum_{m=0}^{\frac{M-2}{2}} x_{k-2mL} + d_l \sum_{m=0}^{\frac{M-2}{2}} x_{k-(2m+1)L} & \text{if } M \text{ is even,} \\ \sum_{m=0}^{\frac{M-1}{2}} x_{k-2mL} + d_l \sum_{m=0}^{\frac{M-3}{2}} x_{k-(2m+1)L} & \text{if } M \text{ is odd,} \end{cases} \tag{5.2.1}$$

where in each case, the first and the second terms represent summation of all the unmodulated and modulated chaotic signals, respectively.

5.2.2 Receiver Structure of the GCDSK system

As in Section 5.1, we consider an AWGN channel and use the same correlator-type detector shown in Fig. 5.1(b). For the lth symbol, the corresponding output of the correlator equals

$$y_l = \sum_{k=2(l-1)\beta+1}^{2l\beta} r_k r_{k-L}. \tag{5.2.2}$$

When M is even and odd, respectively,

$$\begin{aligned} y_l = & \sum_{k=2(l-1)\beta+1}^{2(l-1)\beta+L-2\beta\lfloor\frac{L}{2\beta}\rfloor} \left[\left(\sum_{m=0}^{\frac{M-2}{2}} (x_{k-2mL} + d_l x_{k-(2m+1)L}) + \xi_k\right)\right. \\ & \times \left.\left(\sum_{m=0}^{\frac{M-2}{2}} \left(x_{k-(2m+1)L} + d_{l-\lfloor\frac{L}{2\beta}\rfloor-1} x_{k-(2m+2)L}\right) + \xi_{k-L}\right)\right] \\ & + \sum_{k=2(l-1)\beta+L-2\beta\lfloor\frac{L}{2\beta}\rfloor+1}^{2l\beta} \left[\left(\sum_{m=0}^{\frac{M-2}{2}} (x_{k-2mL} + d_l x_{k-(2m+1)L}) + \xi_k\right)\right. \\ & \times \left.\left(\sum_{m=0}^{\frac{M-2}{2}} \left(x_{k-(2m+1)L} + d_{l-\lfloor\frac{L}{2\beta}\rfloor} x_{k-(2m+2)L}\right) + \xi_{k-L}\right)\right] \\ = & \lambda_{\text{even}} + \sigma_{\text{even}} + \mu_{\text{even}} + \nu_{\text{even}}, \end{aligned} \tag{5.2.3}$$

and

$$
\begin{aligned}
y_l = & \sum_{k=2(l-1)\beta+1}^{2(l-1)\beta+L-2\beta\lfloor \frac{L}{2\beta}\rfloor} \left[\left(\sum_{m=0}^{\frac{M-1}{2}} x_{k-2mL} + d_l \sum_{m=0}^{\frac{M-3}{2}} x_{k-(2m+1)L} + \xi_k \right) \right. \\
& \left. \times \left(\sum_{m=0}^{\frac{M-1}{2}} x_{k-(2m+1)L} + d_{l-\lfloor \frac{L}{2\beta}\rfloor -1} \sum_{m=0}^{\frac{M-3}{2}} x_{k-(2m+2)L} + \xi_{k-L} \right) \right] \\
& + \sum_{k=2(l-1)\beta+L-2\beta\lfloor \frac{L}{2\beta}\rfloor+1}^{2l\beta} \left[\left(\sum_{m=0}^{\frac{M-1}{2}} x_{k-2mL} + d_l \sum_{m=0}^{\frac{M-3}{2}} x_{k-(2m+1)L} + \xi_k \right) \right. \\
& \left. \times \left(\sum_{m=0}^{\frac{M-1}{2}} x_{k-(2m+1)L} + d_{l-\lfloor \frac{L}{2\beta}\rfloor} \sum_{m=0}^{\frac{M-3}{2}} x_{k-(2m+2)L} + \xi_{k-L} \right) \right] \\
= & \lambda_{\text{odd}} + \sigma_{\text{odd}} + \mu_{\text{odd}} + \nu_{\text{odd}},
\end{aligned}
\tag{5.2.4}
$$

where λ_{even}, σ_{even}, μ_{even}, ν_{even}, λ_{odd}, σ_{odd}, μ_{odd} and ν_{odd} are defined as in Appendix 5A. Also $\{\lambda_{\text{even}}, \sigma_{\text{even}}, \mu_{\text{even}}, \nu_{\text{even}}\}$ and $\{\lambda_{\text{odd}}, \sigma_{\text{odd}}, \mu_{\text{odd}}, \nu_{\text{odd}}\}$ denote the sets of the required signal, the inter-symbol interference, the intra-signal interference, and the noise component when M is even and odd, respectively. Based on the value of y_l, the symbol is decoded according to (5.1.7). It can be observed that when L is small compared to 2β, the useful signal component in the GCDSK receiver is approximately $(M-1)$ times larger than that of the CDSK case. Although both the intra-signal interference and the noise component increase compared to the CDSK case, their effect can be compensated by the increase in signal component. Therefore, with appropriate values of M and L, GCDSK can be designed to outperform CDSK.

5.2.3 Gaussian-Approximated Bit Error Rates

In our study, we make use of the Chebyshev map of degree 2 to generate the chaotic signal. The map is given by

$$x_{k+1} = 2x_k^2 - 1, \tag{5.2.5}$$

and its correlation properties can be found in Appendix 3A [see also: Geisel and Fairen, 1984; Lau and Tse, 2003a]. Denote

$$\check{\boldsymbol{d}}_l = \left[d_{l-\lfloor \frac{L}{2\beta}\rfloor} \; d_{l-\lfloor \frac{L}{2\beta}\rfloor -1} \right]. \tag{5.2.6}$$

Assuming that the chaotic signal $\{x_k\}$ is stationary, it is readily shown that for a given transmitted symbol d_l, the mean value of the correlator output, denoted by $\mathrm{E}[y_l \mid (d_l, \check{\boldsymbol{d}}_l)]$, equals

$$\mathrm{E}\big[y_l \mid (d_l, \check{\boldsymbol{d}}_l)\big] = 2d_l\beta(M-1)P_s - d_l L\left\lfloor \frac{M-1}{2} \right\rfloor P_s + d_{l-1}L\left\lfloor \frac{M-1}{2} \right\rfloor P_s \tag{5.2.7}$$

if $L < 2\beta$ and

$$\begin{aligned}\mathrm{E}\big[y_l \mid (d_l, \check{\boldsymbol{d}}_l)\big] &= 2d_l\beta\left[(M-1) - \left\lfloor \frac{M-1}{2} \right\rfloor\right]P_s \\ &+ d_{l-\lfloor \frac{L}{2\beta} \rfloor}\left[2\beta - \left(L - 2\beta\left\lfloor \frac{L}{2\beta} \right\rfloor\right)\right]\left\lfloor \frac{M-1}{2} \right\rfloor P_s \\ &+ d_{l-\lfloor \frac{L}{2\beta} \rfloor - 1}\left(L - 2\beta\left\lfloor \frac{L}{2\beta} \right\rfloor\right)\left\lfloor \frac{M-1}{2} \right\rfloor P_s\end{aligned} \tag{5.2.8}$$

if $L \geqslant 2\beta$, where

$$P_s = \mathrm{E}\big[x_k^2\big] \tag{5.2.9}$$

denotes the average output power of the chaos generator. From (5.2.7), it can be observed that when $M > 2$ and $L < 2\beta$, the useful part of the signal ($2d_l\beta(M-1)P_s - d_l L\lfloor\frac{M-1}{2}\rfloor P_s$) decreases with L whereas the intersymbol interference ($d_{l-1}L\lfloor\frac{M-1}{2}\rfloor P_s$) increases with L. For the case when $M > 2$ and $L \geqslant 2\beta$, (5.2.8) indicates that the useful signal and the intersymbol interference remain more or less unchanged with L. In particular, for $M = 2$, i.e., for the CDSK system, the mean value of y_l does not depend on $d_{l-\lfloor \frac{L}{2\beta} \rfloor}$ and $d_{l-\lfloor \frac{L}{2\beta} \rfloor - 1}$, and both (5.2.7) and (5.2.8) can be simplified to

$$\mathrm{E}[y_l \mid d_l] = 2d_l\beta P_s. \tag{5.2.10}$$

Assuming that the conditional correlator output follows a Gaussian distribution, the approximate bit error rate for the GCDSK system can then be derived analytically and is given by (see Appendix 5B for details)

$$\begin{aligned}\mathrm{BER_{GCDSK}} = \sum_{\check{\boldsymbol{d}}_l \in \{-1,+1\}^2} &\left[\frac{1}{2}\mathrm{Prob}\big(y_l \leqslant 0 \mid (d_l = +1, \check{\boldsymbol{d}}_l)\big)\,\mathrm{Prob}(\check{\boldsymbol{d}}_l)\right. \\ &\left. + \frac{1}{2}\mathrm{Prob}\big(y_l > 0 \mid (d_l = -1, \check{\boldsymbol{d}}_l)\big)\,\mathrm{Prob}(\check{\boldsymbol{d}}_l)\right]\end{aligned}$$

$$= \frac{1}{16} \sum_{\check{\boldsymbol{d}}_l \in \{-1,+1\}^2} \left[\operatorname{erfc}\left(\frac{\mathrm{E}[y_l \mid (d_l = +1, \check{\boldsymbol{d}}_l)]}{\sqrt{2 \operatorname{var}[y_l \mid (d_l = +1, \check{\boldsymbol{d}}_l)]}} \right) + \operatorname{erfc}\left(\frac{\mathrm{E}[y_l \mid (d_l = -1, \check{\boldsymbol{d}}_l)]}{\sqrt{2 \operatorname{var}[y_l \mid (d_l = -1, \check{\boldsymbol{d}}_l)]}} \right) \right] \tag{5.2.11}$$

and the relevant expressions for the variances are shown in (5.B.4) to (5.B.16).

5.2.4 Results and Discussions

In this section, we present our findings on the bit error performance of the GCDSK system. We denote the average bit energy by E_b which can be readily shown equal to

$$E_b = 2\beta M P_s. \tag{5.2.12}$$

For various average bit-energy-to-noise-power-spectral-density ratios (E_b/N_0), we simulate the GCDSK system and record the BERs. Also, we compute the approximate BERs using (5.2.11). We then compare our results with those derived from the CDSK and DCSK systems whenever appropriate.

5.2.4.1 *Effects of Delay L* First, we investigate the effects of the delay L on the bit error performance. Figure 5.3 plots the BER of the CDSK system together with that of the GCDSK system with $M = 4$ and 6. A spreading factor of 100 is used, i.e., $2\beta = 100$. It is observed that the bit error performance for the GCDSK system degrades as the delay L increases. It is readily shown that there are L terms that do not contain any useful component in the correlation process and will only add more uncertainty to the detected signal. Hence, the correlator output becomes more unreliable when $(2\beta - L)$ reduces due to an increase in L, thereby increasing the BER. For the CDSK system, all 2β terms contain the useful component in the correlator block and is independent of the delay L. Therefore, the bit error performance of the CDSK system is found to be unaffected by the delay L. Comparing the CDSK and GCDSK systems, it can be observed that for the same E_b/N_0, the GCDSK system outperforms the CDSK system with delay values up to 60.

In Fig. 5.4, the simulated BER is plotted against E_b/N_0 for the CDSK and GCDSK systems. The spreading factor is kept at 100. Again, it is shown that for a fixed E_b/N_0, the BER for the CDSK system is hardly affected by the delay L whereas the BER for the GCDSK system increases as L increases. As expected, the BER for both systems improves with increasing E_b/N_0.

5.2.4.2 *Effect of the Number of Delay Blocks* Given that the number of delay blocks in the transmitter side equals $(M - 1)$. In Fig. 5.5, we plot the simu-

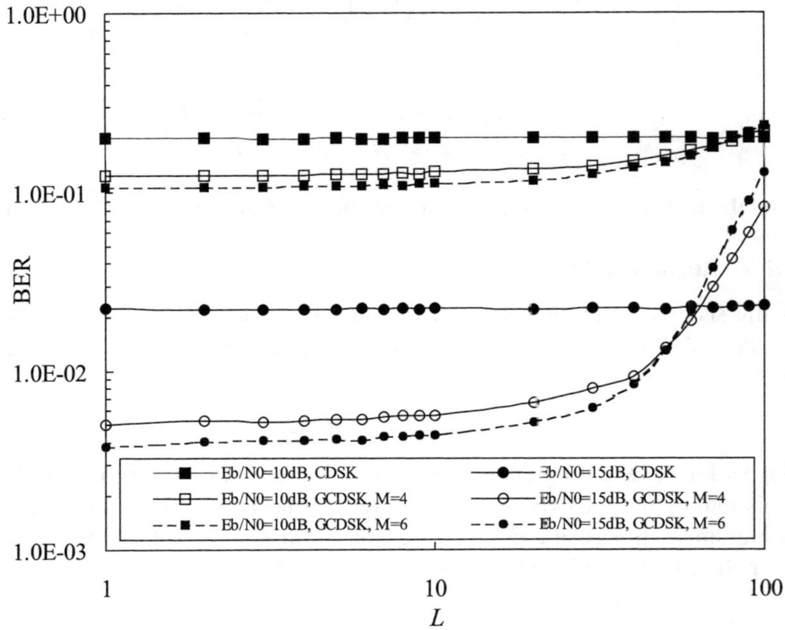

Figure 5.3. Simulated BER versus delay L for CDSK and GCDSK systems. $2\beta = 100$.

lated BER against M. (Note that the CDSK system corresponds to the case where $M = 2$.) A spreading factor of 100 is used. It is shown that for a fixed E_b/N_0 value, the BER reaches an optimum value at a certain value of M. Specifically, when M increases, the average bit energy and the detected signal component given by (5.2.12) and (5.2.7)–(5.2.8), respectively, increase initially. Although the intra-signal interference and the noise power also go up for a given E_b/N_0, there is a net improvement in the signal quality initially, thereby improving the BER. As the value of M is further increased beyond the optimum point, the percentage increase in the detected signal component is overshadowed by the degradation due to intra-signal interference and noise. Therefore, the BER starts to degrade for large values of M.

5.2.4.3 *Effects of Spreading Factor* Next, we study the effects of the spreading factor on the bit error performance. Figure 5.6 plots the simulated BER versus the spreading factor. A delay value $L = 1$ is used. As in other noncoherent chaos-based communication systems, it is observed that the BER improves initially before an optimum point is reached [Sushchik *et al.*, 2000; Tse *et al.*, 2002]. Apparently, the gain in the signal component is significant as the spreading factor

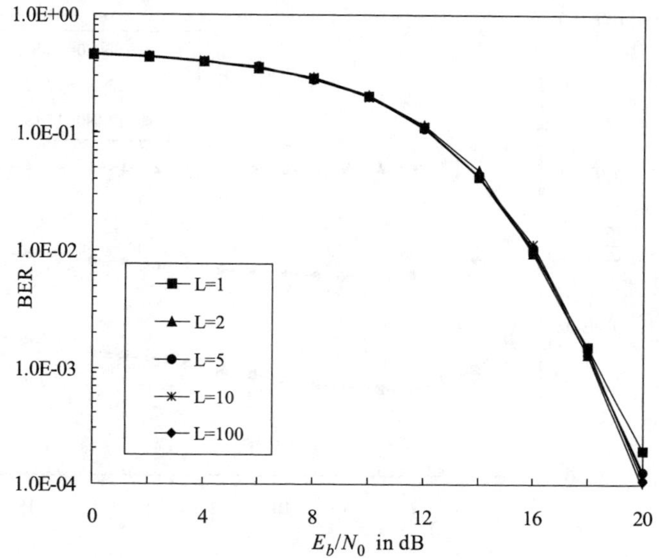

(a)

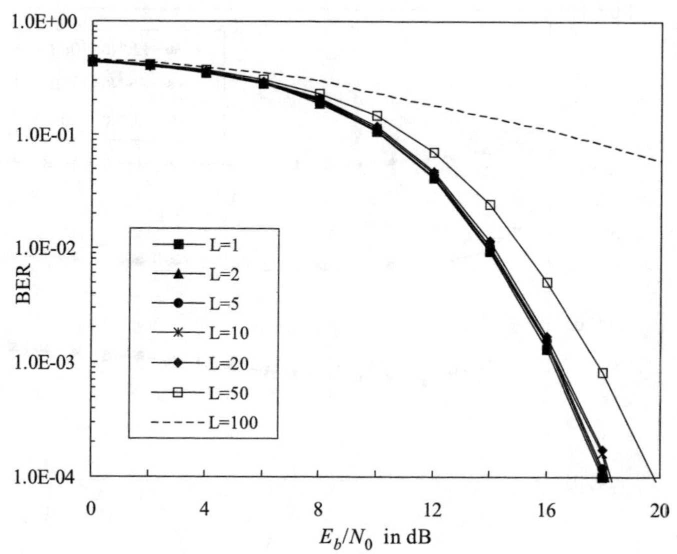

(b)

Figure 5.4. Simulated BER versus E_b/N_0. $2\beta = 100$. (a) CDSK system; (b) GCDSK system with $M = 6$.

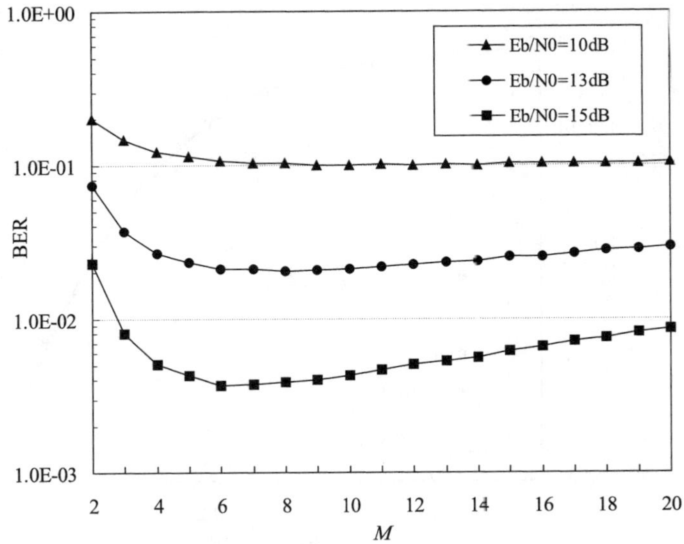

(a)

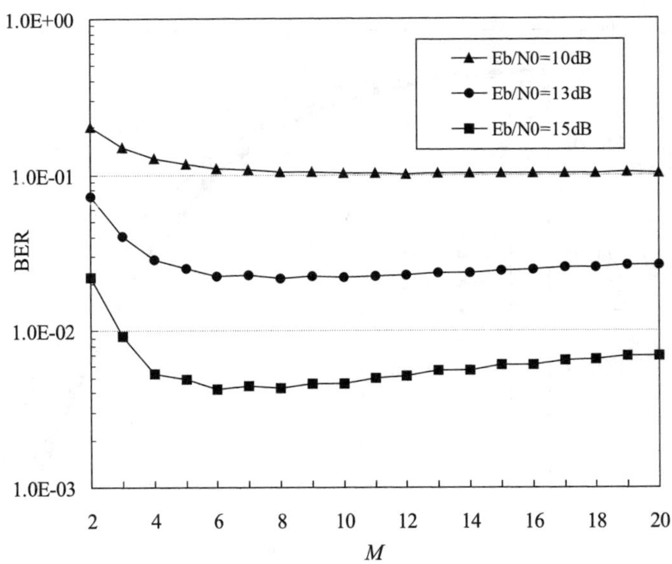

(b)

Figure 5.5. Simulated BER versus M for the CDSK ($M = 2$) and GCDSK systems. $2\beta = 100$. (a) $L = 1$; (b) $L = 5$.

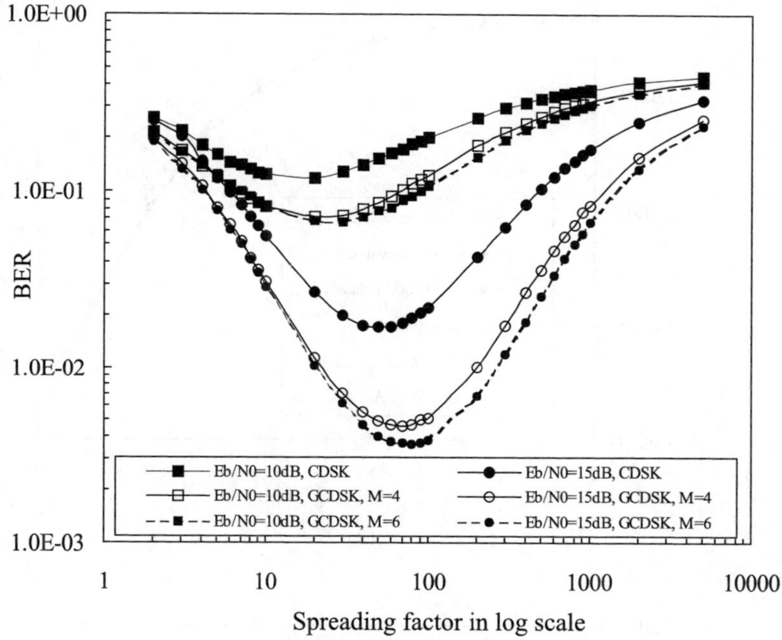

Figure 5.6. Simulated BER versus spreading factor for the CDSK and GCDSK systems. $L = 1$.

first increases. Further increasing the spreading factor beyond the critical point degrades the performance because the increase in noise component become more prominent. It is also found that for different combinations of E_b/N_0 and M, the corresponding optimum values of the spreading factor are different.

5.2.4.4 *Comparison of the Simulated and Gaussian-Approximated BERs* Figure 5.7 compares the simulated and Gaussian-approximated BERs for the CDSK and GCDSK systems (see Appendix 5C for the derivation of the Gaussian-approximated BER for the CDSK system). A spreading factor of 100 is used. In Fig. 5.7(a), it is observed that for the CDSK system, the simulated and Gaussian-approximated BERs are quite close when the delay L is large, which is in good agreement with the results reported by Sushchik *et al.* [2000]. For the GCDSK system with $M = 6$, the discrepancy between the simulated and Gaussian-approximated BERs is quite significant when E_b/N_0 is large. When we analyze the statistics of the correlator output from the simulation results, it is found that the conditional means and variances match with those derived in

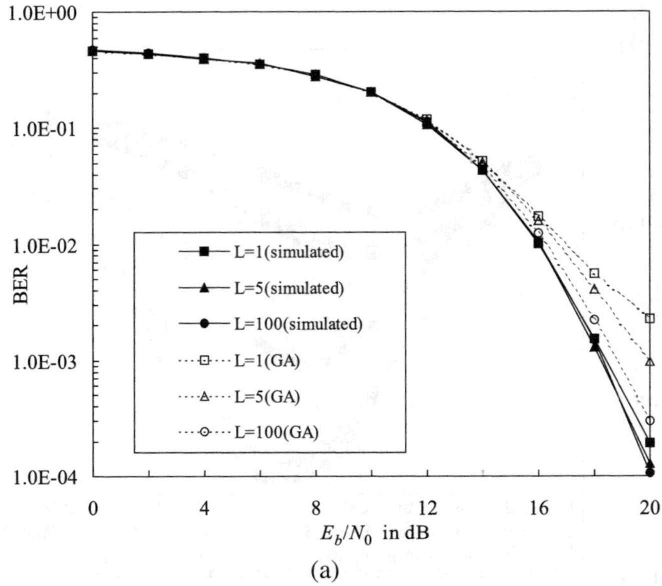

(a)

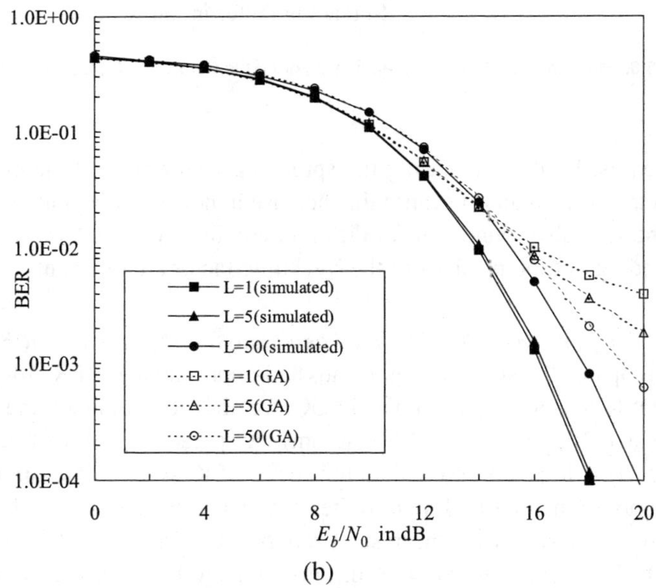

(b)

Figure 5.7. Simulated and Gaussian-approximated BERs versus E_b/N_0. $2\beta = 100$. Simulated results are plotted as solid lines and Gaussian-approximated results are plotted as dotted lines. (a) CDSK system; (b) GCDSK system with $M = 6$.

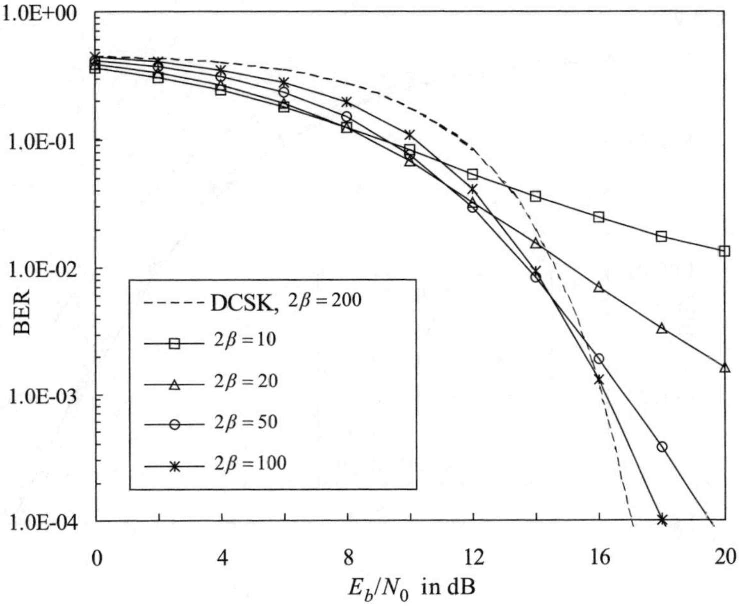

Figure 5.8. Simulated BER versus E_b/N_0 for the GCDSK ($L = 1$, $M = 6$) and DCSK systems.

Appendix 5B. But the conditional correlator output does not exactly follow a Gaussian distribution.

5.2.4.5 *Comparison with DCSK System* Figure 5.8 plots the simulated BERs of the GCDSK and DCSK systems. For the DCSK system, a spreading factor of 200 is used, i.e., $2\beta = 200$. It can be observed that with $L = 1$ and $2\beta = 100$, the GCDSK system can achieve similar BER performance as the DCSK system. In particular, when E_b/N_0 is lower than 16 dB, the GCDSK system slightly outperforms the DCSK one.

In Fig. 5.9, the simulated BERs are plotted again for the CDSK, GCDSK and DCSK systems. A spreading factor of 100 is used for all systems and a delay of 1 is employed for the CDSK and GCDSK systems. To achieve the performance of $\text{BER} = 10^{-3}$, the required E_b/N_0 value of the CDSK system is 2 dB larger than that of the DCSK system. For the GCDSK system, the BERs are about the same for $M = 4, 6, 8$ and 10. Its performance is similar to that of the DCSK system and is better for E_b/N_0 values below 16 dB.

Having evaluated the performance of the GCDSK system for a single-user case, we proceed to study the GCDSK system in a multi-user environment.

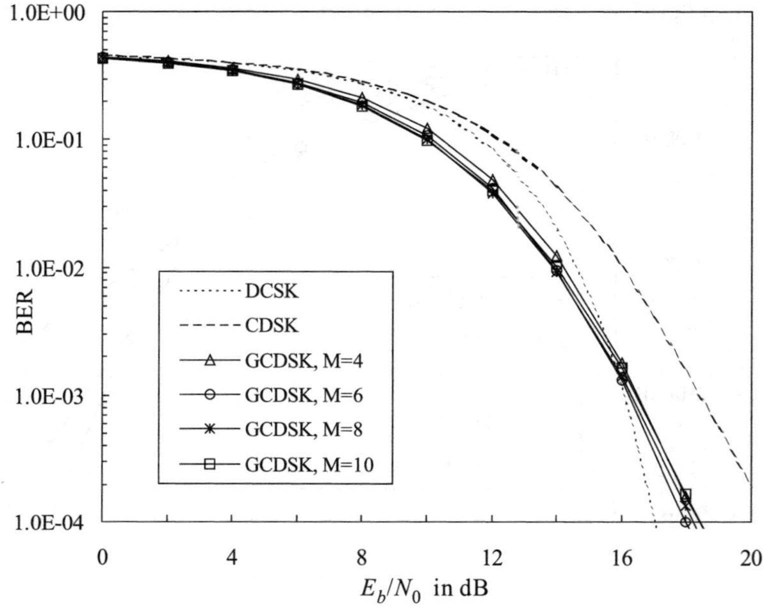

Figure 5.9. Simulated BER versus E_b/N_0 for the CDSK, GCDSK and DCSK systems. $2\beta = 100$ for all systems. $L = 1$ for the CDSK and GCDSK systems.

5.3 MULTIPLE-ACCESS GCDSK SCHEME

5.3.1 Transmitter Structure of the MA-GCDSK System

The transmitter structure of the multiple-access GCDSK (MA-GCDSK) communication system is shown in Fig. 5.10. The chaotic signal and the minimum delay for the ith transmitter are now denoted by $x_k^{(i)}$ and $L^{(i)}$, respectively, which will be used to distinguish different users. To ensure that all the transceivers operate with similar performance, we assume that all users in the system employ the same number of delay blocks in the transmitter, i.e.,

$$M^{(1)} = M^{(2)} = \cdots = M^{(N)} = M. \tag{5.3.1}$$

Denote the spreading factor by 2β. During the lth bit duration, i.e., for $k = 2(l-1)\beta + 1, 2(l-1)\beta + 2, \ldots, 2l\beta$, the transmitted signal for the ith user, $s_k^{(i)}$, is

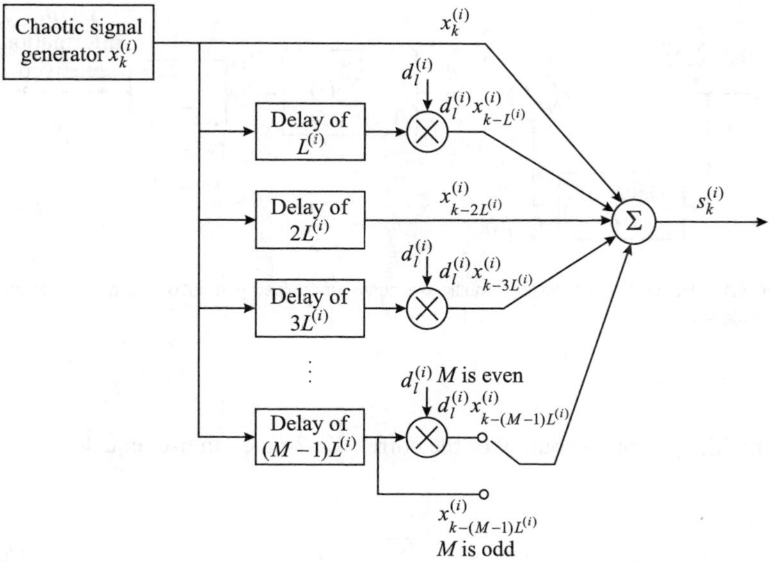

Figure 5.10. Transmitter structure of a multiple-access generalized correlation-delay-shift-keying communication system.

given by

$$s_k^{(i)} = \begin{cases} \sum_{m=0}^{\frac{M-2}{2}} x_{k-2mL^{(i)}}^{(i)} + d_l^{(i)} \sum_{m=0}^{\frac{M-2}{2}} x_{k-(2m+1)L^{(i)}}^{(i)} & \text{if } M \text{ is even,} \\ \sum_{m=0}^{\frac{M-1}{2}} x_{k-2mL^{(i)}}^{(i)} + d_l^{(i)} \sum_{m=0}^{\frac{M-3}{2}} x_{k-(2m+1)L^{(i)}}^{(i)} & \text{if } M \text{ is odd,} \end{cases} \tag{5.3.2}$$

where $d_l^{(i)}$ represents the lth transmitted symbol of the ith user. The overall transmitted signal of the system is then given by

$$s_k = \sum_{i=1}^{N} s_k^{(i)}. \tag{5.3.3}$$

5.3.2 Receiver Structure of the MA-GCDSK System

The receiver structure of the MA-GCDSK system is depicted in Fig. 5.11. Assuming that the transmitted signal is contaminated by an AWGN source, the received signal, denoted by r_k, is given by

$$r_k = s_k + \xi_k. \tag{5.3.4}$$

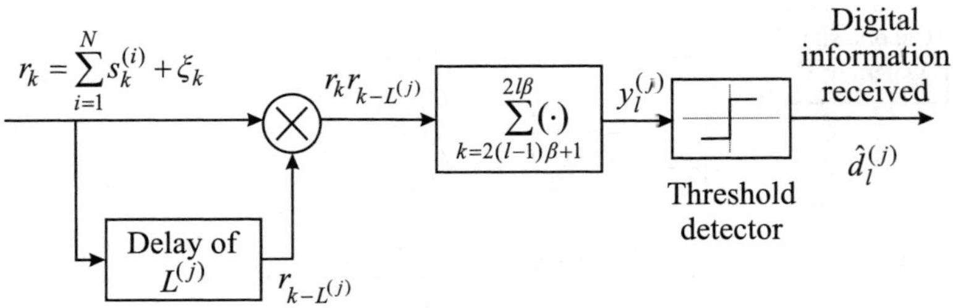

Figure 5.11. Receiver structure of a multiple-access generalized correlation-delay-shift-keying communication system.

For the lth symbol, the output of the correlator for the jth user equals

$$y_l^{(j)} = \sum_{k=2(l-1)\beta+1}^{2l\beta} r_k r_{k-L^{(j)}}. \tag{5.3.5}$$

When M is even and odd, respectively, the expressions for $y_l^{(j)}$ are

$$\begin{aligned}
y_l^{(j)} = & \sum_{k=2(l-1)\beta+1}^{2(l-1)\beta+L^{(j)}-2\beta\lfloor \frac{L^{(j)}}{2\beta} \rfloor} \left[\left(\sum_{i=1}^{N} \sum_{m=0}^{\frac{M-2}{2}} \left(x_{k-2mL^{(i)}}^{(i)} + d_l^{(i)} x_{k-(2m+1)L^{(i)}}^{(i)} \right) + \xi_k \right) \right. \\
& \times \left. \left(\sum_{i=1}^{N} \sum_{m=0}^{\frac{M-2}{2}} \left(x_{k-2mL^{(i)}-L^{(j)}}^{(i)} + d_{l-\lfloor \frac{L^{(i)}}{2\beta} \rfloor -1}^{(i)} x_{k-(2m+1)L^{(i)}-L^{(j)}}^{(i)} \right) + \xi_{k-L^{(j)}} \right) \right] \\
& + \sum_{\substack{k=2(l-1)\beta+L^{(j)} \\ -2\beta\lfloor L^{(j)}/2\beta \rfloor+1}}^{2l\beta} \left[\left(\sum_{i=1}^{N} \sum_{m=0}^{\frac{M-2}{2}} \left(x_{k-2mL^{(i)}}^{(i)} + d_l^{(i)} x_{k-(2m+1)L^{(i)}}^{(i)} \right) + \xi_k \right) \right. \\
& \times \left. \left(\sum_{i=1}^{N} \sum_{m=0}^{\frac{M-2}{2}} \left(x_{k-2mL^{(i)}-L^{(j)}}^{(i)} + d_{l-\lfloor \frac{L^{(i)}}{2\beta} \rfloor}^{(i)} x_{k-(2m+1)L^{(i)}-L^{(j)}}^{(i)} \right) + \xi_{k-L^{(j)}} \right) \right] \\
= & \lambda_{\text{even}}^{(j)} + \sigma_{\text{even}}^{(j)} + \mu_{\text{even}}^{(j)} + \theta_{\text{even}}^{(j)} + \nu_{\text{even}}^{(j)}
\end{aligned} \tag{5.3.6}$$

and

$$
\begin{aligned}
y_l^{(j)} = & \sum_{\substack{k=2(l-1)\beta+1}}^{\substack{2(l-1)\beta+L^{(j)}\\-2\beta\lfloor L^{(j)}/2\beta\rfloor}} \left[\left(\sum_{i=1}^{N} \left(\sum_{m=0}^{\frac{M-1}{2}} x_{k-2mL^{(i)}}^{(i)} + d_l^{(i)} \sum_{m=0}^{\frac{M-3}{2}} x_{k-(2m+1)L^{(i)}}^{(i)} \right) + \xi_k \right) \right. \\
& \times \left(\sum_{i=1}^{N} \left(\sum_{m=0}^{\frac{M-1}{2}} x_{k-2mL^{(i)}-L^{(j)}}^{(i)} \right. \right. \\
& \left. \left. \left. + d_{l-\lfloor \frac{L^{(i)}}{2\beta} \rfloor -1}^{(i)} \sum_{m=0}^{\frac{M-3}{2}} x_{k-(2m+1)L^{(i)}-L^{(j)}}^{(i)} \right) + \xi_{k-L^{(j)}} \right) \right] \\
& + \sum_{\substack{k=2(l-1)\beta+L^{(j)}\\-2\beta\lfloor L^{(j)}/2\beta\rfloor+1}}^{2l\beta} \left[\left(\sum_{i=1}^{N} \left(\sum_{m=0}^{\frac{M-1}{2}} x_{k-2mL^{(i)}}^{(i)} + d_l^{(i)} \sum_{m=0}^{\frac{M-3}{2}} x_{k-(2m+1)L^{(i)}}^{(i)} \right) + \xi_k \right) \right. \\
& \times \left(\sum_{i=1}^{N} \left(\sum_{m=0}^{\frac{M-1}{2}} x_{k-2mL^{(i)}-L^{(j)}}^{(i)} \right. \right. \\
& \left. \left. \left. + d_{l-\lfloor \frac{L^{(i)}}{2\beta} \rfloor}^{(i)} \sum_{m=0}^{\frac{M-3}{2}} x_{k-(2m+1)L^{(i)}-L^{(j)}}^{(i)} \right) + \xi_{k-L^{(j)}} \right) \right] \\
= & \lambda_{\text{odd}}^{(j)} + \sigma_{\text{odd}}^{(j)} + \mu_{\text{odd}}^{(j)} + \theta_{\text{odd}}^{(j)} + \nu_{\text{odd}}^{(j)}. \qquad (5.3.7)
\end{aligned}
$$

In (5.3.6) and (5.3.7), $\{\lambda_{\text{even}}^{(j)}, \sigma_{\text{even}}^{(j)}, \mu_{\text{even}}^{(j)}, \theta_{\text{even}}^{(j)}, \nu_{\text{even}}^{(j)}\}$ and $\{\lambda_{\text{odd}}^{(j)}, \sigma_{\text{odd}}^{(j)}, \mu_{\text{odd}}^{(j)}, \theta_{\text{odd}}^{(j)}, \nu_{\text{odd}}^{(j)}\}$ denote the sets of the required signal, the inter-symbol interference, the intra-user interference, the inter-user interference, and the noise component for the jth user when M is even and odd, respectively. Expressions for the aforementioned variables can be found in Appendix 5D. Finally, note that in the case of a single-user system, the inter-user interference ($\theta_{\text{even}}^{(j)}$ or $\theta_{\text{odd}}^{(j)}$) will disappear.

5.3.3 Gaussian-Approximated Bit Error Rates

In our study, we assume that the Chebyshev map of degree 2 is used as the chaos generator for each user, while each generator starts with a different initial condition. Thus, the mean value of the jth correlator output can be shown equal to

$$\mathrm{E}\big[y_l^{(j)} \mid (d_l^{(j)}, \check{\boldsymbol{d}}_l^{(j)})\big] = \begin{cases} 2d_l^{(j)}\beta(M-1)P_s - d_l^{(j)}L^{(j)}\lfloor\frac{M-1}{2}\rfloor P_s \\ \quad + d_{l-1}^{(j)}L^{(j)}\lfloor\frac{M-1}{2}\rfloor P_s \quad \text{if } L^{(j)} < 2\beta, \\ 2d_l^{(j)}\beta\big[(M-1) - \lfloor\frac{M-1}{2}\rfloor\big]P_s \\ \quad + d^{(j)}_{l-\lfloor\frac{L^{(j)}}{2\beta}\rfloor}\big[2\beta - (L^{(j)} - 2\beta\lfloor\frac{L^{(j)}}{2\beta}\rfloor)\big]\lfloor\frac{M-1}{2}\rfloor P_s \\ \quad + d^{(j)}_{l-\lfloor\frac{L^{(j)}}{2\beta}\rfloor-1}(L^{(j)} - 2\beta\lfloor\frac{L^{(j)}}{2\beta}\rfloor)\lfloor\frac{M-1}{2}\rfloor P_s \quad \text{if } L^{(j)} \geqslant 2\beta, \end{cases} \tag{5.3.8}$$

where

$$\check{\boldsymbol{d}}_l^{(j)} = \big[d^{(j)}_{l-\lfloor\frac{L}{2\beta}\rfloor} \;\; d^{(j)}_{l-\lfloor\frac{L}{2\beta}\rfloor-1}\big]. \tag{5.3.9}$$

Similar to the single-user case, we assume that the correlator output follows a normal distribution. Then the approximate bit error rate for the jth user can be found using

$$\begin{aligned} \mathrm{BER}_{\mathrm{GCDSK}}^{(j)} &= \sum_{\check{\boldsymbol{d}}_l^{(j)} \in \{-1,+1\}^2} \bigg[\frac{1}{2}\mathrm{Prob}\big(y_l^{(j)} \leqslant 0 \,\big|\, (d_l^{(j)} = +1, \check{\boldsymbol{d}}_l^{(j)})\big)\,\mathrm{Prob}\big(\check{\boldsymbol{d}}_l^{(j)}\big) \\ &\quad + \frac{1}{2}\mathrm{Prob}\big(y_l^{(j)} > 0 \,\big|\, (d_l^{(j)} = -1, \check{\boldsymbol{d}}_l^{(j)})\big)\,\mathrm{Prob}\big(\check{\boldsymbol{d}}_l^{(j)}\big)\bigg] \\ &= \frac{1}{16}\sum_{\check{\boldsymbol{d}}_l^{(j)} \in \{-1,+1\}^2} \left[\mathrm{erfc}\left(\frac{\mathrm{E}[y_l^{(j)} \mid (d_l^{(j)} = +1, \check{\boldsymbol{d}}_l^{(j)})]}{\sqrt{2\,\mathrm{var}[y_l^{(j)} \mid (d_l^{(j)} = +1, \check{\boldsymbol{d}}_l^{(j)})]}}\right)\right. \\ &\quad \left. + \,\mathrm{erfc}\left(\frac{\mathrm{E}[y_l^{(j)} \mid (d_l^{(j)} = -1, \check{\boldsymbol{d}}_l^{(j)})]}{\sqrt{2\,\mathrm{var}[y_l^{(j)} \mid (d_l^{(j)} = -1, \check{\boldsymbol{d}}_l^{(j)})]}}\right)\right]. \end{aligned} \tag{5.3.10}$$

Details of the derivation are shown in Appendix 5E.

5.3.4 Results and Discussions

In this section, we present our simulation and analytical results under different scenarios. We assume that all users make use of the Chebyshev map of degree 2 as the chaos generator. We also denote the delay vector by $\boldsymbol{L}$, which is defined as

$$\boldsymbol{L} = \big[L^{(1)} \quad L^{(2)} \quad \cdots \quad L^{(N)}\big]. \tag{5.3.11}$$

5.3.4.1 *Effects of the Number of Users* First, we study the effects of the number of users on the bit error performance. A spreading factor of 200 is used. Figure 5.12 plots the simulated average BER versus E_b/N_0 of the MA-GCDSK system for the cases $M = 2$ (MA-CDSK) and $M = 6$. In order to minimize the inter-user interference and to reduce the effect of noise, $L^{(i)}, i = 1, \ldots, N$, should be selected in such a way that the conditions described in (5.E.1) to (5.E.4) are satisfied. Here, we select $\boldsymbol{L} = [11\ 13]$ and $\boldsymbol{L} = [11\ 13\ 19\ 40]$ for the 2- and 4-user systems, respectively. From the results, it is observed that the case where $M = 6$ outperforms the case where $M = 2$. Moreover, as would be expected, the BERs increase with the number of users N in both cases because of the larger inter-user interference.

5.3.4.2 *Effects of Delay* $L^{(i)}$ Next, we investigate the effects of the delay $L^{(i)}$ on the bit error performance. The delay $L^{(i)}$ affects the BER performance in two different ways. First, for individual users, when $L^{(i)}$ increases, the useful part of the signal diminishes while the intra-user interference increases (except for $M = 2$), causing an increase in BER. Second, if the set of delays $\boldsymbol{L}$ is selected in such a way that (5.E.1) to (5.E.4) are satisfied, the amount of inter-user interference as well as the effect of noise can be reduced, thus lowering the BER. Hence, the BER can be optimized by balancing the effects of these two factors.

Figure 5.13 plots the BERs of individual users in a 2-user MA-GCDSK system with $M = 2$ (MA-CDSK system) and $M = 6$. $2\beta = 200$ and $\boldsymbol{L} = [10\ 51]$ are used. As can be observed, for the case where $M = 2$ (MA-CDSK), the BERs for the users are similar, indicating that the BERs are not very much affected by the choice of $L^{(i)}$. However, for the case where $M = 6$, the BER is found to be higher for a larger $L^{(j)}$. For example, when $E_b/N_0 = 20$ dB, the BERs for Users 1 and 2 are 0.0011 and 0.0031, respectively. As explained before, when $L^{(i)}$ increases, the useful part of the signal diminishes while the intra-user interference increases, causing an increase in BER.

Figure 5.14 plots the simulated average BER versus E_b/N_0 in a 2-user MA-GCDSK system using different sets of delays. $2\beta = 200$ and $M = 6$ are used. From the figure, it can be seen that different sets of delays give rise to different BERs. Consider the BERs corresponding to E_b/N_0 over 12 dB, when the BERs are mainly determined by the intra- and inter-user interference. When the delay values are small, for example using $\boldsymbol{L} = [1\ 2]$ and $\boldsymbol{L} = [2\ 4]$, the intra-signal interference is small. However, due to the fact that the inter-user interference is not optimized, i.e., (5.E.1) to (5.E.4) are not satisfied, the BERs are worse than those in other cases where the delays are larger. Among the sets of delays shown, only $\boldsymbol{L} = [11\ 13]$, $\boldsymbol{L} = [50\ 51]$ and $\boldsymbol{L} = [100\ 101]$ meet the requirements described in (5.E.1) to (5.E.4) and they produce comparatively better BER performance. Moreover, the performance for $\boldsymbol{L} = [11\ 13]$ is better than that for $\boldsymbol{L} = [50\ 51]$,

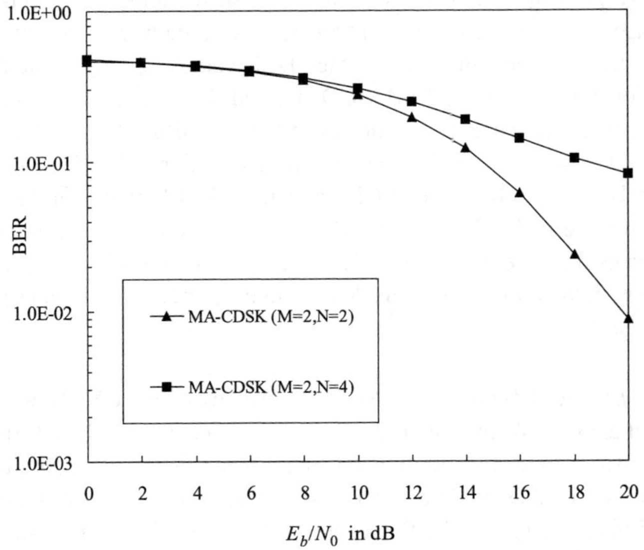

(a)

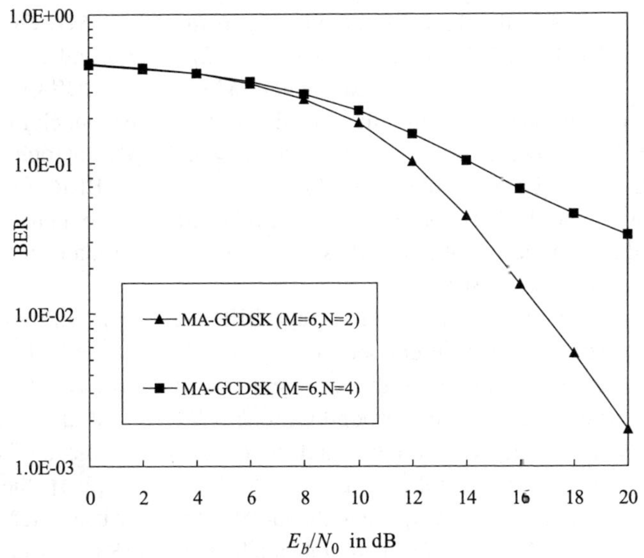

(b)

Figure 5.12. Simulated average BERs versus E_b/N_0 for $N = 2$ and 4. $2\beta = 200$. $\boldsymbol{L} = [11\ 13]$ and $\boldsymbol{L} = [11\ 13\ 19\ 40]$ for the 2- and 4-user MA-GCDSK systems, respectively. (a) $M = 2$ (MA-CDSK system); (b) $M = 6$.

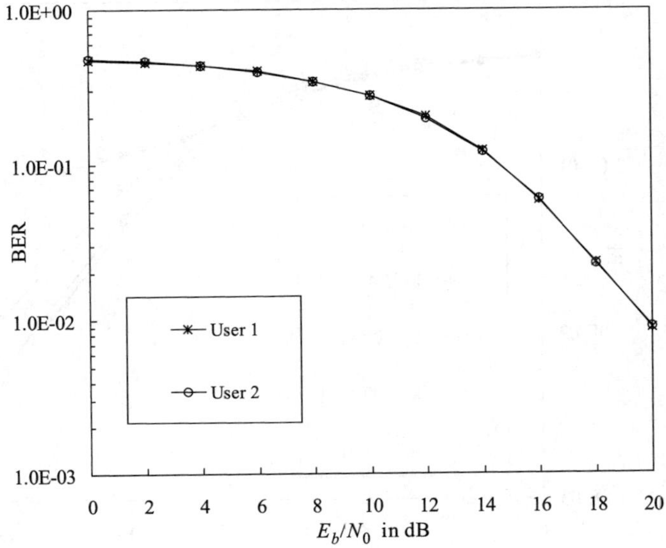

(a)

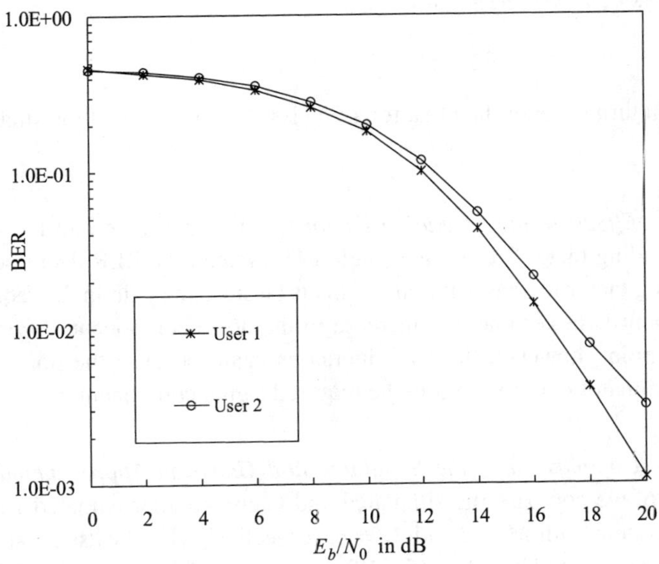

(b)

Figure 5.13. Simulated BERs versus E_b/N_0 in a 2-user MA-GCDSK system. $2\beta = 200$ and $\boldsymbol{L} = [10\ 51]$. (a) $M = 2$ (MA-CDSK system); (b) $M = 6$.

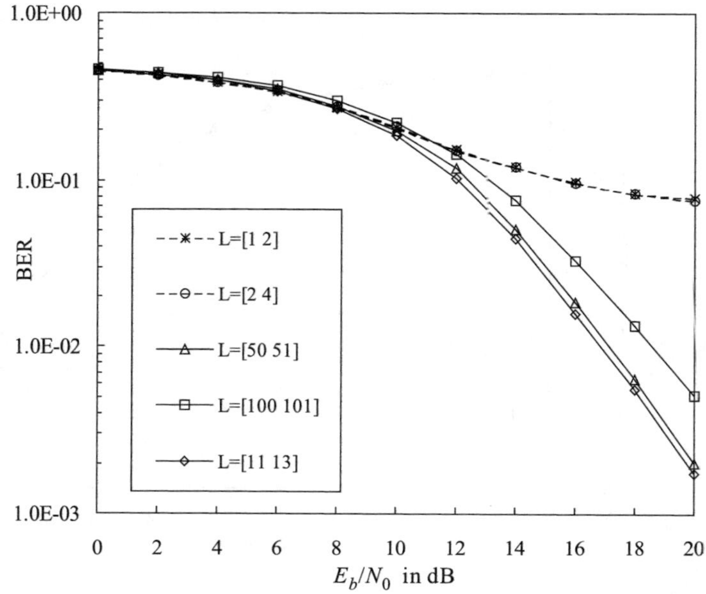

Figure 5.14. Simulated average BER versus E_b/N_0 for the 2-user MA-GCDSK system using different sets of delays. $2\beta = 200$ and $M = 6$.

which in turn is better than that for $\boldsymbol{L} = [100\ 101]$ because of the smaller average delay.

5.3.4.3 *Effects of the Spreading Factor* In Fig. 5.15, we plot the BER versus the spreading factor 2β. As in a single-user system, the BER decreases when the spreading factor increases initially. This is because the gain in the required signal component is larger than the increase in the noise component. After reaching a critical point, however, the BER increases again because the noise component now dominates over the gain in the required signal component.

5.3.4.4 *Comparison of the Simulated and Gaussian-Approximated BERs* In Fig. 5.16, we compare the simulated and Gaussian-approximated BERs for the 2-user system with $M = 2$ and $M = 6$, respectively. The Gaussian-approximated BERs are evaluated based on (5.3.10). A spreading factor of 200 and $\boldsymbol{L} = [10\ 51]$ are used. Since the BERs are different for different users, we plot the BER for each desired user. The curves in Fig. 5.16 indicate that the Gaussian-approximated method can provide a reasonable estimation of the actual system performance.

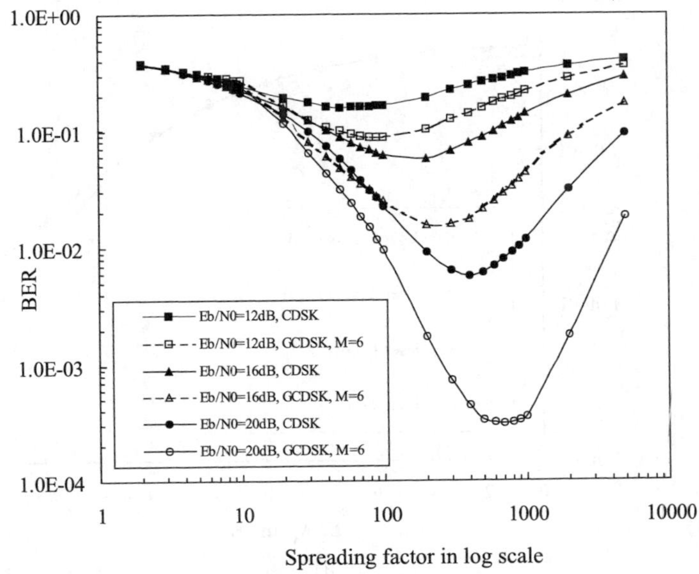

Figure 5.15. Simulated average BERs versus spreading factor for the 2-user MA-GCDSK system using different E_b/N_0 values. $\boldsymbol{L} = [11\ 13]$ and $M = 6$.

5.3.4.5 *Comparison with MA-DCSK System* Figure 5.17 compares the simulated average BERs of the MA-CDSK, MA-GCDSK and MA-DCSK systems. For all systems, the same spreading factor, i.e., 200 is used. For the 2- and 4-user MA-CDSK and MA-GCDSK systems, the sets of delays used are $\boldsymbol{L} = [11\ 13]$ and $\boldsymbol{L} = [11\ 13\ 19\ 40]$, respectively. It can be observed from the figure that the performance of MA-CDSK system is inferior to those of the MA-GCDSK and MA-DCSK systems. Moreover, when E_b/N_0 is below 14 dB, the MA-GCDSK and MA-DCSK schemes have similar performance. Beyond 14 dB, the MA-DCSK scheme slightly outperforms the MA-GCDSK scheme.

5.4 SUMMARY

In this chapter, we have developed and studied in detail a generalized correlation-delay-shift-keying (GCDSK) scheme for noncoherent chaos-based digital communications. We have also compared the bit error rate (BER) of the system with

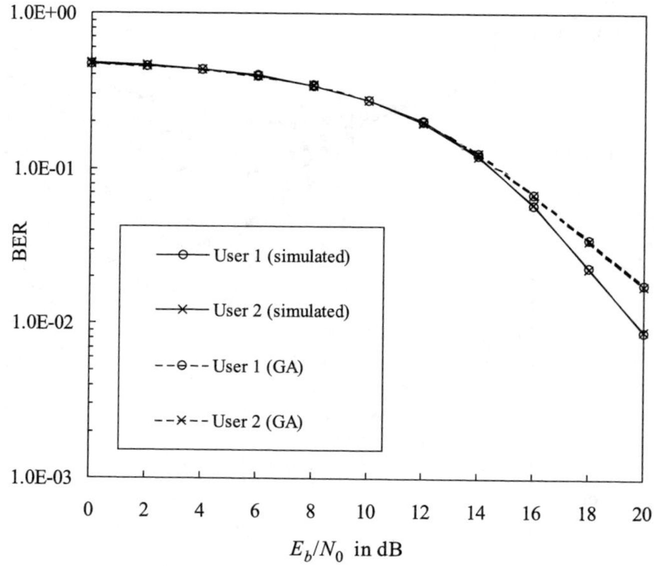

(a)

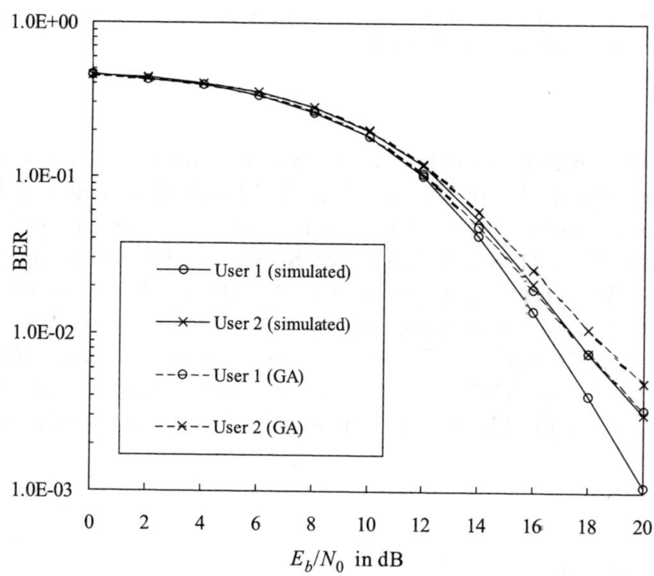

(b)

Figure 5.16. Simulated and Gaussian-approximated (GA) BERs versus E_b/N_0. $2\beta = 200$. Simulated results are plotted as solid lines and Gaussian-approximated results are plotted as dotted lines for the 2-user MA-GCDSK system. $\boldsymbol{L} = [10\ 51]$ is used. (a) $M = 2$ (MA-CDSK system); (b) $M = 6$.

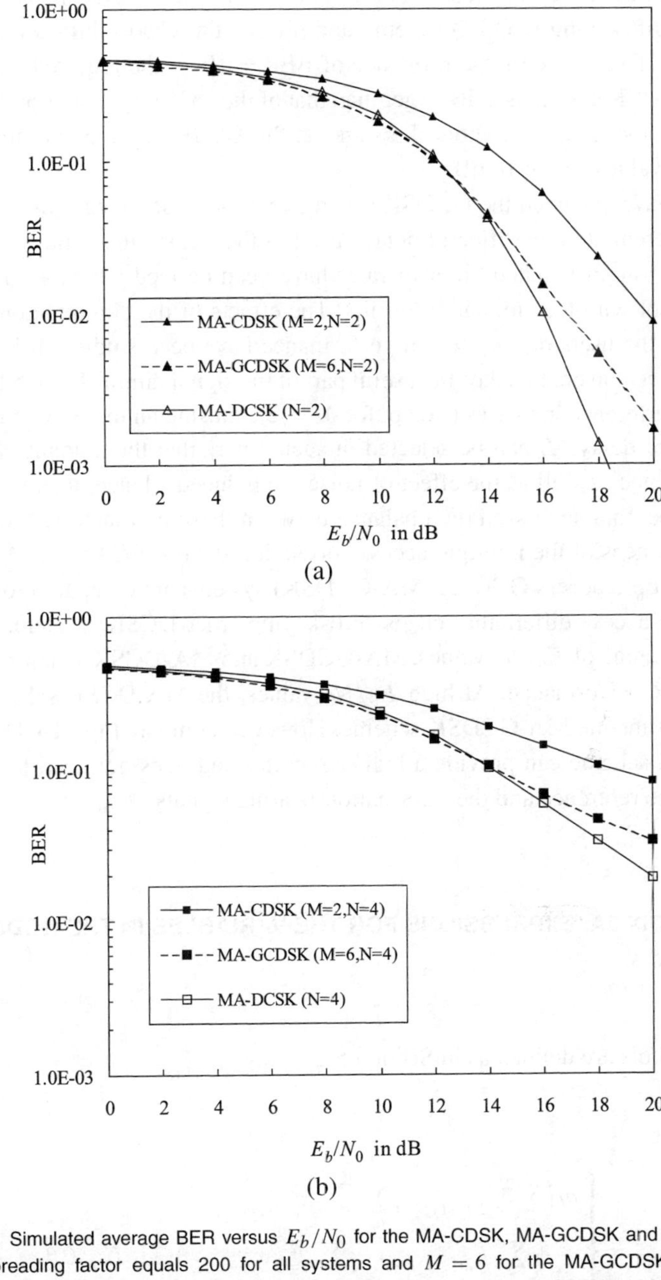

Figure 5.17. Simulated average BER versus E_b/N_0 for the MA-CDSK, MA-GCDSK and MA-DCSK systems. Spreading factor equals 200 for all systems and $M = 6$ for the MA-GCDSK systems. $\boldsymbol{L} = [11\ 13]$ and $\boldsymbol{L} = [11\ 13\ 19\ 40]$, respectively, for the 2- and 4-user MA-CDSK and MA-GCDSK systems. (a) 2-user systems; (b) 4-user systems.

two other noncoherent chaos-based communication schemes, namely correlation-delay-shift-keying (CDSK) scheme and differential chaos-shift-keying (DCSK) scheme. To achieve the performance of BER $= 10^{-3}$, the required E_b/N_0 value of the CDSK system is 2 dB larger than that of the DCSK system. For the GCDSK scheme, the BERs are about the same as the DCSK scheme and are better for E_b/N_0 values below 16 dB.

We have extended the GCDSK scheme to a multiuser environment. The users are differentiated by different delays used in the transmitters and receivers. The Gaussian-approximated bit error rates have been derived which are found to be consistent with the simulation results. The effects of the combination of the delays for the users on the system performance have been studied. It is found that with a larger average delay, the useful part of the signal diminishes while the intra-user interference increases (except for $M = 2$), causing an increase in BER. Also, the set of delays $\boldsymbol{L}$ can be selected in such a way that the amount of inter-user interference as well as the effect of noise are reduced. Hence, the optimum BER would be obtained based on a balance between these two factors. Moreover, the performances of the multiple-access correlation-delay-shift-keying (MA-CDSK) and multiple-access GCDSK (MA-GCDSK) systems are compared to that of the multiple-access differential chaos-shift-keying (MA-DCSK) system. For a reasonable range of E_b/N_0 values, MA-GCDSK and MA-DCSK schemes have similar BER performance. At high E_b/N_0 values, the MA-DCSK scheme slightly outperforms the MA-GCDSK scheme. However, compared to MA-DCSK, MA-GCDSK scheme can provide a higher security and does not need to switch between the reference and the information-bearing signals.

APPENDIX 5A. EXPRESSIONS FOR THE VARIABLES IN THE GCDSK SYSTEM

All symbols are defined as in Section 5.2.

1. λ_{even}

$$\lambda_{\text{even}} = \begin{cases} d_l \Big(\sum_{k=2(l-1)\beta+1}^{2l\beta} \sum_{m=0}^{\frac{M-2}{2}} x_{k-(2m+1)L}^2 \\ \quad + \sum_{k=2(l-1)\beta+L+1}^{2l\beta} \sum_{m=1}^{\frac{M-2}{2}} x_{k-2mL}^2 \Big) & \text{if } L < 2\beta, \\ d_l \sum_{k=2(l-1)\beta+1}^{2l\beta} \sum_{m=0}^{\frac{M-2}{2}} x_{k-(2m+1)L}^2 & \text{if } L \geqslant 2\beta. \end{cases} \tag{5.A.1}$$

2. σ_{even}

$$\sigma_{\text{even}} = \begin{cases} d_{l-1} \sum_{k=2(l-1)\beta+1}^{2(l-1)\beta+L} \left[\sum_{m=0}^{\frac{M-2}{2}} (x_{k-2mL} + d_l x_{k-(2m+1)L}) \right. \\ \quad \left. \times \sum_{m=0}^{\frac{M-2}{2}} x_{k-(2m+2)L} \right] \quad \text{if } L < 2\beta, \\ d_{l-\lfloor \frac{L}{2\beta} \rfloor -1} \sum_{k=2(l-1)\beta+1}^{2(l-1)\beta+L-2\beta\lfloor \frac{L}{2\beta} \rfloor} \left[\sum_{m=0}^{\frac{M-2}{2}} (x_{k-2mL} + d_l x_{k-(2m+1)L}) \right. \\ \quad \left. \times \sum_{m=0}^{\frac{M-2}{2}} x_{k-(2m+2)L} \right] \\ \quad + d_{l-\lfloor \frac{L}{2\beta} \rfloor} \sum_{k=2(l-1)\beta+L-2\beta\lfloor \frac{L}{2\beta} \rfloor+1}^{2l\beta} \left[\sum_{m=0}^{\frac{M-2}{2}} (x_{k-2mL} + d_l x_{k-(2m+1)L}) \right. \\ \quad \left. \times \sum_{m=0}^{\frac{M-2}{2}} x_{k-(2m+2)L} \right] \quad \text{if } L \geqslant 2\beta. \end{cases} \tag{5.A.2}$$

3. μ_{even}

$$\begin{aligned} \mu_{\text{even}} = & \sum_{k=2(l-1)\beta+1}^{2(l-1)\beta+L-2\beta\lfloor \frac{L}{2\beta} \rfloor} \left[\sum_{m=0}^{\frac{M-2}{2}} (x_{k-2mL} + d_l x_{k-(2m+1)L}) \right. \\ & \left. \times \sum_{m=0}^{\frac{M-2}{2}} \left(x_{k-(2m+1)L} + d_{l-\lfloor \frac{L}{2\beta} \rfloor -1} x_{k-(2m+2)L} \right) \right] \\ & + \sum_{k=2(l-1)\beta+L-2\beta\lfloor \frac{L}{2\beta} \rfloor+1}^{2l\beta} \left[\sum_{m=0}^{\frac{M-2}{2}} (x_{k-2mL} + d_l x_{k-(2m+1)L}) \right. \\ & \left. \times \sum_{m=0}^{\frac{M-2}{2}} \left(x_{k-(2m+1)L} + d_{l-\lfloor \frac{L}{2\beta} \rfloor} x_{k-(2m+2)L} \right) \right] \\ & - \lambda_{\text{even}} - \sigma_{\text{even}}. \end{aligned} \tag{5.A.3}$$

4. ν_{even}

$$\begin{aligned} \nu_{\text{even}} = & \sum_{k=2(l-1)\beta+1}^{2l\beta} \sum_{m=0}^{\frac{M-2}{2}} (x_{k-2mL} + d_l x_{k-(2m+1)L}) \xi_{k-L} \\ & + \sum_{k=2(l-1)\beta+1}^{2(l-1)\beta+L-2\beta\lfloor \frac{L}{2\beta} \rfloor} \sum_{m=0}^{\frac{M-2}{2}} \left(x_{k-(2m+1)L} + d_{l-\lfloor \frac{L}{2\beta} \rfloor -1} x_{k-(2m+2)L} \right) \xi_k \end{aligned}$$

$$+\sum_{k=2(l-1)\beta+L-2\beta\lfloor\frac{L}{2\beta}\rfloor+1}^{2l\beta}\sum_{m=0}^{\frac{M-2}{2}}\left(x_{k-(2m+1)L}+d_{l-\lfloor\frac{L}{2\beta}\rfloor}x_{k-(2m+2)L}\right)\xi_k$$

$$+\sum_{k=2(l-1)\beta+1}^{2l\beta}\xi_k\xi_{k-L}. \tag{5.A.4}$$

5. λ_{odd}

$$\lambda_{\text{odd}}=\begin{cases} d_l\Big(\sum_{k=2(l-1)\beta+1}^{2l\beta}\sum_{m=0}^{\frac{M-3}{2}}x_{k-(2m+1)L}^2 \\ \quad+\sum_{k=2(l-1)\beta+L+1}^{2l\beta}\sum_{m=1}^{\frac{M-1}{2}}x_{k-2mL}^2\Big) & \text{if } L<2\beta, \\ d_l\sum_{k=2(l-1)\beta+1}^{2l\beta}\sum_{m=0}^{\frac{M-3}{2}}x_{k-(2m+1)L}^2 & \text{if } L\geqslant 2\beta. \end{cases} \tag{5.A.5}$$

6. σ_{odd}

$$\sigma_{\text{odd}}=\begin{cases} d_{l-1}\sum_{k=2(l-1)\beta+1}^{2(l-1)\beta+L}\Big[\Big(\sum_{m=0}^{\frac{M-1}{2}}x_{k-2mL}+d_l\sum_{m=0}^{\frac{M-3}{2}}x_{k-(2m+1)L}\Big) \\ \quad\times\sum_{m=0}^{\frac{M-3}{2}}x_{k-(2m+2)L}\Big] & \text{if } L<2\beta, \\ d_{l-\lfloor\frac{L}{2\beta}\rfloor-1}\sum_{k=2(l-1)\beta+1}^{2(l-1)\beta+L-2\beta\lfloor\frac{L}{2\beta}\rfloor}\Big[\Big(\sum_{m=0}^{\frac{M-1}{2}}x_{k-2mL} \\ \quad+d_l\sum_{m=0}^{\frac{M-3}{2}}x_{k-(2m+1)L}\Big)\Big(\sum_{m=0}^{\frac{M-3}{2}}x_{k-(2m+2)L}\Big)\Big] \\ \quad+d_{l-\lfloor\frac{L}{2\beta}\rfloor}\sum_{k=2(l-1)\beta+L-2\beta\lfloor\frac{L}{2\beta}\rfloor+1}^{2l\beta}\Big[\Big(\sum_{m=0}^{\frac{M-1}{2}}x_{k-2mL} \\ \quad+d_l\sum_{m=0}^{\frac{M-3}{2}}x_{k-(2m+1)L}\Big) \\ \quad\times\Big(\sum_{m=0}^{\frac{M-3}{2}}x_{k-(2m+2)L}\Big)\Big] & \text{if } L\geqslant 2\beta. \end{cases} \tag{5.A.6}$$

7. μ_{odd}

$$\mu_{\text{odd}}=\sum_{k=2(l-1)\beta+1}^{2(l-1)\beta+L-2\beta\lfloor\frac{L}{2\beta}\rfloor}\left[\left(\sum_{m=0}^{\frac{M-1}{2}}x_{k-2mL}+d_l\sum_{m=0}^{\frac{M-3}{2}}x_{k-(2m+1)L}\right)\right.$$

$$\left.\times\left(\sum_{m=0}^{\frac{M-1}{2}}x_{k-(2m+1)L}+d_{l-\lfloor\frac{L}{2\beta}\rfloor-1}\sum_{m=0}^{\frac{M-3}{2}}x_{k-(2m+2)L}\right)\right]$$

$$
\begin{aligned}
&+ \sum_{k=2(l-1)\beta+L-2\beta\lfloor\frac{L}{2\beta}\rfloor+1}^{2l\beta} \left[\left(\sum_{m=0}^{\frac{M-1}{2}} x_{k-2mL} + d_l \sum_{m=0}^{\frac{M-3}{2}} x_{k-(2m+1)L}\right)\right.\\
&\times \left.\left(\sum_{m=0}^{\frac{M-1}{2}} x_{k-(2m+1)L} + d_{l-\lfloor\frac{L}{2\beta}\rfloor} \sum_{m=0}^{\frac{M-3}{2}} x_{k-(2m+2)L}\right)\right]\\
&- \lambda_{\text{odd}} - \sigma_{\text{odd}}.
\end{aligned}
\tag{5.A.7}
$$

8. ν_{odd}

$$
\begin{aligned}
\nu_{\text{odd}} = &\sum_{k=2(l-1)\beta+1}^{2l\beta} \left(\sum_{m=0}^{\frac{M-1}{2}} x_{k-2mL} + d_l \sum_{m=0}^{\frac{M-3}{2}} x_{k-(2m+1)L}\right)\xi_{k-L}\\
&+ \sum_{k=2(l-1)\beta+1}^{2(l-1)\beta+L-2\beta\lfloor\frac{L}{2\beta}\rfloor} \left(\sum_{m=0}^{\frac{M-1}{2}} x_{k-(2m+1)L} + d_{l-\lfloor\frac{L}{2\beta}\rfloor-1} \sum_{m=0}^{\frac{M-3}{2}} x_{k-(2m+2)L}\right)\xi_k\\
&+ \sum_{k=2(l-1)\beta+L-2\beta\lfloor\frac{L}{2\beta}\rfloor+1}^{2l\beta} \left(\sum_{m=0}^{\frac{M-1}{2}} x_{k-(2m+1)L} + d_{l-\lfloor\frac{L}{2\beta}\rfloor} \sum_{m=0}^{\frac{M-3}{2}} x_{k-(2m+2)L}\right)\xi_k\\
&+ \sum_{k=2(l-1)\beta+1}^{2l\beta} \xi_k \xi_{k-L}.
\end{aligned}
\tag{5.A.8}
$$

APPENDIX 5B. DERIVATION OF BIT ERROR RATES FOR GCDSK SYSTEM BASED ON SIMPLE GAUSSIAN APPROXIMATION

All symbols are defined as in Sections 5.1 and 5.2. Without loss of generality, we consider the probability of error for the lth symbol when all delay units at the transmitter side begin generating chaotic signals. Denote

$$
\check{\boldsymbol{d}}_l = \left[d_{l-\lfloor\frac{L}{2\beta}\rfloor} \; d_{l-\lfloor\frac{L}{2\beta}\rfloor-1} \right].
\tag{5.B.1}
$$

Based on (5.2.3), (5.2.4) and (5.A.1)–(5.A.8), it is readily shown that for the given transmitted symbols d_l and $\check{\boldsymbol{d}}_l$, the mean value of y_l equals

$$\mathrm{E}\big[y_l \mid (d_l, \check{\boldsymbol{d}}_l)\big] = \begin{cases} 2d_l\beta(M-1)P_s - d_l L\lfloor\frac{M-1}{2}\rfloor P_s + d_{l-1}L\lfloor\frac{M-1}{2}\rfloor P_s \\ \quad \text{if } L < 2\beta, \\ 2d_l\beta\big[(M-1) - \lfloor\frac{M-1}{2}\rfloor\big]P_s \\ \quad + d_{l-\lfloor\frac{L}{2\beta}\rfloor}\big[2\beta - \big(L - 2\beta\lfloor\frac{L}{2\beta}\rfloor\big)\big]\lfloor\frac{M-1}{2}\rfloor P_s \\ \quad + d_{l-\lfloor\frac{L}{2\beta}\rfloor-1}\big(L - 2\beta\lfloor\frac{L}{2\beta}\rfloor\big)\lfloor\frac{M-1}{2}\rfloor P_s \\ \quad \text{if } L \geqslant 2\beta. \end{cases} \tag{5.B.2}$$

We investigate the system under the assumption that the Chebyshev map of degree 2 is used to generate the chaotic signals. Using such a generator, it is readily shown that

$$\mathrm{E}\big[x_k^2 x_{k+L} x_{k+2L}\big]\begin{cases} \neq 0 & \text{if } L = 1, \\ = 0 & \text{if } L > 1. \end{cases} \tag{5.B.3}$$

Next, we evaluate the variance of $y_l \mid (d_l, \check{\boldsymbol{d}}_l)$ under three different scenarios: $L = 1$, $2 \leqslant L < 2\beta$ and $L \geqslant 2\beta$.

Case I: $L = 1$
When $L = 1$, the variance of $y_l \mid (d_l, \check{\boldsymbol{d}}_l)$ can be shown equal to

$$\begin{aligned} \operatorname{var}\big[y_l \mid (d_l, \check{\boldsymbol{d}}_l)\big] &= V_1(2\beta + L, M)\operatorname{var}\big[x_k^2\big] + V_2(2\beta + L, M){P_s}^2 \\ &\quad + V_3(2\beta + L, M)P_s\frac{N_0}{2} + \beta\frac{N_0^2}{2} \\ &\quad + 2P_c\big[C_1(2\beta + L, M, d_l, d_{l-1}) + C_2(2\beta + L, M, d_l, d_{l-1})\big], \end{aligned} \tag{5.B.4}$$

where

$$P_c = E\big[x_k^2 x_{k+1} x_{k+2}\big], \tag{5.B.5}$$

$$V_1(\psi, M) = \begin{cases} \sum_{u=1}^{M-1} u^2 + (\psi - M)(M-1)^2 + \sum_{u=1}^{M-2} u^2 & \text{if } \psi \geqslant M, \\ \sum_{u=1}^{\psi-1} u^2 + (M - \psi)(\psi - 1)^2 + \sum_{u=1}^{\psi-2} u^2 & \text{if } \psi < M, \end{cases} \tag{5.B.6}$$

$$V_2(\psi, M) = \begin{cases} (\psi-1) + \sum_{u=0}^{M-2} 2(M-u-1)(2u+1)^2 \\ \quad + \sum_{u=1}^{M-1} (2u)^2(\psi-u-1) \quad \text{if } \psi \geqslant M, \\ (\psi-1) + \sum_{u=0}^{\psi-2} 2(M-u-1)(2u+1)^2 \\ \quad + \sum_{u=1}^{\psi-2} (2u)^2(\psi-u-1) \\ \quad + 2(M-\psi+1)(M-\psi)(\psi-1)^2 \quad \text{if } \psi < M, \end{cases} \tag{5.B.7}$$

$$V_3(\psi, M) = 2(2M\psi - 3M - 2\psi + 4). \tag{5.B.8}$$

Also, when M is even,

$$C_1(\psi, M, d_l, d_{l-1}) = \begin{cases} d_l(2\psi-3) \quad \text{if } \psi > M = 2, \\ d_l + \sum_{u=1}^{\frac{M-2}{2}} (4u-2+d_l d_{l-1})(4u-1)d_l \\ \quad + \sum_{u=1}^{\frac{M-4}{2}} (4u+d_l d_{l-1})\big[(4u-1)d_l + 2d_{l-1}\big] \\ \quad + d_l \sum_{u=1}^{M-2} (2u-1)(2u+1) \\ \quad + (2M-4+d_l d_{l-1})\big[(2M-5)d_l + d_{l-1}\big] \\ \quad + d_l(\psi-M)(2M-2)(2M-4) \quad \text{if } \psi \geqslant M > 2, \\ d_l + \sum_{u=1}^{\frac{\psi-2}{2}} \big[(4u-2+d_l d_{l-1})(4u-1)d_l \\ \quad + (4u+d_l d_{l-1})((4u-1)d_l + 2d_{l-1})\big] \\ \quad + d_l \sum_{u=1}^{\psi-2} (2u-1)(2u+1) \\ \quad + (2\psi-3+d_l d_{l-1})\big[(2\psi-4)d_l + d_{l-1}\big] \\ \quad + \frac{M-\psi}{2}(2\psi-3+d_l d_{l-1})(2\psi-2)d_l \\ \quad + \frac{M-\psi-2}{2}(2\psi-3+d_l d_{l-1})\big[(2\psi-4)d_l + 2d_{l-1}\big] \\ \quad \text{if } \psi < M \text{ and } \psi \text{ is even}, \\ d_l + \sum_{u=1}^{\frac{\psi-1}{2}} (4u-2+d_l d_{l-1})(4u-1)d_l \\ \quad + \sum_{u=1}^{\frac{\psi-3}{2}} (4u+d_l d_{l-1})\big[(4u-1)d_l + 2d_{l-1}\big] \\ \quad + d_i \sum_{u=1}^{\psi-2} (2u-1)(2u+1) \\ \quad + (2\psi-3+d_l d_{l-1})\big[(2\psi-4)d_l + d_{l-1}\big] \\ \quad + \frac{M-\psi-1}{2}(2\psi-3+d_l d_{l-1}) \\ \quad \times \big[(2\psi-2)d_l + ((2\psi-4)d_l + 2d_{l-1})\big] \\ \quad \text{if } \psi < M \text{ and } \psi \text{ is odd}, \end{cases} \tag{5.B.9}$$

$$C_2(\phi, M, d_l, d_{l-1}) = \begin{cases} d_l(2\phi-5) \quad \text{if } \phi > M = 2, \\ \sum_{u=1}^{\frac{M-2}{2}} \big[(2u-1)d_l\big](4u + d_l d_{l-1}) \\ \quad + \sum_{u=1}^{\frac{M-4}{2}} \big[(2u-1)d_l + d_{l-1}\big](4u+2+d_l d_{l-1}) \\ \quad + d_l \sum_{u=1}^{M-3}(2u-1)(u+1) \\ \quad + \big[(M-3)d_l + d_{l-1}\big](2M-2) \\ \quad + d_l(M-1)\big[(2M-3)+(2M-5)\big] \\ \quad + 2d_l(\phi-M-1)(M-1)^2 \quad \text{if } \phi > M \geqslant 4, \\ (5M-12)d_l + (3M-8)d_{l-1} \quad \text{if } 3 = \phi < M, \\ \sum_{u=1}^{\frac{\phi-2}{2}} \big[(2u-1)d_l\big](4u + d_l d_{l-1}) \\ \quad + \sum_{u=1}^{\frac{\phi-4}{2}} \big[(2u-1)d_l + d_{l-1}\big](4u+2+d_l d_{l-1}) \\ \quad + d_l \sum_{u=1}^{\phi-3}(2u-1)(u+1) \\ \quad + \big[(\phi-3)d_l + d_{l-1}\big](2\phi-3+d_l d_{l-1}) \\ \quad + \big[(\phi-2)d_l + d_{l-1}\big](2\phi-3) \\ \quad + d_l(\phi-1)(2\phi-5) \\ \quad + \frac{M-\phi}{2}(2\phi-3+d_l d_{l-1})(\phi-1)d_l \\ \quad + \frac{M-\phi-2}{2}(2\phi-3+d_l d_{l-1})\big[(\phi-2)d_l + d_{l-1}\big] \\ \qquad \text{if } 4 \leqslant \phi < M \text{ and } \phi \text{ is even}, \\ \sum_{u=1}^{\frac{\phi-3}{2}} \Big[\big((2u-1)d_l\big)(4u + d_l d_{l-1}) \\ \quad + \big((2u-1)d_l + d_{l-1}\big)(4u+2+d_l d_{l-1})\Big] \\ \quad + d_l \sum_{u=1}^{\phi-3}(2u-1)(u+1) \\ \quad + \big[(\phi-2)d_l\big](2\phi-3+d_l d_{l-1}) \\ \quad + \big[(\phi-2)d_l + d_{l-1}\big](2\phi-3) \\ \quad + d_l(\phi-1)(2\phi-5) \\ \quad + \frac{M-\phi-1}{2}(2\phi-3+d_l d_{l-1}) \\ \quad \times\big[(\phi-1)d_l + \big((\phi-2)d_l + d_{l-1}\big)\big] \\ \qquad \text{if } 4 \leqslant \phi < M \text{ and } \phi \text{ is odd}, \\ \sum_{u=1}^{\frac{\phi-2}{2}} \big[(2u-1)d_l\big](4u + d_l d_{l-1}) \\ \quad + \sum_{u=1}^{\frac{\phi-4}{2}} \big[(2u-1)d_l + d_{l-1}\big](4u+2+d_l d_{l-1}) \\ \quad + d_l \sum_{u=1}^{\phi-3}(2u-1)(u+1) \\ \quad + \big[(\phi-3)d_l + d_{l-1}\big](2\phi-3) \\ \quad + d_l(\phi-1)(2\phi-5) \quad \text{if } \phi = M. \end{cases} \tag{5.B.10}$$

When M is odd,

$$C_1(\psi, M, d_l, d_{l-1}) = \begin{cases} d_l + \sum_{u=1}^{\frac{M-3}{2}} \big[(4u-2+d_l d_{l-1})\big((4u-3)d_l + 2d_{l-1}\big) \\ \quad + (4u + d_l d_{l-1})(4u+1)d_l\big] \\ \quad + d_l \sum_{u=1}^{M-2} (2u-1)(2u+1) \\ \quad + (2M-4+d_l d_{l-1})\big[(2M-5)d_l + d_{l-1}\big] \\ \quad + d_l(\psi - M)(2M-2)(2M-4) \\ \quad\ \text{if } \psi \geqslant M > 2, \\ d_l + \sum_{u=1}^{\frac{\psi-2}{2}} (4u-2+d_l d_{l-1})\big[(4u-3)d_l + 2d_{l-1}\big] \\ \quad + (4u + d_l d_{l-1})(4u+1)d_l \\ \quad + d_l \sum_{u=1}^{\psi-2} (2u-1)(2u+1) \\ \quad + (2\psi-3+d_l d_{l-1})\big[(2\psi-4)d_l + d_{l-1}\big] \\ \quad + \frac{M-\psi-1}{2}(2\psi-3+d_l d_{l-1}) \\ \quad \times\big[(2\psi-2)d_l + \big((2\psi-4)d_l + 2d_{l-1}\big)\big] \\ \quad\ \text{if } \psi < M \text{ and } \psi \text{ is even}, \\ d_l + \sum_{u=1}^{\frac{\psi-1}{2}} (4u-2+d_l d_{l-1})\big[(4u-3)d_l + 2d_{l-1}\big] \\ \quad + \sum_{u=1}^{\frac{\psi-3}{2}} (4u + d_l d_{l-1})(4u+1)d_l \\ \quad + d_l \sum_{u=1}^{\psi-2} (2u-1)(2u+1) \\ \quad + (2\psi-3+d_l d_{l-1})\big[(2\psi-4)d_l + d_{l-1}\big] \\ \quad + \frac{M-\psi}{2}(2\psi-3+d_l d_{l-1})(2\psi-2)d_l \\ \quad + \frac{M-\psi-2}{2}(2\psi-3+d_l d_{l-1})\big[(2\psi-4)d_l + 2d_{l-1}\big] \\ \quad\ \text{if } \psi < M \text{ and } \psi \text{ is odd}, \end{cases} \tag{5.B.11}$$

$$
C_2(\psi, M, d_l, d_{l-1}) =
\begin{cases}
d_l(8\psi - 24) + 4d_{l-1} \quad \text{if } \psi > M = 3, \\
\sum_{u=1}^{\frac{M-3}{2}} \big[\big(2(u-1)d_l + d_{l-1}\big)(4u + d_l d_{l-1}) \\
\quad + 2ud_l(4u + 2 + d_l d_{l-1})\big] \\
\quad + d_l \sum_{u=1}^{M-3} (2u - 1)(u + 1) \\
\quad + \big[(M-3)d_l + d_{l-1}\big](2M - 2) \\
\quad + d_l(M-1)\big[(2M-3) + (2M-5)\big] \\
\quad + 2d_l(\psi - M - 1)(M-1)^2 \quad \text{if } \psi > M \geqslant 4, \\
(5M - 13)d_l + (3M - 7)d_{l-1} \quad \text{if } 3 = \psi < M, \\
\sum_{u=1}^{\frac{\psi-2}{2}} \big[(2u-2)d_l + d_{l-1}\big](4u + d_l d_{l-1}) \\
\quad + \sum_{u=1}^{\frac{\psi-4}{2}} 2ud_l(4u + 2 + d_l d_{l-1}) \\
\quad + d_l \sum_{u=1}^{\psi-3} (2u-1)(u+1) \\
\quad + \big[(\psi-2)d_l\big](2\psi - 3 + d_l d_{l-1}) \\
\quad + \big[(\psi-2)d_l + d_{l-1}\big](2\psi - 3) \\
\quad + d_l(\psi - 1)(2\psi - 5) \\
\quad + \frac{M-\psi-1}{2}(2\psi - 3 + d_l d_{l-1}) \\
\quad \times\big[(\psi-1)d_l + \big((\psi-2)d_l + d_{l-1}\big)\big] \\
\qquad \text{if } 4 \leqslant \phi < M \text{ and } \phi \text{ is even}, \\
\sum_{u=1}^{\frac{\psi-3}{2}} \big[\big((2u-2)d_l + d_{l-1}\big)(4u + d_l d_{l-1}) \\
\quad + 2ud_l(4u + 2 + d_l d_{l-1})\big] \\
\quad + d_l \sum_{u=1}^{\psi-3} (2u-1)(u+1) \\
\quad + \big[(\psi-3)d_l + d_{l-1}\big](2\psi - 3 + d_l d_{l-1}) \\
\quad + \big[(\psi-2)d_l + d_{l-1}\big](2\psi - 3) \\
\quad + d_l(\psi - 1)(2\psi - 5) \\
\quad + \frac{M-\psi}{2}(2\psi - 3 + d_l d_{l-1})(\psi - 1)d_l \\
\quad + \frac{M-\psi-2}{2}(2\psi - 3 + d_l d_{l-1})\big[(\psi-2)d_l + d_{l-1}\big] \\
\qquad \text{if } 4 \leqslant \phi < M \text{ and } \phi \text{ is odd}, \\
\sum_{u=1}^{\frac{\psi-3}{2}} \big[\big((2u-2)d_l + d_{l-1}\big)(4u + d_l d_{l-1}) \\
\quad + 2ud_l(4u + 2 + d_l d_{l-1})\big] \\
\quad + d_l \sum_{u=1}^{\psi-3} (2u-1)(u+1) \\
\quad + \big[(\psi-3)d_l + d_{l-1}\big](2\psi - 3) \\
\quad + d_l(\psi - 1)(2\psi - 5) \quad \text{if } \psi = M.
\end{cases}
\tag{5.B.12}
$$

Case II: $2 \leqslant L < 2\beta$

For the case where $2 \leqslant L < 2\beta$, we define

$$\gamma = \left\lfloor \frac{2\beta}{L} \right\rfloor + 1, \tag{5.B.13}$$

$$\alpha = 2\beta - \left\lfloor \frac{2\beta}{L} \right\rfloor L. \tag{5.B.14}$$

Also, it can be shown that the variance of $y_l \mid (d_l, \check{\boldsymbol{d}}_l)$ is independent of d_l and $\check{\boldsymbol{d}}_l$, and is equal to

$$\begin{aligned}
&\operatorname{var}\big[y_l \mid (d_l, \check{\boldsymbol{d}}_l)\big] \\
&\quad = \operatorname{var}[y_l] \\
&\quad = \alpha\left[V_1(\gamma+1, M)\operatorname{var}\big[x_k^2\big] + V_2(\gamma+1, M)P_s^2 + V_3(\gamma+1, M)P_s\frac{N_0}{2}\right] \\
&\qquad + (L-\alpha)\left[V_1(\gamma, M)\operatorname{var}\big[x_k^2\big] + V_2(\gamma, M)P_s^2 + V_3(\gamma, M)P_s\frac{N_0}{2}\right] \\
&\qquad + \beta\frac{N_0^2}{2}.
\end{aligned} \tag{5.B.15}$$

Case III: $L \geqslant 2\beta$

When $L \geqslant 2\beta$, the variance of $y_l \mid (d_l, \check{\boldsymbol{d}}_l)$ is also independent of d_l and $\check{\boldsymbol{d}}_l$, and its expression is given by

$$\begin{aligned}
\operatorname{var}\big[y_l \mid (d_l, \check{\boldsymbol{d}}_l)\big] &= \operatorname{var}[y_l] \\
&= 2\beta \operatorname{var}\big[x_k^2\big] + 2\big(M^2-1\big)\beta P_s^2 + 2M\beta P_s N_0 + \beta\frac{N_0^2}{2}.
\end{aligned} \tag{5.B.16}$$

Assuming that $y_l \mid (d_l, \check{\boldsymbol{d}}_l)$ follows a normal distribution for the given transmitted symbols d_l and $\check{\boldsymbol{d}}_l$, the approximate BER of the GCDSK system can be computed using

$$\begin{aligned}
\mathrm{BER}_{\mathrm{GCDSK}} = \sum_{\check{\boldsymbol{d}}_l \in \{-1,+1\}^2} &\left[\frac{1}{2}\operatorname{Prob}\big(y_l \leqslant 0 \,\big|\, (d_l = +1, \check{\boldsymbol{d}}_l)\big)\operatorname{Prob}(\check{\boldsymbol{d}}_l)\right. \\
&\left. + \frac{1}{2}\operatorname{Prob}\big(y_l > 0 \,\big|\, (d_l = -1, \check{\boldsymbol{d}}_l)\big)\operatorname{Prob}(\check{\boldsymbol{d}}_l)\right]
\end{aligned}$$

$$= \frac{1}{16} \sum_{\check{\boldsymbol{d}}_l \in \{-1,+1\}^2} \left[\operatorname{erfc}\left(\frac{\mathrm{E}[y_l \mid (d_l = +1, \check{\boldsymbol{d}}_l)]}{\sqrt{2 \operatorname{var}[y_l \mid (d_l = +1, \check{\boldsymbol{d}}_l)]}} \right) + \operatorname{erfc}\left(\frac{\mathrm{E}[y_l \mid (d_l = -1, \check{\boldsymbol{d}}_l)]}{\sqrt{2 \operatorname{var}[y_l \mid (d_l = -1, \check{\boldsymbol{d}}_l)]}} \right) \right] \tag{5.B.17}$$

where erfc(·) represents the complementary error function given by (3.3.5) [Proakis and Salehi, 1994].

APPENDIX 5C. DERIVATION OF BIT ERROR RATES FOR CDSK SYSTEM BASED ON SIMPLE GAUSSIAN APPROXIMATION

All symbols are defined as in Section 5.1. Without loss of generality, the probability of error for the lth symbol is considered. The derivation is similar to that of the GCDSK system and will not be shown here. In summary, we have

$$\mathrm{E}[y_l \mid d_l] = 2 d_l \beta P_s \tag{5.C.1}$$

and

$$\operatorname{var}[y_l \mid d_l] = \begin{cases} 2\beta \operatorname{var}\left[x_k^2\right] + (10\beta - 2) {P_s}^2 + 2\beta M P_s N_0 \\ \quad + \beta \frac{N_0^2}{2} + 8 d_l (2\beta - 1) P_c \quad \text{when } L = 1, \\ \alpha f_1(\gamma + 1) + (L - \alpha) f_1(\gamma) + 2\beta M P_s N_0 \\ \quad + \beta \frac{N_0^2}{2} \quad \text{when } 2 \leqslant L < 2\beta, \\ 2\beta \operatorname{var}\left[x_k^2\right] + 6\beta P_s^2 + 4\beta P_s N_0 + \beta \frac{N_0^2}{2} \quad \text{when } L \geqslant 2\beta, \end{cases} \tag{5.C.2}$$

where

$$\gamma = \left\lfloor \frac{2\beta}{L} \right\rfloor + 1, \tag{5.C.3}$$

$$\alpha = 2\beta - \left\lfloor \frac{2\beta}{L} \right\rfloor L, \tag{5.C.4}$$

$$f_1(\psi) = (\psi - 1) \operatorname{var}\left[x_k^2\right] + (5\psi - 7) {P_s}^2. \tag{5.C.5}$$

Note that for the case $L \geqslant 2\beta$, the variance of $y_l \mid d_l$ can be re-written as

$$\operatorname{var}[y_l \mid d_l] = 2\beta \operatorname{var}\left[x_k^2\right] + \frac{3 E_b^2}{8\beta} + E_b N_0 + \beta \frac{N_0^2}{2}, \tag{5.C.6}$$

which is independent of d_l and can be shown equal to that obtained by Sushchik *et al.* [2000]. The only difference is that the value of var$[x_k^2]$ is different for different maps (symmetric tent map used by Sushchik *et al.* and Chebyshev map used here). The Gaussian-approximated BER can then be obtained similarly as in (5.B.17).

APPENDIX 5D. EXPRESSIONS FOR THE VARIABLES IN THE MA-GCDSK SYSTEM

All symbols are defined as in Section 5.3.

1. $\lambda_{\text{even}}^{(j)}$

$$\lambda_{\text{even}}^{(j)} = \begin{cases} d_l^{(j)}\Big(\sum_{k=2(l-1)\beta+1}^{2l\beta} \sum_{m=0}^{\frac{M-2}{2}} \big(x_{k-(2m+1)L^{(j)}}^{(j)}\big)^2 \\ \quad + \sum_{k=2(l-1)\beta+L^{(j)}+1}^{2l\beta} \sum_{m=1}^{\frac{M-2}{2}} \big(x_{k-2mL^{(j)}}^{(j)}\big)^2\Big) & \text{if } L^{(j)} < 2\beta, \\ d_l^{(j)} \sum_{k=2(l-1)\beta+1}^{2l\beta} \sum_{m=0}^{\frac{M-2}{2}} \big(x_{k-(2m+1)L^{(j)}}^{(j)}\big)^2 & \text{if } L^{(j)} \geqslant 2\beta. \end{cases} \tag{5.D.1}$$

2. $\sigma_{\text{even}}^{(j)}$

$$\sigma_{\text{even}}^{(j)} = \begin{cases} d_{l-1}^{(j)} \sum_{k=2(l-1)\beta+1}^{2(l-1)\beta+L^{(j)}} \Big[\sum_{m=0}^{\frac{M-2}{2}} \big(x_{k-2mL^{(j)}}^{(j)} + d_l^{(j)} x_{k-(2m+1)L^{(j)}}^{(j)}\big) \\ \quad \times \sum_{m=0}^{\frac{M-2}{2}} x_{k-(2m+2)L^{(j)}}^{(j)}\Big] \quad \text{if } L^{(j)} < 2\beta, \\ d_{l-\lfloor \frac{L^{(j)}}{2\beta} \rfloor -1}^{(j)} \sum_{k=2(l-1)\beta+1}^{2(l-1)\beta+L^{(j)}-2\beta\lfloor \frac{L^{(j)}}{2\beta}\rfloor} \\ \quad \times \Big[\sum_{m=0}^{\frac{M-2}{2}} \big(x_{k-2mL^{(j)}}^{(j)} + d_l^{(j)} x_{k-(2m+1)L^{(j)}}^{(j)}\big) \sum_{m=0}^{\frac{M-2}{2}} x_{k-(2m+2)L^{(j)}}^{(j)}\Big] \\ \quad + d_{l-\lfloor \frac{L^{(j)}}{2\beta} \rfloor}^{(j)} \sum_{k=2(l-1)\beta+L^{(j)}-2\beta\lfloor \frac{L^{(j)}}{2\beta}\rfloor+1}^{2l\beta} \\ \quad \times \Big[\sum_{m=0}^{\frac{M-2}{2}} \big(x_{k-2mL^{(j)}}^{(j)} + d_l^{(j)} x_{k-(2m+1)L^{(j)}}^{(j)}\big) \sum_{m=0}^{\frac{M-2}{2}} x_{k-(2m+2)L^{(j)}}^{(j)}\Big] \\ \quad \text{if } L^{(j)} \geqslant 2\beta. \end{cases} \tag{5.D.2}$$

3. $\mu_{\text{even}}^{(j)}$

$$
\begin{aligned}
\mu_{\text{even}}^{(j)} = & \sum_{k=2(l-1)\beta+1}^{2(l-1)\beta+L^{(j)}-2\beta\lfloor\frac{L^{(j)}}{2\beta}\rfloor} \left[\sum_{m=0}^{\frac{M-2}{2}} \left(x_{k-2mL^{(j)}}^{(j)} + d_l^{(j)} x_{k-(2m+1)L^{(j)}}^{(j)} \right) \right. \\
& \left. \times \sum_{m=0}^{\frac{M-2}{2}} \left(x_{k-(2m+1)L^{(j)}}^{(j)} + d_{l-\lfloor\frac{L^{(j)}}{2\beta}\rfloor-1}^{(j)} x_{k-(2m+2)L^{(j)}}^{(j)} \right) \right] \\
& + \sum_{k=2(l-1)\beta+L^{(j)}-2\beta\lfloor\frac{L^{(j)}}{2\beta}\rfloor+1}^{2l\beta} \left[\sum_{m=0}^{\frac{M-2}{2}} \left(x_{k-2mL^{(j)}}^{(j)} + d_l^{(j)} x_{k-(2m+1)L^{(j)}}^{(j)} \right) \right. \\
& \left. \times \sum_{m=0}^{\frac{M-2}{2}} \left(x_{k-(2m+1)L^{(j)}}^{(j)} + d_{l-\lfloor\frac{L^{(j)}}{2\beta}\rfloor}^{(j)} x_{k-(2m+2)L^{(j)}}^{(j)} \right) \right] \\
& - \lambda_{\text{even}}^{(j)} - \sigma_{\text{even}}^{(j)}.
\end{aligned}
\tag{5.D.3}
$$

4. $\theta_{\text{even}}^{(j)}$

$$
\begin{aligned}
\theta_{\text{even}}^{(j)} = & \sum_{k=2(l-1)\beta+1}^{2(l-1)\beta+L^{(j)}-2\beta\lfloor\frac{L^{(j)}}{2\beta}\rfloor} \left[\sum_{i=1}^{N} \sum_{m=0}^{\frac{M-2}{2}} \left(x_{k-2mL^{(i)}}^{(i)} + d_l^{(i)} x_{k-(2m+1)L^{(i)}}^{(i)} \right) \right. \\
& \left. \times \sum_{u=1,(i,u)\neq(j,j)}^{N} \sum_{m=0}^{\frac{M-2}{2}} \left(x_{k-2mL^{(u)}-L^{(j)}}^{(u)} + d_{l-\lfloor\frac{L^{(u)}}{2\beta}\rfloor-1}^{(u)} x_{k-(2m+1)L^{(u)}-L^{(j)}}^{(u)} \right) \right] \\
& + \sum_{k=2(l-1)\beta+L^{(j)}-2\beta\lfloor\frac{L^{(j)}}{2\beta}\rfloor+1}^{2l\beta} \left[\sum_{i=1}^{N} \sum_{m=0}^{\frac{M-2}{2}} \left(x_{k-2mL^{(i)}}^{(i)} + d_l^{(i)} x_{k-(2m+1)L^{(i)}}^{(i)} \right) \right. \\
& \left. \times \sum_{u=1,(i,u)\neq(j,j)}^{N} \sum_{m=0}^{\frac{M-2}{2}} \left(x_{k-2mL^{(u)}-L^{(j)}}^{(u)} + d_{l-\lfloor\frac{L^{(u)}}{2\beta}\rfloor}^{(u)} x_{k-(2m+1)L^{(u)}-L^{(j)}}^{(u)} \right) \right]
\end{aligned}
\tag{5.D.4}
$$

5. $\nu_{\text{even}}^{(j)}$

$$\nu_{\text{even}}^{(j)} = \sum_{k=2(l-1)\beta+1}^{2l\beta} \left[\sum_{i=1}^{N} \sum_{m=0}^{\frac{M-2}{2}} \left(x_{k-2mL^{(i)}}^{(i)} + d_l^{(i)} x_{k-(2m+1)L^{(i)}}^{(i)} \right) \right] \xi_{k-L^{(j)}}$$

$$+ \sum_{k=2(l-1)\beta+1}^{2(l-1)\beta+L^{(j)}-2\beta\lfloor\frac{L^{(j)}}{2\beta}\rfloor} \left[\sum_{i=1}^{N} \sum_{m=0}^{\frac{M-2}{2}} \left(x_{k-2mL^{(i)}-L^{(j)}}^{(i)} \right.\right.$$

$$\left.\left. + d_{l-\lfloor\frac{L^{(i)}}{2\beta}\rfloor-1}^{(i)} x_{k-(2m+1)L^{(i)}-L^{(j)}}^{(i)} \right) \right] \xi_k$$

$$+ \sum_{k=2(l-1)\beta+L^{(j)}-2\beta\lfloor\frac{L^{(j)}}{2\beta}\rfloor+1}^{2l\beta} \left[\sum_{i=1}^{N} \sum_{m=0}^{\frac{M-2}{2}} \left(x_{k-2mL^{(i)}-L^{(j)}}^{(i)} \right.\right.$$

$$\left.\left. + d_{l-\lfloor\frac{L^{(i)}}{2\beta}\rfloor}^{(i)} x_{k-(2m+1)L^{(i)}-L^{(j)}}^{(i)} \right) \xi_k \right]$$

$$+ \sum_{k=2(l-1)\beta+1}^{2l\beta} \xi_k \xi_{k-L^{(j)}}. \tag{5.D.5}$$

6. $\lambda_{\text{odd}}^{(j)}$

$$\lambda_{\text{odd}}^{(j)} = \begin{cases} d_l^{(j)} \left(\sum_{k=2(l-1)\beta+1}^{2l\beta} \sum_{m=0}^{\frac{M-3}{2}} \left(x_{k-(2m+1)L^{(j)}}^{(j)} \right)^2 \right. \\ \quad \left. + \sum_{k=2(l-1)\beta+L^{(j)}+1}^{2l\beta} \sum_{m=1}^{\frac{M-1}{2}} \left(x_{k-2mL^{(j)}}^{(j)} \right)^2 \right) & \text{if } L^{(j)} < 2\beta, \\ d_l^{(j)} \sum_{k=2(l-1)\beta+1}^{2l\beta} \sum_{m=0}^{\frac{M-3}{2}} \left(x_{k-(2m+1)L^{(j)}}^{(j)} \right)^2 & \text{if } L^{(j)} \geqslant 2\beta. \end{cases} \tag{5.D.6}$$

7. $\sigma_{\text{odd}}^{(j)}$

$$\sigma_{\text{odd}}^{(j)} = \begin{cases} d_{l-1}^{(j)} \sum_{k=2(l-1)\beta+1}^{2(l-1)\beta+L^{(j)}} \Bigl[\Bigl(\sum_{m=0}^{\frac{M-1}{2}} x_{k-2mL^{(j)}}^{(j)} + d_l^{(j)} \sum_{m=0}^{\frac{M-3}{2}} x_{k-(2m+1)L^{(j)}}^{(j)} \Bigr) \\ \quad \times \sum_{m=0}^{\frac{M-3}{2}} x_{k-(2m+2)L^{(j)}}^{(j)} \Bigr] \quad \text{if } L^{(j)} < 2\beta, \\ d_{l-\lfloor \frac{L^{(j)}}{2\beta} \rfloor -1}^{(j)} \sum_{k=2(l-1)\beta+1}^{2(l-1)\beta+L^{(j)}-2\beta\lfloor \frac{L^{(j)}}{2\beta} \rfloor} \\ \quad \times \Bigl[\Bigl(\sum_{m=0}^{\frac{M-1}{2}} x_{k-2mL^{(j)}}^{(j)} + d_l^{(j)} \sum_{m=0}^{\frac{M-3}{2}} x_{k-(2m+1)L^{(j)}}^{(j)} \Bigr) \\ \quad \times \sum_{m=0}^{\frac{M-3}{2}} x_{k-(2m+2)L^{(j)}}^{(j)} \Bigr] \\ \quad + d_{l-\lfloor \frac{L^{(j)}}{2\beta} \rfloor}^{(j)} \sum_{k=2(l-1)\beta+L^{(j)}-2\beta\lfloor \frac{L^{(j)}}{2\beta} \rfloor+1}^{2l\beta} \\ \quad \times \Bigl[\Bigl(\sum_{m=0}^{\frac{M-1}{2}} x_{k-2mL^{(j)}}^{(j)} + d_l^{(j)} \sum_{m=0}^{\frac{M-3}{2}} x_{k-(2m+1)L^{(j)}}^{(j)} \Bigr) \\ \quad \times \sum_{m=0}^{\frac{M-3}{2}} x_{k-(2m+2)L^{(j)}}^{(j)} \Bigr] \quad \text{if } L^{(j)} \geqslant 2\beta. \end{cases} \tag{5.D.7}$$

8. $\mu_{\text{odd}}^{(j)}$

$$\begin{aligned} \mu_{\text{odd}}^{(j)} = & \sum_{k=2(l-1)\beta+1}^{2(l-1)\beta+L^{(j)}-2\beta\lfloor \frac{L^{(j)}}{2\beta} \rfloor} \left[\left(\sum_{m=0}^{\frac{M-1}{2}} x_{k-2mL^{(j)}}^{(j)} + d_l^{(j)} \sum_{m=0}^{\frac{M-3}{2}} x_{k-(2m+1)L^{(j)}}^{(j)} \right) \right. \\ & \left. \times \left(\sum_{m=0}^{\frac{M-1}{2}} x_{k-(2m+1)L^{(j)}}^{(j)} + d_{l-\lfloor \frac{L^{(j)}}{2\beta} \rfloor -1}^{(j)} \sum_{m=0}^{\frac{M-3}{2}} x_{k-(2m+2)L^{(j)}}^{(j)} \right) \right] \\ & + \sum_{k=2(l-1)\beta+L^{(j)}-2\beta\lfloor \frac{L^{(j)}}{2\beta} \rfloor+1}^{2l\beta} \left[\left(\sum_{m=0}^{\frac{M-1}{2}} x_{k-2mL^{(j)}}^{(j)} + d_l^{(j)} \sum_{m=0}^{\frac{M-3}{2}} x_{k-(2m+1)L^{(j)}}^{(j)} \right) \right. \\ & \left. \times \left(\sum_{m=0}^{\frac{M-1}{2}} x_{k-(2m+1)L^{(j)}}^{(j)} + d_{l-\lfloor \frac{L^{(j)}}{2\beta} \rfloor}^{(j)} \sum_{m=0}^{\frac{M-3}{2}} x_{k-(2m+2)L^{(j)}}^{(j)} \right) \right] \\ & - \lambda_{\text{odd}}^{(j)} - \sigma_{\text{odd}}^{(j)}. \end{aligned} \tag{5.D.8}$$

9. $\theta_{\text{odd}}^{(j)}$

$$\theta_{\text{odd}}^{(j)} = \sum_{k=2(l-1)\beta+1}^{2(l-1)\beta+L^{(j)}-2\beta\lfloor\frac{L^{(j)}}{2\beta}\rfloor} \left[\sum_{i=1}^{N}\left(\sum_{m=0}^{\frac{M-1}{2}} x_{k-2mL^{(i)}}^{(i)} + d_l^{(i)} \sum_{m=0}^{\frac{M-3}{2}} x_{k-(2m+1)L^{(i)}}^{(i)}\right)\right.$$
$$\times \left.\sum_{\substack{u=1\\(i,u)\neq(j,j)}}^{N} \left(\sum_{m=0}^{\frac{M-1}{2}} x_{k-2mL^{(u)}-L^{(j)}}^{(u)} + d_{l-\lfloor\frac{L^{(u)}}{2\beta}\rfloor-1}^{(u)} \sum_{m=0}^{\frac{M-3}{2}} x_{k-(2m+1)L^{(u)}-L^{(j)}}^{(u)}\right)\right]$$
$$+ \sum_{\substack{k=2(l-1)\beta+L^{(j)}\\-2\beta\lfloor L^{(j)}/2\beta\rfloor+1}}^{2l\beta} \left[\sum_{i=1}^{N}\left(\sum_{m=0}^{\frac{M-1}{2}} x_{k-2mL^{(i)}}^{(i)} + d_l^{(i)} \sum_{m=0}^{\frac{M-3}{2}} x_{k-(2m+1)L^{(i)}}^{(i)}\right)\right.$$
$$\times \left.\sum_{\substack{u=1\\(i,u)\neq(j,j)}}^{N} \left(\sum_{m=0}^{\frac{M-1}{2}} x_{k-2mL^{(u)}-L^{(j)}}^{(u)} + d_{l-\lfloor\frac{L^{(u)}}{2\beta}\rfloor}^{(u)} \sum_{m=0}^{\frac{M-3}{2}} x_{k-(2m+1)L^{(u)}-L^{(j)}}^{(u)}\right)\right]. \tag{5.D.9}$$

10. $\nu_{\text{odd}}^{(j)}$

$$\nu_{\text{odd}}^{(j)} = \sum_{k=2(l-1)\beta+1}^{2l\beta} \left[\sum_{i=1}^{N}\left(\sum_{m=0}^{\frac{M-1}{2}} x_{k-2mL^{(i)}}^{(i)} + d_l^{(i)} \sum_{m=0}^{\frac{M-3}{2}} x_{k-(2m+1)L^{(i)}}^{(i)}\right)\right]\xi_{k-L^{(j)}}$$
$$+ \sum_{k=2(l-1)\beta+1}^{2(l-1)\beta+L^{(j)}-2\beta\lfloor\frac{L^{(j)}}{2\beta}\rfloor} \left[\sum_{i=1}^{N}\left(\sum_{m=0}^{\frac{M-1}{2}} x_{k-2mL^{(i)}-L^{(j)}}^{(i)}\right.\right.$$
$$\left.\left.+ d_{l-\lfloor\frac{L^{(i)}}{2\beta}\rfloor-1}^{(i)} \sum_{m=0}^{\frac{M-3}{2}} x_{k-(2m+1)L^{(i)}-L^{(j)}}^{(i)}\right)\right]\xi_k$$
$$+ \sum_{k=2(l-1)\beta+L^{(j)}-2\beta\lfloor\frac{L^{(j)}}{2\beta}\rfloor+1}^{2l\beta} \left[\sum_{i=1}^{N}\left(\sum_{m=0}^{\frac{M-1}{2}} x_{k-2mL^{(i)}-L^{(j)}}^{(i)}\right.\right.$$
$$\left.\left.+ d_{l-\lfloor\frac{L^{(i)}}{2\beta}\rfloor}^{(i)} \sum_{m=0}^{\frac{M-3}{2}} x_{k-(2m+1)L^{(i)}-L^{(j)}}^{(i)}\right)\right]\xi_k + \sum_{k=2(l-1)\beta+1}^{2l\beta} \xi_k\xi_{k-L^{(j)}}. \tag{5.D.10}$$

APPENDIX 5E. DERIVATION OF BIT ERROR RATES FOR MA-GCDSK SYSTEM

All symbols are defined as in Section 5.3. In the evaluation of the performance for the jth user in a MA-GCDSK system, in order to reduce the inter-user interference and the effect of noise, we need to reduce the variance of the conditional decision variable $y_l^{(j)} \mid (d_l^{(j)}, \check{\boldsymbol{d}}_l^{(j)})$. Hence, it can be shown that the set of delays $L^{(i)}, i = 1, \ldots, N$, should be selected such that they satisfy the following conditions:

$$\zeta^{(i)} L^{(i)} \neq \zeta^{(j)} L^{(j)},$$
$$\text{where } i, j = 1, \ldots, N,\ i \neq j,\ \zeta^{(j)} = 1, \ldots, (M+1),\ \zeta^{(i)} = 1, \ldots, (M-1) \tag{5.E.1}$$

$$\zeta^{(i)} L^{(i)} \neq \zeta^{(u)} L^{(u)},$$
$$\text{where } i, u = 1, \ldots, N,\ i \neq j,\ u \neq j,\ i \neq u,\ \zeta^{(i)}, \zeta^{(j)},\ \zeta^{(u)} = 1, \ldots, (M-1) \tag{5.E.2}$$

$$\zeta^{(i)} L^{(i)} \neq 2L^{(j)} + \zeta^{(u)} L^{(u)},$$
$$\text{where } i, u = 1, \ldots, N,\ i \neq j,\ u \neq j,\ i \neq u,\ \zeta^{(i)}, \zeta^{(j)}, \zeta^{(u)} = 1, \ldots, (M-1) \tag{5.E.3}$$

$$2L^{(j)} \neq \zeta^{(i)} L^{(i)} + \zeta^{(u)} L^{(u)},$$
$$\text{where } i, u = 1, \ldots, N,\ i \neq j,\ u \neq j,\ i \neq u,\ \zeta^{(i)}, \zeta^{(j)}, \zeta^{(u)} = 1, \ldots, (M-1). \tag{5.E.4}$$

For the variance of $y_l^{(j)} \mid (d_l^{(j)}, \check{\boldsymbol{d}}_l^{(j)})$, we evaluate its value under three different scenarios: $L^{(j)} = 1$, $2 \leqslant L^{(j)} < 2\beta$ and $L^{(j)} \geqslant 2\beta$. Before proceeding further, we first define the following functions:

$$\gamma^{(i)} = \left\lfloor \frac{2\beta}{L^{(i)}} \right\rfloor + 1, \tag{5.E.5}$$

$$\alpha^{(i)} = 2\beta - \left\lfloor \frac{2\beta}{L^{(i)}} \right\rfloor L^{(i)}, \tag{5.E.6}$$

$$q\left(\alpha^{(i)}, \gamma^{(i)}\right) = \begin{cases} \alpha^{(i)} f_2\left(\gamma^{(i)}\right) + \left(L^{(i)} - \alpha^{(i)}\right) f_2\left(\gamma^{(i)} - 1\right) & \text{if } 0 < L^{(i)} < 2\beta, \\ 0 & \text{if } L^{(i)} \geqslant 2\beta, \end{cases} \tag{5.E.7}$$

$$f_2(\psi) = \begin{cases} \sum_{u=0}^{M-1}(2M-2u-1)(u+1)^2 + (\psi - M)\Big(2\sum_{u=1}^{M-1}u^2 + M^2\Big) \\ \quad + \sum_{v=1}^{M-1}\Big[2\sum_{u=1}^{v}u^2 + (2M-2v-1)v^2\Big] \quad \text{if } \psi \geqslant M, \\ \sum_{u=0}^{\psi-1}(2M-2u-1)(u+1)^2 + \sum_{u=\psi}^{M-1}(2M-2u-1)\psi^2 \\ \quad + \sum_{v=1}^{\psi-1}\Big[2\sum_{u=1}^{v}u^2 + (2M-2v-1)v^2\Big] \quad \text{if } \psi < M. \end{cases} \tag{5.E.8}$$

When $L^{(j)} = 1$, the variance of $y_l^{(j)} \mid (d_l^{(j)}, \check{\boldsymbol{d}}_l^{(j)})$ can be shown equal to

$$\begin{aligned} &\operatorname{var}\big[y_l^{(j)} \mid \big(d_l^{(j)}, \check{\boldsymbol{d}}_l^{(j)}\big)\big] \\ &\quad = V_1(2\beta + L^{(j)}, M)\operatorname{var}\big[\big(x_k^{(j)}\big)^2\big] + V_2\big(2\beta + L^{(j)}, M\big)P_s{}^2 \\ &\qquad + V_3\big(2\beta + L^{(j)}, M\big)P_s\frac{N_0}{2} \\ &\qquad + 2P_c\big[C_1\big(2\beta + L^{(j)}, M, d_l, d_{l-1}\big) + C_2\big(2\beta + L^{(j)}, M, d_l, d_{l-1}\big)\big] \\ &\qquad + \sum_{i=1, i\neq j}^{N} q\big(\alpha^{(i)}, \gamma^{(i)}\big)P_s^2 + 2\beta M^2\big(N^2 - \tilde{N} - 1\big)P_s^2 \\ &\qquad + 2(N-1)\big(2\beta - L^{(j)}\big)M(M-2)P_s^2 \\ &\qquad + 2\beta M(N-1)P_s N_0 + \beta\frac{N_0^2}{2}. \end{aligned} \tag{5.E.9}$$

For the case where $2 \leqslant L^{(j)} < 2\beta$, the variance of $y_l^{(j)} \mid (d_l^{(j)}, \check{\boldsymbol{d}}_l^{(j)})$ can be found equal to

$$\begin{aligned} &\operatorname{var}\big[y_l^{(j)} \mid \big(d_l^{(j)}, \check{\boldsymbol{d}}_l^{(j)}\big)\big] \\ &\quad = \alpha^{(j)}\Big[V_1(\gamma^{(j)} + 1, M)\operatorname{var}\big[\big(x_k^{(j)}\big)^2\big] + V_2\big(\gamma^{(j)} + 1, M\big)P_s^2 \\ &\qquad + V_3\big(\gamma^{(j)} + 1, M\big)P_s\frac{N_0}{2}\Big] \\ &\qquad + \big(L^{(j)} - \alpha^{(j)}\big)\Big[V_1\big(\gamma^{(j)}, M\big)\operatorname{var}\big[\big(x_k^{(j)}\big)^2\big] + V_2\big(\gamma^{(j)}, M\big)P_s^2 \\ &\qquad + V_3\big(\gamma^{(j)}, M\big)P_s\frac{N_0}{2}\Big] \end{aligned}$$

$$+\sum_{i=1,i\neq j}^{N} q\left(\alpha^{(i)},\gamma^{(i)}\right)P_s^2+2\beta M^2\left(N^2-\tilde{N}-1\right)P_s^2$$
$$+2(N-1)\left(2\beta-L^{(j)}\right)M(M-2)P_s^2$$
$$+2\beta M(N-1)P_sN_0+\beta\frac{N_0^2}{2}, \tag{5.E.10}$$

where $(\tilde{N}-1)$ represents the number of users with a delay $L^{(i)}$ smaller than 2β. Hence, if $L^{(i)}<2\beta$ for all i, $\tilde{N}=N-1$. Finally, when $L^{(j)}\geqslant 2\beta$, we have

$$\begin{aligned}\operatorname{var}\left[y_l^{(j)} \mid \left(d_l^{(j)},\check{\boldsymbol{d}}_l^{(j)}\right)\right] &= 2\beta\operatorname{var}\left[\left(x_k^{(j)}\right)^2\right]+2\left(M^2-1\right)\beta P_s^2\\ &\quad+\sum_{i=1,i\neq j}^{N} q\left(\alpha^{(i)},\gamma^{(i)}\right)P_s^2+2\beta M^2\left(N^2-\tilde{N}-1\right)P_s^2\\ &\quad+2\beta MNP_sN_0+\beta\frac{N_0^2}{2}.\end{aligned} \tag{5.E.11}$$

For the cases $2\leqslant L^{(j)}<2\beta$ and $L^{(j)}\geqslant 2\beta$, it can be observed from (5.E.10) and (5.E.11) that the variance of $y_l^{(j)} \mid (d_l^{(j)},\check{\boldsymbol{d}}_l^{(j)})$ is independent of the transmitted symbols $d_l^{(j)}$ and $\check{\boldsymbol{d}}_l^{(j)}$.

Assuming that $y_l^{(j)} \mid (d_l^{(j)},\check{\boldsymbol{d}}_l^{(j)})$ follows a normal distribution for the given transmitted symbols $d_l^{(j)}$ and $\check{\boldsymbol{d}}_l^{(j)}$, the approximate BER for the jth user will be given by

$$\begin{aligned}&\mathrm{BER}_{\mathrm{GCDSK}}^{(j)}\\ &=\sum_{\check{\boldsymbol{d}}_l^{(j)}\in\{-1,+1\}^2}\left[\frac{1}{2}\operatorname{Prob}\left(y_l^{(j)}\leqslant 0 \,\middle|\, \left(d_l^{(j)}=+1,\check{\boldsymbol{d}}_l^{(j)}\right)\right)\operatorname{Prob}\left(\check{\boldsymbol{d}}_l^{(j)}\right)\right.\\ &\quad\left.+\frac{1}{2}\operatorname{Prob}\left(y_l^{(j)}>0 \,\middle|\, \left(d_l^{(j)}=-1,\check{\boldsymbol{d}}_l^{(j)}\right)\right)\operatorname{Prob}\left(\check{\boldsymbol{d}}_l^{(j)}\right)\right]\\ &=\frac{1}{16}\sum_{\check{\boldsymbol{d}}_l^{(j)}\in\{-1,+1\}^2}\left[\operatorname{erfc}\left(\frac{\mathrm{E}[y_l^{(j)} \mid (d_l^{(j)}=+1,\check{\boldsymbol{d}}_l^{(j)})]}{\sqrt{2\operatorname{var}[y_l^{(j)} \mid (d_l^{(j)}=+1,\check{\boldsymbol{d}}_l^{(j)})]}}\right)\right.\\ &\quad\left.+\operatorname{erfc}\left(\frac{\mathrm{E}[y_l^{(j)} \mid (d_l^{(j)}=-1,\check{\boldsymbol{d}}_l^{(j)})]}{\sqrt{2\operatorname{var}[y_l^{(j)} \mid (d_l^{(j)}=-1,\check{\boldsymbol{d}}_l^{(j)})]}}\right)\right].\end{aligned} \tag{5.E.12}$$

6

ENHANCEMENTS OF CORRELATION-DELAY-SHIFT-KEYING SCHEME

In Chapter 5, we have reviewed the operations of the correlation-delay-shift-keying (CDSK) scheme. Due to the correlations between the delay versions of the chaotic signals, intra-signal interference occurs, limiting the performance of CDSK systems. In the generalized CDSK (GCDSK) scheme described in Section 5.2, the signal component has been enhanced, and so is the intra-signal interference. In this chapter, we modify the modulation scheme of CDSK with an aim to diminishing the intra-signal interference, thus enhancing the system performance. Two enhancement methods for CDSK will be investigated. In both cases, the spreading factor is set to be the same as the delay, which means that the receiver detects the symbol by correlating the received signal in the current bit duration with that in the previous bit duration.

In the first method, when the symbol "$+1$" is to be sent, a delayed signal from the chaos generator will be transmitted. The chaotic signal not only carries the information of the current bit, but also serves as the reference signal for the next information bit. The same chaotic signal will be re-sent in the following bit duration as long as the next symbol is also "$+1$". If the next transmitted symbol is "-1", the chaotic reference signal transmitted in the previous bit will be modulated by the data symbol (i.e., -1). It is then added to the output of the chaos generator, and the resultant signal is amplified before transmission. If two consecutive "-1"s are to be sent, the transmitted signal in the second symbol duration will have a similar structure as the CDSK signal, which is the summation of a chaotic signal and the delayed and modulated version of it. But the signal here is further amplified before transmission. Compared to the CDSK scheme, the first enhancement method combines two chaotic signals for transmission only for the symbol "-1". For the symbol "$+1$", we only need to transmit one (delayed) chaotic signal. The advantage of such an arrangement is twofold. Firstly, it can reduce the intra-signal interference. Secondly, the average bit energy is reduced because a smaller signal is being sent. It also implies that a lower E_b/N_0 value will be sufficient to achieve a desired performance level.

For the first enhancement method, it is found that the intra-signal interference remains the same as that for the CDSK scheme if both the current and the previous symbols are "-1". This is the worst case scenario and it dominates the system performance, especially when the E_b/N_0 value is small. To further alleviate the intra-signal interference problem, we study another enhancement method which has a slightly more complicated transmitter structure. In particular, the method simply sends an inverted copy of the transmitted signal in the previous bit duration if both the current and the previous symbols are "-1". The method can enhance the useful signal component while keeping the average bit energy unchanged. Hence, the performance should be improved compared to the first method.

The organization of this chapter is as follows. First, the operation and the performance analysis of the enhancement methods are presented in Sections 6.1 and 6.2. In Section 6.3, based on a permutation transformation, the multiple-access capabilities of the enhancement methods as well as CDSK are explored. Also, the analytical bit error rate (BER) of the permutation-based multiple-access CDSK system is derived. Finally, the simulated and Gaussian approximated BER results are reported in Section 6.4. In addition, the performances of the enhancement methods are compared with those of the DCSK and CDSK schemes.

6.1 ENHANCED CORRELATION-DELAY-SHIFT-KEYING SYSTEM WITH METHOD ONE

6.1.1 Transmitter Structure

Figure 6.1 shows the transmitter structure of the enhanced correlation-delay-shift-keying system with Method 1 (ECDSK-M1). Suppose the previous symbol sent is "$+1$" ("-1"). If the binary symbol "$+1$" is to be sent in the current bit duration, the chaotic signal sent (the chaotic signal produced by the chaos generator) during the previous bit duration will be transmitted. When a "-1" is to be sent, the chaotic signal sent (the chaotic signal produced by the chaos generator) in the previous bit duration is multiplied by -1 and added to the output of the chaos generator. The output of the adder is then amplified by a factor of $\sqrt{G}$ before

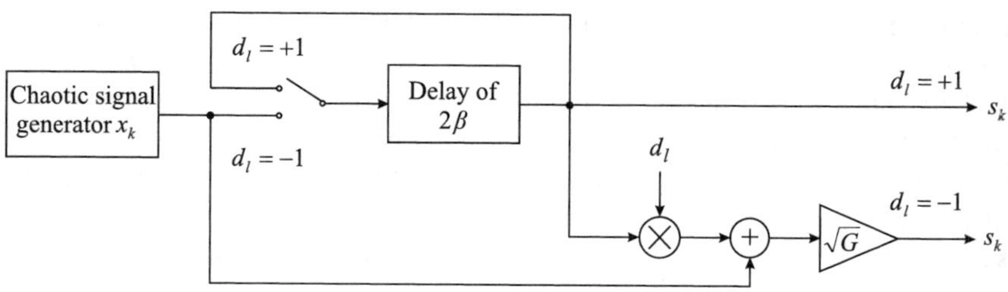

Figure 6.1. Transmitter structure of the enhanced correlation-delay-shift-keying system with Method 1 (ECDSK-M1).

transmission. The value of the gain $\sqrt{G}$ can be adjusted such that the optimum bit error rate (BER) is obtained.

Denote the lth transmitted symbol by $d_l \in \{-1, +1\}$ and assume that the probabilities of occurrence for "+1" and "−1" are the same. Denoting the spreading factor by 2β and considering the lth transmitted symbol, i.e., for $k = 2(l-1)\beta + 1, 2(l-1)\beta + 2, \ldots, 2l\beta$, the transmitted signal can be represented by

$$s_k = \begin{cases} s_{k-2\beta} & \text{if } d_l = \text{“}+1\text{” and } d_{l-1} = \text{“}+1\text{”}, \\ x_{k-2\beta} & \text{if } d_l = \text{“}+1\text{” and } d_{l-1} = \text{“}-1\text{”}, \\ \sqrt{G}(-s_{k-2\beta} + x_k) & \text{if } d_l = \text{“}-1\text{” and } d_{l-1} = \text{“}+1\text{”}, \\ \sqrt{G}(-x_{k-2\beta} + x_k) & \text{if } d_l = \text{“}-1\text{” and } d_{l-1} = \text{“}-1\text{”}. \end{cases} \tag{6.1.1}$$

6.1.2 Receiver Structure

After passing through the additive white Gaussian noise channel with two-sided power spectral density $N_0/2$, the received signal is given by

$$r_k = s_k + \xi_k, \tag{6.1.2}$$

where ξ_k is a Gaussian random variable with zero mean and variance $N_0/2$. A simple correlator-based receiver, as in Fig. 6.2, is employed to correlate the incoming signal in the current bit duration with that in the previous bit duration. The output of the correlator for the lth symbol duration equals

$$y_l = \sum_{k=2(l-1)\beta+1}^{2l\beta} r_k r_{k-2\beta}. \tag{6.1.3}$$

We define $\hat{n}$ as the number of consecutive "+1"s immediately before and excluding the $(l-1)$th bit and we also define, for convenience, a new variable ω

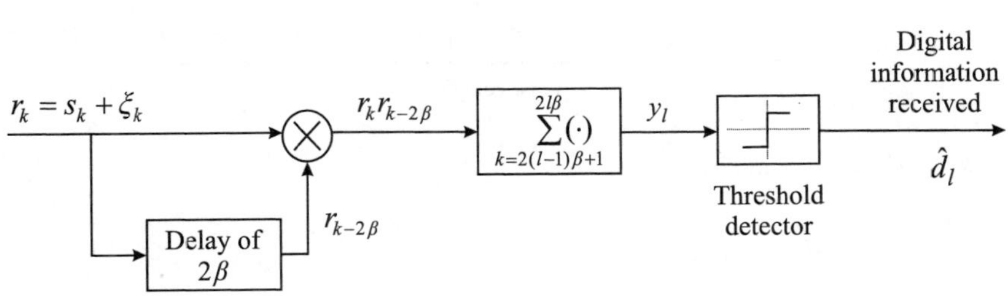

Figure 6.2. Receiver structure of the enhanced correlation-delay-shift-keying system with Method 1 (ECDSK-M1).

as

$$\omega = \hat{n} + 2. \tag{6.1.4}$$

Next, we evaluate the output of the correlator under four different scenarios.

Case 1: $d_l = +1$ and $d_{l-1} = +1$

$$\begin{aligned} y_l \mid (d_l = +1, d_{l-1} = +1) &= \sum_{k=2(l-1)\beta+1}^{2l\beta} (x_{k-2\omega\beta} + \xi_k)(x_{k-2\omega\beta} + \xi_{k-2\beta}) \\ &= \sum_{k=2(l-1)\beta+1}^{2l\beta} \Bigg[\underbrace{x_{k-2\omega\beta}^2}_{\text{required signal}} + \underbrace{x_{k-2\omega\beta}\xi_{k-2\beta} + x_{k-2\omega\beta}\xi_k + \xi_k\xi_{k-2\beta}}_{\text{noise components}} \Bigg]. \end{aligned} \tag{6.1.5}$$

Case 2: $d_l = +1$ and $d_{l-1} = -1$

$$\begin{aligned} y_l \mid (d_l = +1, d_{l-1} = -1) &= \sum_{k=2(l-1)\beta+1}^{2l\beta} [x_{k-2\beta} + \xi_k]\big[\sqrt{G}(-x_{k-2\omega\beta} + x_{k-2\beta}) + \xi_{k-2\beta}\big] \\ &= \sum_{k=2(l-1)\beta+1}^{2l\beta} \Bigg[\underbrace{\sqrt{G}x_{k-2\beta}^2}_{\text{required signal}} - \underbrace{\sqrt{G}x_{k-2\beta}x_{k-2\omega\beta}}_{\text{intra-signal interference}} \\ &\quad + \underbrace{x_{k-2\beta}\xi_{k-2\beta} + \sqrt{G}(-x_{k-2\omega\beta} + x_{k-2\beta})\xi_k + \xi_k\xi_{k-2\beta}}_{\text{noise components}} \Bigg]. \end{aligned} \tag{6.1.6}$$

Case 3: $d_l = -1$ and $d_{l-1} = +1$

$$\begin{aligned} y_l \mid (d_l = -1, d_{l-1} = +1) &= \sum_{k=2(l-1)\beta+1}^{2l\beta} \big[\sqrt{G}(-x_{k-2\omega\beta} + x_k) + \xi_k\big][x_{k-2w\beta} + \xi_{k-2\beta}] \end{aligned}$$

$$= \sum_{k=2(l-1)\beta+1}^{2l\beta} \Bigg[-\underbrace{\sqrt{G}x_{k-2w\beta}^2}_{\text{required signal}} + \underbrace{\sqrt{G}x_k x_{k-2\omega\beta}}_{\text{intra-signal interference}}$$

$$+ \underbrace{\sqrt{G}(-x_{k-2\omega\beta} + x_k)\xi_{k-2\beta} + x_{k-2w\beta}\xi_k + \xi_k\xi_{k-2\beta}}_{\text{noise components}} \Bigg] \quad (6.1.7)$$

Case 4: $d_l = -1$ and $d_{l-1} = -1$

$$y_l \mid (d_l = -1, d_{l-1} = -1)$$

$$= \sum_{k=2(l-1)\beta+1}^{2l\beta} \left[\sqrt{G}(-x_{k-2\beta} + x_k) + \xi_k\right]\left[\sqrt{G}(-x_{k-2\omega\beta} + x_{k-2\beta}) + \xi_{k-2\beta}\right]$$

$$= \sum_{k=2(l-1)\beta+1}^{2l\beta} \Bigg[- \underbrace{Gx_{k-2\beta}^2}_{\text{required signal}}$$

$$+ \underbrace{Gx_{k-2\beta}x_{k-2\omega\beta} - Gx_k x_{k-2\omega\beta} + Gx_k x_{k-2\beta}}_{\text{intra-signal interference}}$$

$$+ \underbrace{\sqrt{G}(-x_{k-2\beta} + x_k)\xi_{k-2\beta}}_{\text{noise components}}$$

$$+ \underbrace{\sqrt{G}(-x_{k-2\omega\beta} + x_{k-2\beta})\xi_k + \xi_k\xi_{k-2\beta}}_{\text{noise components}} \Bigg] \quad (6.1.8)$$

Based on different combinations of d_l and d_{l-1}, we compare the correlator output in the ECDSK-M1 system with that in the CDSK system (see (5.1.3) to (5.1.6)) and have made the following observations for the ECDSK-M1 system.

- When $d_l =$ "+1" and $d_{l-1} =$ "+1", the intra-signal interference disappears and the noise components diminish.
- When $d_l =$ "+1" and $d_{l-1} =$ "−1" or $d_l =$ "−1" and $d_{l-1} =$ "+1", both the intra-signal interference and the noise components are reduced.
- When $d_l =$ "−1" and $d_{l-1} =$ "−1", only the noise components diminish.

Because of the reduction of the intra-signal interference and the noise components, we expect the bit error performance of the ECDSK-M1 system to be better compared to that of the CDSK system.

6.1.3 Gaussian-Approximated Bit Error Rates for ECDSK-M1 System

Given the transmitted symbols d_l and d_{l-1}, the conditional mean value of the correlator output can be shown equal to

$$\mathrm{E}\big[y_l \mid (d_l = +1, d_{l-1} = +1)\big] = 2\beta P_s, \tag{6.1.9}$$

$$\mathrm{E}\big[y_l \mid (d_l = +1, d_{l-1} = -1)\big] = 2\sqrt{G}\beta P_s, \tag{6.1.10}$$

$$\mathrm{E}\big[y_l \mid (d_l = -1, d_{l-1} = +1)\big] = -2\sqrt{G}\beta P_s, \tag{6.1.11}$$

$$\mathrm{E}\big[y_l \mid (d_l = -1, d_{l-1} = -1)\big] = -2G\beta P_s, \tag{6.1.12}$$

where

$$P_s = \mathrm{E}\big[x_k^2\big]. \tag{6.1.13}$$

Also, the conditional variance of the correlator output can be computed by

$$\begin{aligned}
&\mathrm{var}\big[y_l \mid (d_l = +1, d_{l-1} = +1)\big] \\
&\quad = 2\beta\,\mathrm{var}\big[x_k^2\big] + 2\beta P_s N_0 + \beta\frac{N_0^2}{2}, && (6.1.14) \\
&\mathrm{var}\big[y_l \mid (d_l = +1, d_{l-1} = -1)\big] \\
&\quad = 2G\beta\,\mathrm{var}\big[x_k^2\big] + 2G\beta P_s^2 + (1+2G)\beta P_s N_0 + \beta\frac{N_0^2}{2}, && (6.1.15) \\
&\mathrm{var}\big[y_l \mid (d_l = -1, d_{l-1} = +1)\big] \\
&\quad = 2G\beta\,\mathrm{var}\big[x_k^2\big] + 2G\beta P_s^2 + (1+2G)\beta P_s N_0 + \beta\frac{N_0^2}{2}, && (6.1.16) \\
&\mathrm{var}\big[y_l \mid (d_l = -1, d_{l-1} = -1)\big] \\
&\quad = 2G^2\beta\,\mathrm{var}\big[x_k^2\big] + 6G^2\beta P_s^2 + 4G\beta P_s N_0 + \beta\frac{N_0^2}{2}. && (6.1.17)
\end{aligned}$$

Using the Gaussian approximation method, the analytical bit error rate for the ECDSK-M1 system can then be derived and is given by

$$\begin{aligned}
&\mathrm{BER}_{\text{ECDSK-M1}} \\
&\quad = \frac{1}{4}\,\mathrm{Prob}\big(y_l \leqslant 0 \mid (d_l = +1, d_{l-1} = +1)\big) \\
&\qquad + \frac{1}{4}\,\mathrm{Prob}\big(y_l \leqslant 0 \mid (d_l = +1, d_{l-1} = -1)\big)
\end{aligned}$$

$$
\begin{aligned}
&+\frac{1}{4}\,\mathrm{Prob}\big(y_l > 0 \,\big|\, (d_l = -1, d_{l-1} = +1)\big) \\
&+\frac{1}{4}\,\mathrm{Prob}\big(y_l > 0 \,\big|\, (d_l = -1, d_{l-1} = -1)\big) \\
&= \frac{1}{8}\,\mathrm{erfc}\left(\frac{\mathrm{E}[y_l \mid (d_l = +1, d_{l-1} = +1)]}{\sqrt{2\,\mathrm{var}[y_l \mid (d_l = +1, d_{l-1} = +1)]}}\right) \\
&+\frac{1}{8}\,\mathrm{erfc}\left(\frac{\mathrm{E}[y_l \mid (d_l = +1, d_{l-1} = -1)]}{\sqrt{2\,\mathrm{var}[y_l \mid (d_l = +1, d_{l-1} = -1)]}}\right) \\
&+\frac{1}{8}\,\mathrm{erfc}\left(\frac{-\mathrm{E}[y_l \mid (d_l = -1, d_{l-1} = +1)]}{\sqrt{2\,\mathrm{var}[y_l \mid (d_l = -1, d_{l-1} = +1)]}}\right) \\
&+\frac{1}{8}\,\mathrm{erfc}\left(\frac{-\mathrm{E}[y_l \mid (d_l = -1, d_{l-1} = -1)]}{\sqrt{2\,\mathrm{var}[y_l \mid (d_l = -1, d_{l-1} = -1)]}}\right) \\
&= \frac{1}{8}\,\mathrm{erfc}\left(\left[\frac{\Psi}{\beta} + (1+2G)\left(\frac{E_b}{N_0}\right)^{-1} + \frac{(1+2G)^2}{4}\beta\left(\frac{E_b}{N_0}\right)^{-2}\right]^{-1/2}\right) \\
&+\frac{1}{8}\,\mathrm{erfc}\left(\left[\frac{\Psi}{\beta} + \frac{1}{\beta} + \frac{(1+2G)^2}{2G}\left(\frac{E_b}{N_0}\right)^{-1}\right.\right. \\
&\left.\left. + \frac{(1+2G)^2}{4G}\beta\left(\frac{E_b}{N_0}\right)^{-2}\right]^{-1/2}\right) \\
&+\frac{1}{8}\,\mathrm{erfc}\left(\left[\frac{\Psi}{\beta} + \frac{1}{\beta} + \frac{(1+2G)^2}{2G}\left(\frac{E_b}{N_0}\right)^{-1}\right.\right. \\
&\left.\left. + \frac{(1+2G)^2}{4G}\beta\left(\frac{E_b}{N_0}\right)^{-2}\right]^{-1/2}\right) \\
&+\frac{1}{8}\,\mathrm{erfc}\left(\left[\frac{\Psi}{\beta} + \frac{3}{\beta} + \frac{2(1+2G)}{G}\left(\frac{E_b}{N_0}\right)^{-1}\right.\right. \\
&\left.\left. + \frac{(1+2G)^2}{4G^2}\beta\left(\frac{E_b}{N_0}\right)^{-2}\right]^{-1/2}\right),
\end{aligned}
\tag{6.1.18}
$$

where

$$\Psi = \frac{\text{var}[x_k^2]}{P_s^2} = \frac{\text{var}[x_k^2]}{\text{E}^2[x_k^2]}. \tag{6.1.19}$$

Denote the average bit energy by E_b. When $G = 3/2$, the average bit energy for the ECDSK-M1 scheme equals

$$E_b = E_{b_1} + E_{b_2} = \frac{1+2G}{2} 2\beta P_s = 4\beta P_s, \tag{6.1.20}$$

where E_{b_1} and E_{b_2} represent the average bit energy when the symbols "+1" and "−1" are transmitted, respectively. The average bit energy is now same as that for the CDSK system. Substituting $G = 3/2$ into (6.1.18), we obtain

$$\begin{aligned}
&\text{BER}_{\text{ECDSK-M1}} \\
&= \frac{1}{8}\text{erfc}\left(\left[\frac{\Psi}{\beta} + 4\left(\frac{E_b}{N_0}\right)^{-1} + 4\beta\left(\frac{E_b}{N_0}\right)^{-2}\right]^{-1/2}\right) \\
&+ \frac{1}{8}\text{erfc}\left(\left[\frac{\Psi}{\beta} + \frac{1}{\beta} + \frac{16}{3}\left(\frac{E_b}{N_0}\right)^{-1} + \frac{8}{3}\beta\left(\frac{E_b}{N_0}\right)^{-2}\right]^{-1/2}\right) \\
&+ \frac{1}{8}\text{erfc}\left(\left[\frac{\Psi}{\beta} + \frac{1}{\beta} + \frac{16}{3}\left(\frac{E_b}{N_0}\right)^{-1} + \frac{8}{3}\beta\left(\frac{E_b}{N_0}\right)^{-2}\right]^{-1/2}\right) \\
&+ \frac{1}{8}\text{erfc}\left(\left[\frac{\Psi}{\beta} + \frac{3}{\beta} + \frac{16}{3}\left(\frac{E_b}{N_0}\right)^{-1} + \frac{16}{9}\beta\left(\frac{E_b}{N_0}\right)^{-2}\right]^{-1/2}\right).
\end{aligned} \tag{6.1.21}$$

6.2 ENHANCED CORRELATION-DELAY-SHIFT-KEYING SYSTEM WITH METHOD TWO

6.2.1 Transmitter Structure

Figure 6.3 shows the transmitter structure of the enhanced correlation-delay-shift-keying system with Method 2 (ECDSK-M2). Consider the lth bit duration. If both d_l and d_{l-1} are "+1", the chaotic signal sent in the previous bit duration will be re-transmitted. Conversely, if both symbols are "−1", an inverted copy of the chaotic signal sent in the previous bit duration will be transmitted. When $d_l =$ "+1" and $d_{l-1} =$ "−1", the chaotic signal generated during the $(l - \check{n})$th bit duration will

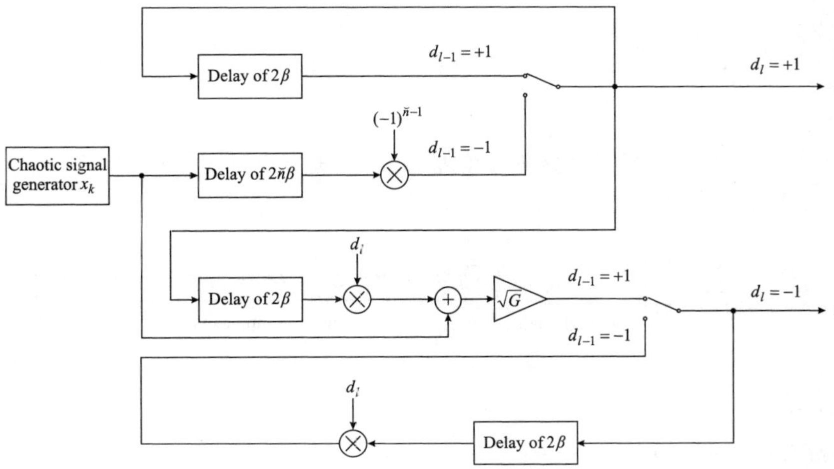

Figure 6.3. Transmitter structure of the enhanced correlation-delay-shift-keying system with Method 2 (ECDSK-M2).

be sent after multiplied by $(-1)^{(\check{n}-1)}$, where $\check{n}$ denotes the number of consecutive "-1"s sent immediately before the lth bit. For the last case when $d_l =$ "-1" and $d_{l-1} =$ "$+1$", the chaotic signal sent during the previous symbol duration is inverted and then added to the output of the chaos generator. The output of the adder is further scaled by the factor $\sqrt{G}$ before transmission. In summary, the transmitted signal is expressed as

$$s_k = \begin{cases} s_{k-2\beta} & \text{if } d_l = \text{“}+1\text{” and } d_{l-1} = \text{“}+1\text{”}, \\ (-1)^{\check{n}-1} x_{k-2\check{n}\beta} & \text{if } d_l = \text{“}+1\text{” and } d_{l-1} = \text{“}-1\text{”}, \\ \sqrt{G}(-s_{k-2\beta} + x_k) & \text{if } d_l = \text{“}-1\text{” and } d_{l-1} = \text{“}+1\text{”}, \\ -s_{k-2\beta} & \text{if } d_l = \text{“}-1\text{” and } d_{l-1} = \text{“}-1\text{”}. \end{cases} \tag{6.2.1}$$

6.2.2 Receiver Structure

The receiver structure is the same as that for the ECDSK-M1 system. Define α and λ as the smallest and second smallest positive integers such that $d_{l-\alpha} = d_{l-\lambda} = -1$ and $d_{l-\alpha-1} = d_{l-\lambda-1} = +1$.

Referring to Fig. 6.2, we evaluate the output of the correlator for the ECDSK-M2 system under four different cases. In the following, $\hat{D}_l, \check{D}_l \in \{-1, +1\}$.

Case 1: $d_l = +1$ *and* $d_{l-1} = +1$

$$y_l \mid (d_l = +1, d_{l-1} = +1)$$

$$= \sum_{k=2(l-1)\beta+1}^{2l\beta} r_k r_{k-2\beta}$$

$$= \sum_{k=2(l-1)\beta+1}^{2l\beta} (\hat{D}_l x_{k-2\alpha\beta} + \xi_k)(\hat{D}_l x_{k-2\alpha\beta} + \xi_{k-2\beta})$$

$$= \sum_{k=2(l-1)\beta+1}^{2l\beta} \left[\underbrace{x_{k-2\alpha\beta}^2}_{\text{required signal}} + \underbrace{\hat{D}_l x_{k-2\alpha\beta}\xi_{k-2\beta} + \hat{D}_l x_{k-2\alpha\beta}\xi_k + \xi_k\xi_{k-2\beta}}_{\text{noise components}} \right]. \tag{6.2.2}$$

Case 2: $d_l = +1$ *and* $d_{l-1} = -1$

$$y_l \mid (d_l = +1, d_{l-1} = -1)$$
$$= \sum_{k=2(l-1)\beta+1}^{2l\beta} (\hat{D}_l x_{k-2\alpha\beta} + \xi_k)\left[\sqrt{G}(\check{D}_l x_{k-2\lambda\beta} + \hat{D}_l x_{k-2\alpha\beta}) + \xi_{k-2\beta}\right]$$

$$= \sum_{k=2(l-1)\beta+1}^{2l\beta} \left[\underbrace{\sqrt{G}x_{k-2\alpha\beta}^2}_{\text{required signal}} + \underbrace{\sqrt{G}\hat{D}_l\check{D}_l x_{k-2\alpha\beta}x_{k-2\lambda\beta}}_{\text{intra-signal interference}} \right.$$
$$\left. + \underbrace{\hat{D}_l x_{k-2\alpha\beta}\xi_{k-2\beta} + \xi_k\xi_{k-2\beta}}_{\text{noise components}} + \underbrace{\sqrt{G}(\check{D}_l x_{k-2\lambda\beta} + \hat{D}_l x_{k-2\alpha\beta})\xi_k}_{\text{noise components}} \right]. \tag{6.2.3}$$

Case 3: $d_l = -1$ *and* $d_{l-1} = +1$

$$y_l \mid (d_l = -1, d_{l-1} = +1)$$
$$= \sum_{k=2(l-1)\beta+1}^{2l\beta} \left[\sqrt{G}(\check{D}_l x_{k-2\alpha\beta} + x_k) + \xi_k\right]\left[-\check{D}_l x_{k-2\alpha\beta} + \xi_{k-2\beta}\right]$$

$$= \sum_{k=2(l-1)\beta+1}^{2l\beta} \left[-\underbrace{\sqrt{G}x_{k-2\alpha\beta}^2}_{\text{required signal}} - \underbrace{\sqrt{G}\check{D}_l x_k x_{k-2\alpha\beta}}_{\text{intra-signal interference}} \right.$$

$$+ \underbrace{\sqrt{G}(\check{D}_l x_{k-2\alpha\beta} + x_k)\xi_{k-2\beta}}_{\text{noise components}} - \underbrace{\check{D}_l x_{k-2\alpha\beta}\xi_k + \xi_k\xi_{k-2\beta}}_{\text{noise components}} \Bigg]. \quad (6.2.4)$$

Case 4: $d_l = -1$ and $d_{l-1} = -1$

$$\begin{aligned} & y_l \mid (d_l = -1, d_{l-1} = -1) \\ &= \sum_{k=2(l-1)\beta+1}^{2l\beta} \left[\sqrt{G}(\check{D}_l x_{k-2\lambda\beta} + \hat{D}_l x_{k-2\alpha\beta}) + \xi_k\right] \\ &\quad \times \left[-\sqrt{G}(\check{D}_l x_{k-2\lambda\beta} + \hat{D}_l x_{k-2\alpha\beta}) + \xi_{k-2\beta}\right] \\ &= \sum_{k=2(l-1)\beta+1}^{2l\beta} \Bigg[- \underbrace{G(\check{D}_l x_{k-2\lambda\beta} + \hat{D}_l x_{k-2\alpha\beta})^2}_{\text{required signal}} \\ &\quad + \underbrace{\sqrt{G}(\check{D}_l x_{k-2\lambda\beta} + \hat{D}_l x_{k-2\alpha\beta})\xi_{k-2\beta}}_{\text{noise components}} \\ &\quad - \underbrace{\sqrt{G}(\check{D}_l x_{k-2\lambda\beta} + \hat{D}_l x_{k-2\alpha\beta})\xi_k + \xi_k\xi_{k-2\beta}}_{\text{noise components}} \Bigg]. \end{aligned} \quad (6.2.5)$$

Comparing (6.2.2)–(6.2.5) to (6.1.5)–(6.1.8), we have the following observations.

- For the cases when $d_l =$ "+1" and $d_{l-1} =$ "+1", $d_l =$ "+1" and $d_{l-1} =$ "−1", and $d_l =$ "−1" and $d_{l-1} =$ "+1", the correlator output for the ECDSK-M2 scheme is similar to that for the ECDSK-M1 scheme.
- When $d_l =$ "−1" and $d_{l-1} =$ "−1", the intra-signal interference has been removed for the ECDSK-M2 scheme.

Hence, we expect that the ECDSK-M2 scheme can achieve a better performance than the ECDSK-M1 scheme.

6.2.3 Gaussian-Approximated Bit Error Rates for ECDSK-M2 System

Given the transmitted symbols d_l and d_{l-1}, the conditional mean values of the correlator output can be obtained using

$$\mathrm{E}\big[y_l \mid (d_l = +1, d_{l-1} = +1)\big] = 2\beta P_s, \quad (6.2.6)$$

$$\mathrm{E}\big[y_l \mid (d_l = +1, d_{l-1} = -1)\big] = 2\sqrt{G}\beta P_s, \quad (6.2.7)$$

$$E\left[y_l \mid (d_l = -1, d_{l-1} = +1)\right] = -2\sqrt{G}\beta P_s, \tag{6.2.8}$$

$$E\left[y_l \mid (d_l = -1, d_{l-1} = -1)\right] = -4G\beta P_s. \tag{6.2.9}$$

Correspondingly, the conditional variances of the correlator output are

$$\begin{aligned}
&\mathrm{var}\left[y_l \mid (d_l = +1, d_{l-1} = +1)\right] \\
&\quad = 2\beta\,\mathrm{var}\left[x_k^2\right] + 2\beta P_s N_0 + \beta\frac{N_0^2}{2}, &(6.2.10)\\
&\mathrm{var}\left[y_l \mid (d_l = +1, d_{l-1} = -1)\right] \\
&\quad = 2G\beta\,\mathrm{var}\left[x_k^2\right] + 2G\beta P_s^2 + (1+2G)\beta P_s N_0 + \beta\frac{N_0^2}{2}, &(6.2.11)\\
&\mathrm{var}\left[y_l \mid (d_l = -1, d_{l-1} = +1)\right] \\
&\quad = 2G\beta\,\mathrm{var}\left[x_k^2\right] + 2G\beta P_s^2 + (1+2G)\beta P_s N_0 + \beta\frac{N_0^2}{2}, &(6.2.12)\\
&\mathrm{var}\left[y_l \mid (d_l = -1, d_{l-1} = -1)\right] \\
&\quad = 4G^2\beta\,\mathrm{var}\left[x_k^2\right] + 8G^2\beta P_s^2 + 4G\beta P_s N_0 + \beta\frac{N_0^2}{2}. &(6.2.13)
\end{aligned}$$

Using a similar method as in Section 6.1.3, the BER for the ECDSK-M2 scheme is readily shown equal to

$$\begin{aligned}
&\mathrm{BER}_{\text{ECDSK-M2}} \\
&= \frac{1}{8}\,\mathrm{erfc}\left(\left[\frac{\Psi}{\beta} + (1+2G)\left(\frac{E_b}{N_0}\right)^{-1} + \frac{(1+2G)^2}{4}\beta\left(\frac{E_b}{N_0}\right)^{-2}\right]^{-1/2}\right) \\
&\quad + \frac{1}{8}\,\mathrm{erfc}\left(\left[\frac{\Psi}{\beta} + \frac{1}{\beta} + \frac{(1+2G)^2}{2G}\left(\frac{E_b}{N_0}\right)^{-1} + \frac{(1+2G)^2}{4G}\beta\left(\frac{E_b}{N_0}\right)^{-2}\right]^{-1/2}\right) \\
&\quad + \frac{1}{8}\,\mathrm{erfc}\left(\left[\frac{\Psi}{\beta} + \frac{1}{\beta} + \frac{(1+2G)^2}{2G}\left(\frac{E_b}{N_0}\right)^{-1} + \frac{(1+2G)^2}{4G}\beta\left(\frac{E_b}{N_0}\right)^{-2}\right]^{-1/2}\right) \\
&\quad + \frac{1}{8}\,\mathrm{erfc}\left(\left[\frac{\Psi}{2\beta} + \frac{1}{\beta} + \frac{2G+1}{2G}\left(\frac{E_b}{N_0}\right)^{-1} + \frac{(1+2G)^2}{16G^2}\beta\left(\frac{E_b}{N_0}\right)^{-2}\right]^{-1/2}\right).
\end{aligned} \tag{6.2.14}$$

Substituting $G = 3/2$ into Eq. (6.2.14), we further obtain

$$\begin{aligned}
&\mathrm{BER}_{\mathrm{ECDSK\text{-}M2}} \\
&= \frac{1}{8}\operatorname{erfc}\left(\left[\frac{\Psi}{\beta} + 4\left(\frac{E_b}{N_0}\right)^{-1} + 4\beta\left(\frac{E_b}{N_0}\right)^{-2}\right]^{-1/2}\right) \\
&+ \frac{1}{8}\operatorname{erfc}\left(\left[\frac{\Psi}{\beta} + \frac{1}{\beta} + \frac{16}{3}\left(\frac{E_b}{N_0}\right)^{-1} + \frac{8}{3}\beta\left(\frac{E_b}{N_0}\right)^{-2}\right]^{-1/2}\right) \\
&+ \frac{1}{8}\operatorname{erfc}\left(\left[\frac{\Psi}{\beta} + \frac{1}{\beta} + \frac{16}{3}\left(\frac{E_b}{N_0}\right)^{-1} + \frac{8}{3}\beta\left(\frac{E_b}{N_0}\right)^{-2}\right]^{-1/2}\right) \\
&+ \frac{1}{8}\operatorname{erfc}\left(\left[\frac{\Psi}{2\beta} + \frac{1}{\beta} + \frac{4}{3}\left(\frac{E_b}{N_0}\right)^{-1} + \frac{4}{9}\beta\left(\frac{E_b}{N_0}\right)^{-2}\right]^{-1/2}\right). \qquad (6.2.15)
\end{aligned}$$

6.3 PERMUTATION-BASED MULTIPLE-ACCESS CDSK SYSTEMS

In Chapter 5, multiple access of the CDSK system has been accomplished by assigning different minimum delay values to different users. In this chapter, the delay values for the CDSK and the enhanced systems are all fixed at 2β, i.e., the spreading factor. To provide multiple-access capability, we introduce a transformation block to each of the transmitters and receivers.

6.3.1 Transmitter Structure

In Fig. 6.4, we depict the ith transmitter of a simple CDSK system with multiple-access capability. Compared to the single-user CDSK system shown in Fig. 5.1, only a transformation block and a switch have been added here. The same transformation block and switch are also included in the ECDSK-M1 and ECDSK-M2 systems to form the permutation-based multiple-access ECDSK-M1 (PMA-ECDSK-M1) and PMA-ECDSK-M2 systems, respectively. The corresponding transmitter structures are shown in Fig. 6.5 and Fig. 6.6.

Consider the lth bit duration and define the signal vector

$$\mathbf{s}_l^{(i)} = \left[s_{2(l-1)\beta+1}^{(i)} \quad s_{2(l-1)\beta+2}^{(i)} \quad \cdots \quad s_{2l\beta}^{(i)} \right]. \qquad (6.3.1)$$

Also, we denote the output vector of the transformation block by

$$\hat{\mathbf{s}}_l^{(i)} = \left[\hat{s}_{2(l-1)\beta+1}^{(i)} \quad \hat{s}_{2(l-1)\beta+2}^{(i)} \quad \cdots \quad \hat{s}_{2l\beta}^{(i)} \right] = F^{(i)}\left(\mathbf{s}_l^{(i)}\right). \qquad (6.3.2)$$

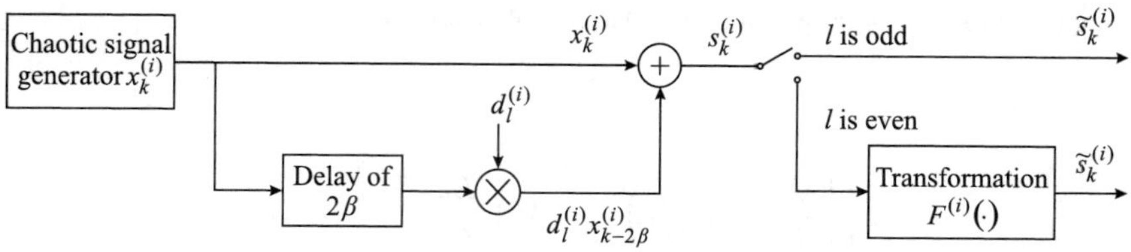

Figure 6.4. Transmitter structure of the ith user in the permutation-based multiple-access CDSK (PMA-CDSK) communication system.

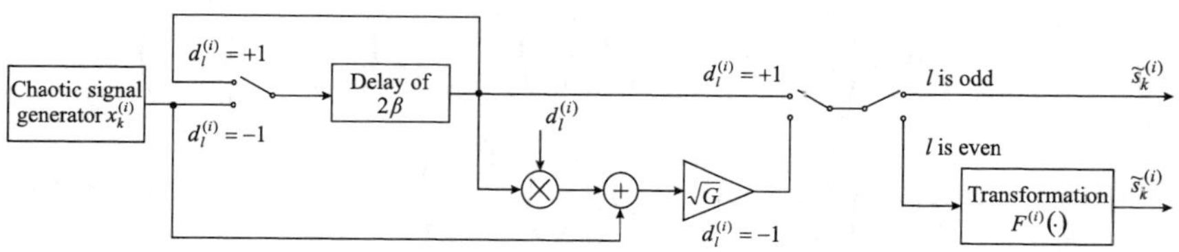

Figure 6.5. Transmitter structure of the permutation-based multiple-access ECDSK-M1 (PMA-ECDSK-M1) communication system.

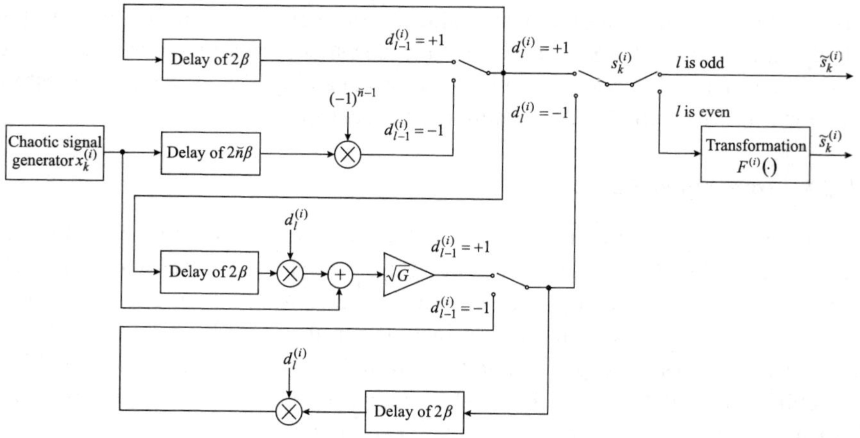

Figure 6.6. Transmitter structure of the permutation-based multiple-access ECDSK-M2 (PMA-ECDSK-M2) communication system.

Here, the transformation block, represented by $F^{(i)}(\cdot)$, is a simple permutation function that exchanges the locations of the chaotic signals within the same bit duration [Lau *et al.*, 2002a]. Depending on l being odd or even, the signal vector

$\boldsymbol{s}_l^{(i)}$ or $\hat{\boldsymbol{s}}_l^{(i)}$ is sent. Therefore, the transmitted signal vector, denoted by $\tilde{\boldsymbol{s}}_l^{(i)}$, equals

$$\tilde{\boldsymbol{s}}_l^{(i)} = \begin{cases} \boldsymbol{s}_l^{(i)} & \text{if } l \text{ is odd,} \\ \hat{\boldsymbol{s}}_l^{(i)} & \text{if } l \text{ is even.} \end{cases} \tag{6.3.3}$$

The overall transmitted signal vector of the system is found by summing all vectors from individual transmitters, i.e.,

$$\tilde{\boldsymbol{s}}_l = \sum_{i=1}^{N} \tilde{\boldsymbol{s}}_l^{(i)}, \tag{6.3.4}$$

where N denotes the number of users in the system.

6.3.2 Receiver Structure

The receiver structure of the permutation-based multiple-access CDSK/ECDSK-M1/ECDSK-M2 system is illustrated in Fig. 6.7. The received signal vector is given by

$$\boldsymbol{r}_l = \tilde{\boldsymbol{s}}_l + \boldsymbol{\xi}_l, \tag{6.3.5}$$

where

$$\boldsymbol{\xi}_l = \begin{bmatrix} \xi_{2(l-1)\beta+1} & \xi_{2(l-1)\beta+2} & \cdots & \xi_{2l\beta} \end{bmatrix} \tag{6.3.6}$$

denotes the noise vector during the lth symbol duration. When l is odd, the incoming signal vector is directly fed into the correlator of the jth user. But when l is even, the received signal vector needs to undergo an inverse transformation $\tilde{F}^{(j)}(\cdot) = [F^{(j)}(\cdot)]^{-1}$ before being sent to the correlator. At the end of each bit duration, the sign of the correlator output will be used to determine whether the transmitted bit is "+1" or "−1".

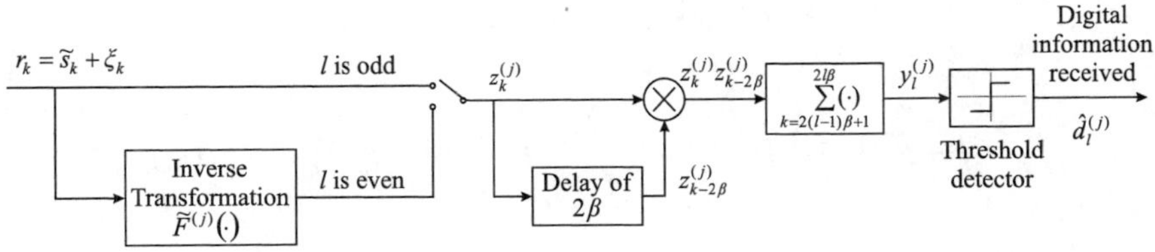

Figure 6.7. Receiver structure of the jth user in a permutation-based multiple-access CDSK/ECDSK-M1/ECDSK-M2 communication system.

6.3.3 Gaussian-Approximated Bit Error Rates

In the following, we derive the analytical BER of PMA-CDSK system based on Gaussian approximation. For the BERs of ECDSK-M1 and ECDSK-M2 systems, they can also be derived accordingly using a similar technique.

Denote the signal vector feeding into the correlator during the lth symbol duration by $z_l^{(j)}$ where

$$z_l^{(j)} = \left[z_{2(l-1)\beta+1}^{(j)} \quad z_{2(l-1)\beta+2}^{(j)} \quad \cdots \quad z_{2l\beta}^{(j)} \right]. \tag{6.3.7}$$

When l is odd, we have

$$\begin{aligned} z_l^{(j)} &= \boldsymbol{r}_l \\ &= \tilde{\boldsymbol{s}}_l + \boldsymbol{\xi}_l \\ &= \sum_{i=1}^{N} \tilde{\boldsymbol{s}}_l^{(i)} + \boldsymbol{\xi}_l \\ &= \boldsymbol{s}_l^{(j)} + \sum_{i=1, i \neq j}^{N} \boldsymbol{s}_l^{(i)} + \boldsymbol{\xi}_l. \end{aligned} \tag{6.3.8}$$

When l is even, the signal vector at each transmitter has to undergo a transformation. Moreover, the incoming signal vector at the receiver side has to go through another transformation. The signal vector going to the correlator is therefore equal to

$$\begin{aligned} z_l^{(j)} &= \tilde{F}^{(j)}(\boldsymbol{r}_l) \\ &= \tilde{F}^{(j)}(\tilde{\boldsymbol{s}}_l + \boldsymbol{\xi}_l) \\ &= \tilde{F}^{(j)}\left(\sum_{i=1}^{N} \hat{\boldsymbol{s}}_l^{(i)} + \boldsymbol{\xi}_l \right) \\ &= \tilde{F}^{(j)}\left(\sum_{i=1}^{N} F^{(i)}\left(\boldsymbol{s}_l^{(i)}\right) + \boldsymbol{\xi}_l \right) \\ &= \tilde{F}^{(j)}\left(F^{(j)}\left(\boldsymbol{s}_l^{(j)}\right)\right) + \sum_{i=1, i \neq j}^{N} \tilde{F}^{(j)}\left(F^{(i)}\left(\boldsymbol{s}_l^{(i)}\right)\right) + \tilde{F}^{(j)}(\boldsymbol{\xi}_l) \\ &= \boldsymbol{s}_l^{(j)} + \sum_{i=1, i \neq j}^{N} \tilde{F}^{(j)}\left(F^{(i)}\left(\boldsymbol{s}_l^{(i)}\right)\right) + \tilde{F}^{(j)}(\boldsymbol{\xi}_l). \end{aligned} \tag{6.3.9}$$

Based on (6.3.8) and (6.3.9), we can derive the output of the correlator, which is denoted by $y_l^{(j)}$ and equals $z_l^{(j)}(z_{l-1}^{(j)})^{\mathrm{T}}$. When l is odd, we have

$$
\begin{aligned}
y_l^{(j)} &= \left(\boldsymbol{s}_l^{(j)} + \sum_{i=1,i\neq j}^{N} \boldsymbol{s}_l^{(i)} + \boldsymbol{\xi}_l\right) \\
&\quad \times \left(\boldsymbol{s}_{l-1}^{(j)} + \sum_{i=1,i\neq j}^{N} \tilde{F}^{(j)}\left(F^{(i)}\left(\boldsymbol{s}_{l-1}^{(i)}\right)\right) + \tilde{F}^{(j)}(\boldsymbol{\xi}_{l-1})\right)^{\mathrm{T}} \\
&= \underbrace{\boldsymbol{s}_l^{(j)}\left(\boldsymbol{s}_{l-1}^{(j)}\right)^{\mathrm{T}}}_{\substack{\text{required signal and intra-signal}\\ \text{interference of the } j\text{th user}}} \\
&\quad \underbrace{+\boldsymbol{s}_l^{(j)}\left(\sum_{i=1,i\neq j}^{N} \tilde{F}^{(j)}\left(F^{(i)}\left(\boldsymbol{s}_{l-1}^{(i)}\right)\right)\right)^{\mathrm{T}} + \left(\sum_{i=1,i\neq j}^{N} \boldsymbol{s}_l^{(i)}\right)\left(\boldsymbol{s}_{l-1}^{(j)}\right)^{\mathrm{T}}}_{\text{inter-user interference}} \\
&\quad \underbrace{+\left(\sum_{i=1,i\neq j}^{N} \boldsymbol{s}_l^{(i)}\right)\left(\sum_{i=1,i\neq j}^{N} \tilde{F}^{(j)}\left(F^{(i)}\left(\boldsymbol{s}_{l-1}^{(i)}\right)\right)\right)^{\mathrm{T}}}_{\text{inter-user interference}} \\
&\quad \underbrace{+\left(\boldsymbol{s}_l^{(j)} + \sum_{i=1,i\neq j}^{N} \boldsymbol{s}_l^{(i)}\right)\left(\tilde{F}^{(j)}(\boldsymbol{\xi}_{l-1})\right)^{\mathrm{T}}}_{\text{noise components}} \\
&\quad \underbrace{+\boldsymbol{\xi}_l\left(\boldsymbol{s}_{l-1}^{(j)} + \sum_{i=1,i\neq j}^{N} \tilde{F}^{(j)}\left(F^{(i)}\left(\boldsymbol{s}_{l-1}^{(i)}\right)\right) + \tilde{F}^{(j)}(\boldsymbol{\xi}_{l-1})\right)^{\mathrm{T}}}_{\text{noise components}}. \qquad (6.3.10)
\end{aligned}
$$

The first term in (6.3.10) can further be expanded into the required signal and the intra-signal interference components, i.e.,

$$
\begin{aligned}
&\boldsymbol{s}_l^{(j)}\left(\boldsymbol{s}_{l-1}^{(j)}\right)^{\mathrm{T}} \\
&= \sum_{k=2(l-1)\beta+1}^{2l\beta} \left(x_k^{(j)} + d_l^{(j)} x_{k-2\beta}^{(j)}\right)\left(x_{k-2\beta}^{(j)} + d_{l-1}^{(j)} x_{k-4\beta}^{(j)}\right)
\end{aligned}
$$

$$= \underbrace{d_l^{(j)} \sum_{k=2(l-1)\beta+1}^{2l\beta} \left(x_{k-2\beta}^{(j)}\right)^2}_{\text{required signal}}$$

$$+ \underbrace{\sum_{k=2(l-1)\beta+1}^{2l\beta} \left(x_k^{(j)} x_{k-2\beta}^{(j)} + d_{l-1}^{(j)} x_k^{(j)} x_{k-4\beta}^{(j)} + d_l^{(j)} d_{l-1}^{(j)} x_{k-2\beta}^{(j)} x_{k-4\beta}^{(j)}\right)}_{\text{intra-signal interference of the } j\text{th user}}. \tag{6.3.11}$$

Similarly, when l is even, it is readily shown that

$$y_l^{(j)} = \left(\boldsymbol{s}_l^{(j)} + \sum_{i=1,i\neq j}^{N} \tilde{F}^{(j)}\left(F^{(i)}\left(\boldsymbol{s}_l^{(i)}\right)\right) + \tilde{F}^{(j)}(\boldsymbol{\xi}_l)\right)$$

$$\times \left(\boldsymbol{s}_{l-1}^{(j)} + \sum_{i=1,i\neq j}^{N} \boldsymbol{s}_{l-1}^{(i)} + \boldsymbol{\xi}_{l-1}\right)^{\mathrm{T}}$$

$$= \underbrace{\boldsymbol{s}_l^{(j)}\left(\boldsymbol{s}_{l-1}^{(j)}\right)^{\mathrm{T}}}_{\substack{\text{required signal and intra-signal}\\ \text{interference of the } j\text{th user}}}$$

$$+ \underbrace{\boldsymbol{s}_l^{(j)}\left(\sum_{i=1,i\neq j}^{N} \boldsymbol{s}_{l-1}^{(i)}\right)^{\mathrm{T}} + \left(\sum_{i=1,i\neq j}^{N} \tilde{F}^{(j)}\left(F^{(i)}\left(\boldsymbol{s}_l^{(i)}\right)\right)\right)\left(\boldsymbol{s}_{l-1}^{(j)}\right)^{\mathrm{T}}}_{\text{inter-user interference}}$$

$$+ \underbrace{\left(\sum_{i=1,i\neq j}^{N} \tilde{F}^{(j)}\left(F^{(i)}\left(\boldsymbol{s}_l^{(i)}\right)\right)\right)\left(\sum_{i=1,i\neq j}^{N} \boldsymbol{s}_{l-1}^{(i)}\right)^{\mathrm{T}}}_{\text{inter-user interference}}$$

$$+ \underbrace{\left(\boldsymbol{s}_l^{(j)} + \sum_{i=1,i\neq j}^{N} \tilde{F}^{(j)}\left(F^{(i)}\left(\boldsymbol{s}_l^{(i)}\right)\right)\right)\boldsymbol{\xi}_{l-1}^{\mathrm{T}}}_{\text{noise components}}$$

$$+ \underbrace{\tilde{F}^{(j)}(\boldsymbol{\xi}_l)\left(\boldsymbol{s}_{l-1}^{(j)} + \sum_{i=1,i\neq j}^{N} \boldsymbol{s}_{l-1}^{(i)} + \boldsymbol{\xi}_{l-1}\right)^{\mathrm{T}}}_{\text{noise components}}, \tag{6.3.12}$$

where the first term in (6.3.12) is identical to that given in (6.3.11). Moreover, it is readily shown that the probability density function of the correlator output is the same regardless of whether l being odd or even.

Assuming that the chaotic signals of different users are generated by the same chaotic function, the mean square value (power) of the chaotic signals are therefore the same. Denoting the mean square value by P_s, we have

$$P_s = \mathrm{E}\big[\big(x_k^{(1)}\big)^2\big] = \mathrm{E}\big[\big(x_k^{(2)}\big)^2\big] = \cdots = \mathrm{E}\big[\big(x_k^{(N)}\big)^2\big]. \tag{6.3.13}$$

Define

$$\Psi = \frac{\mathrm{var}[(x_k^{(1)})^2]}{\mathrm{E}^2[(x_k^{(1)})^2]} = \frac{\mathrm{var}[(x_k^{(2)})^2]}{\mathrm{E}^2[(x_k^{(2)})^2]} = \cdots = \frac{\mathrm{var}[(x_k^{(N)})^2]}{\mathrm{E}^2[(x_k^{(N)})^2]}. \tag{6.3.14}$$

The conditional mean and variance of the jth correlator output are, respectively, given by

$$\mathrm{E}\big[y_l^{(j)} \mid d_l^{(j)} = +1\big] = -\mathrm{E}\big[y_l^{(j)} \mid d_l^{(j)} = -1\big] = 2\beta P_s \tag{6.3.15}$$

and

$$\begin{aligned}\mathrm{var}\big[y_l^{(j)} \mid d_l^{(j)} = +1\big] &= \mathrm{var}\big[y_l^{(j)} \mid d_l^{(j)} = -1\big] \\ &= 2\beta\,\mathrm{var}\big[x_k^2\big] + 2\big(4N^2 - 1\big)\beta P_s^2 + 4N\beta P_s N_0 + \beta\frac{N_0^2}{2}.\end{aligned} \tag{6.3.16}$$

Then the approximate bit error rate for the jth user can be calculated by

$$\begin{aligned}&\mathrm{BER}_{\text{PMA-CDSK}}^{(j)} \\ &= \frac{1}{2}\,\mathrm{Prob}\big(y_l^{(j)} \leqslant 0 \mid d_l^{(j)} = +1\big) + \frac{1}{2}\,\mathrm{Prob}\big(y_l^{(j)} > 0 \mid d_l^{(j)} = -1\big) \\ &= \frac{1}{2}\,\mathrm{erfc}\left(\frac{\mathrm{E}[y_l^{(j)} \mid d_l^{(j)} = +1]}{\sqrt{2\,\mathrm{var}[y_l^{(j)} \mid d_l^{(j)} = +1]}}\right) \\ &= \frac{1}{2}\,\mathrm{erfc}\left(\left[\frac{\Psi}{\beta} + \frac{(4N^2-1)}{\beta} + 8N\left(\frac{E_b}{N_0}\right)^{-1} + 4\beta\left(\frac{E_b}{N_0}\right)^{-2}\right]^{-1/2}\right).\end{aligned} \tag{6.3.17}$$

6.4 RESULTS AND DISCUSSIONS

In our study, we use Chebyshev map of degree 2 to generate the chaotic signals. For the CDSK scheme, E_b is given by

$$E_b = 4\beta P_s. \tag{6.4.1}$$

For both ECDSK-M1 and ECDSK-M2 schemes, E_b is equal to

$$E_b = (1+2G)\beta P_s. \tag{6.4.2}$$

6.4.1 Effect of Gain G

First, we study the effect of G for the ECDSK-M1 and ECDSK-M2 schemes in a single-user system. Figure 6.8 plots the Gaussian-approximated BERs versus G for the ECDSK-M1 and ECDSK-M2 schemes. A spreading factor of 100 is used. When E_b/N_0 is small, say 4 dB, the effect of G is minimal for both schemes because the BERs are already quite high. When E_b/N_0 increases, it can be seen that the optimum G changes for the ECDSK-M1 scheme. But for the ECDSK-M2 scheme, the optimum G is maintained at $1/2$.

In Fig. 6.9, the simulated BER is plotted against E_b/N_0 in the single-user ECDSK-M1 and ECDSK-M2 systems for $G = 1$ and $1/2$. The spreading factor is 100. The results indicate that the BER for $G = 1$ is lower than that for $G = 1/2$ in the ECDSK-M1 system, whereas the converse is true in the ECDSK-M2 system.

Figure 6.10 shows the Gaussian-approximated BERs against E_b/N_0 when $G = 1$ and when the optimum G have been used, respectively, in a single-user ECDSK-M1 system. The spreading factor of 100 is used. It can be found that the performance for $G = 1$ can achieve similar performance as that when the optimum G is used. Therefore, in the following subsections, we used $G = 1$ for the ECDSK-M1 scheme and $G = 1/2$ for the ECDSK-M2 scheme.

6.4.2 Effect of Spreading Factor

Next, we study another important factor that affects the performance of the enhanced systems — the spreading factor. Figure 6.11 plots the simulated BERs versus the spreading factor in the single-user ECDSK-M1 and the ECDSK-M2 systems. As the spreading factor increases, the BER first decreases to a minimum value. Due to the effect of the noise, further increasing the spreading factor will not lower, but increase the BER. The phenomenon is very typical in CDSK systems.

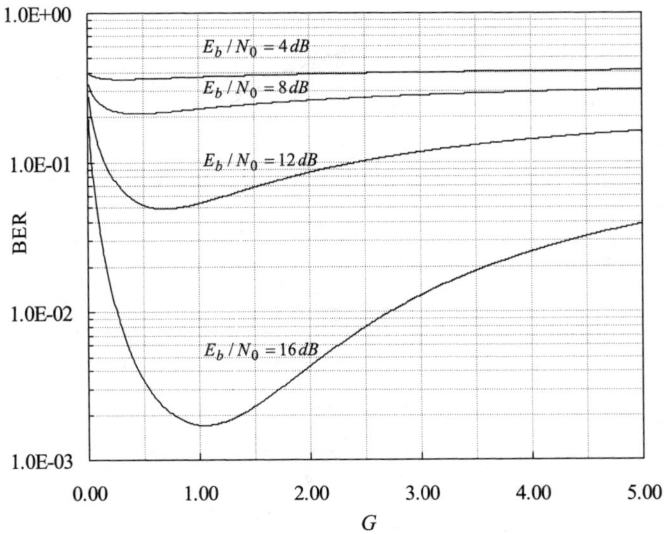

(a)

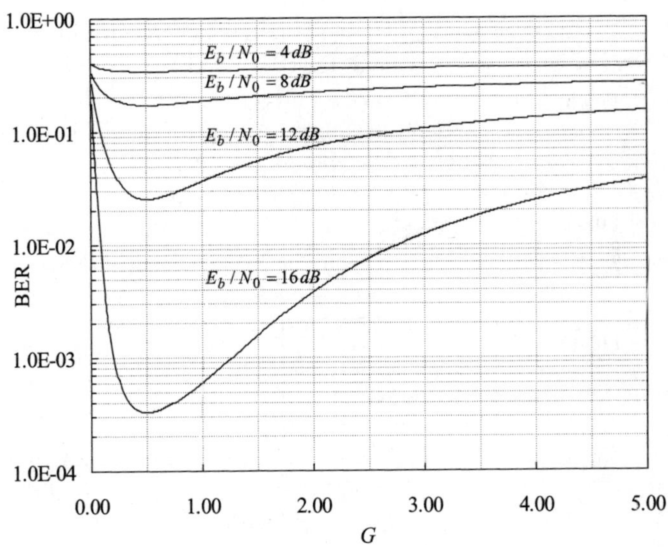

(b)

Figure 6.8. Gaussian-approximated BER versus G in a single-user system. $2\beta = 100$ is used. (a) ECDSK-M1 system; (b) ECDSK-M2 system.

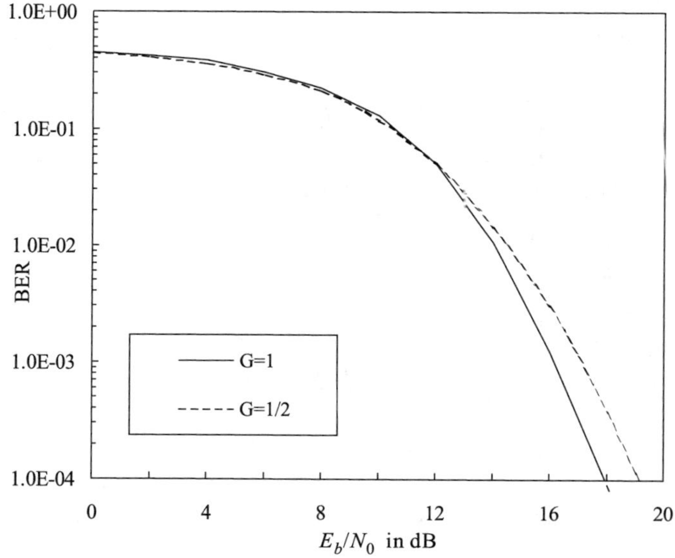

(a)

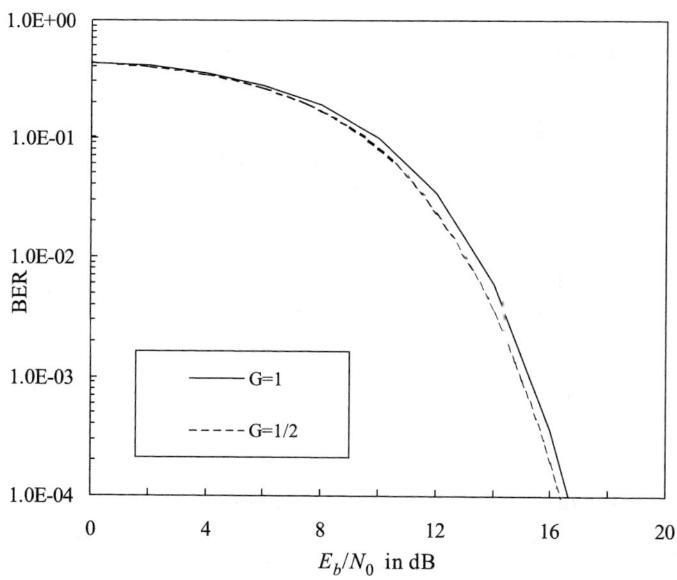

(b)

Figure 6.9. Simulated BER versus E_b/N_0 when $G = 1$ and $1/2$ are used in a single-user system. $2\beta = 100$ is used. (a) ECDSK-M1 system; (b) ECDSK-M2 system.

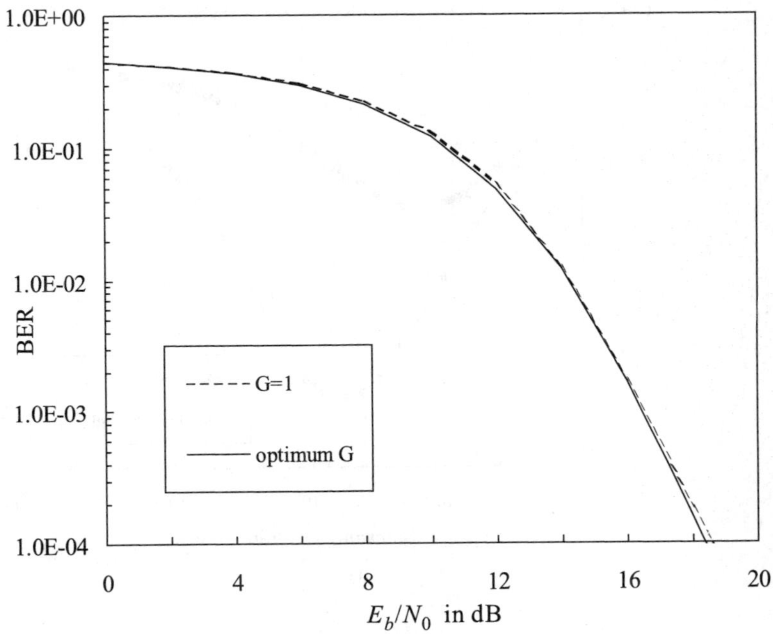

Figure 6.10. Gaussian-approximated BERs versus E_b/N_0 when $G = 1$ and when the optimum G have been used, respectively, in a single-user ECDSK-M1 system. $2\beta = 100$ is used.

6.4.3 Comparison of the Simulated and Gaussian-Approximated Bit Error Rates

Figure 6.12 compares the simulated and Gaussian-approximated BERs in the single-user ECDSK-M1 and the ECDSK-M2 systems. A spreading factor of 100 is used. Results show that the simulated and Gaussian-approximated BERs are consistent with each other. Next, we plot the results again for the 2-, 4- and 6-user PMA-ECDSK-M1 and the PMA-ECDSK-M2 systems in Fig. 6.13. The spreading factor equals 100. For all cases, it can be seen that the Gaussian-approximated BERs are close to the simulated results.

6.4.4 Comparison with PMA-CDSK and PMA-DCSK Systems

Figure 6.14 compares the simulated BERs of the single-user ECDSK-M1, ECDSK-M2, CDSK and DCSK systems. For the DCSK system, a spreading factor of 200 is used. For other systems, a spreading factor of 100 is used. Among the curves, it can be shown that the ECDSK-M2 gives the best results, while the BER of the CDSK system is the worst. It is because the intra-signal interference

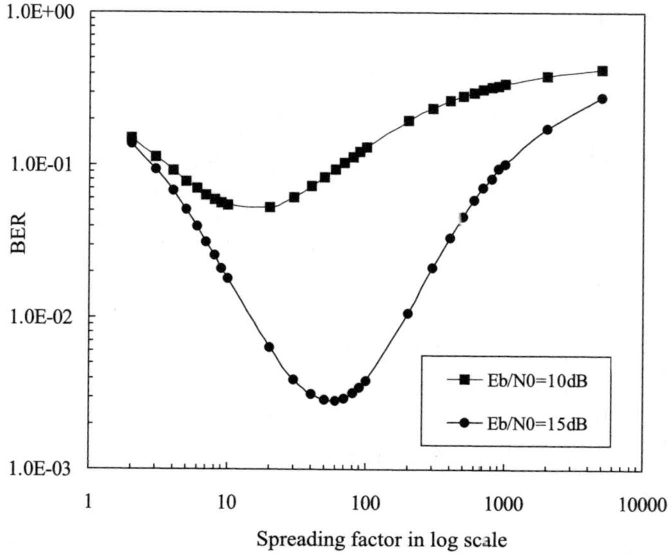

(a)

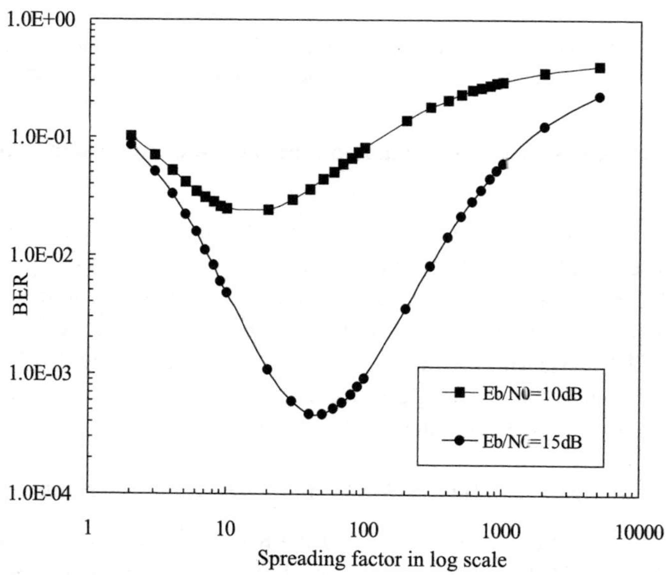

(b)

Figure 6.11. Simulated BER versus spreading factor for (a) ECDSK-M1 system ($G = 1$); (b) ECDSK-M2 system ($G = 1/2$).

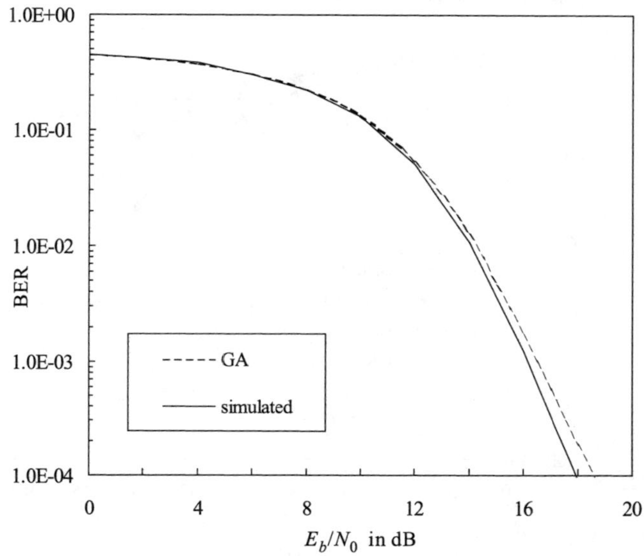

(a)

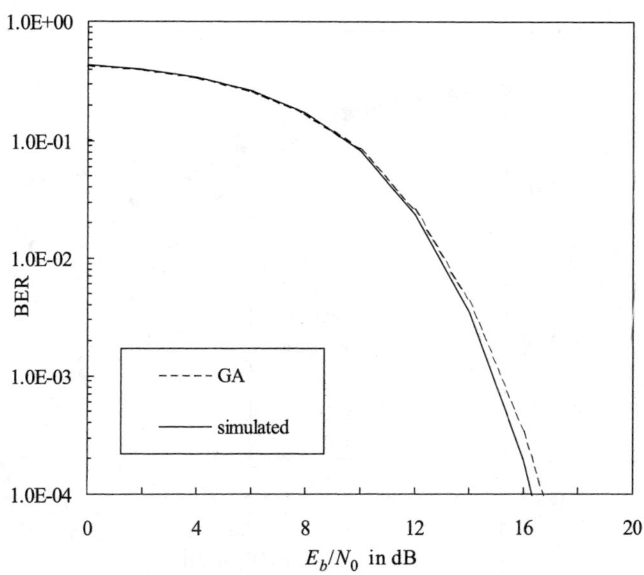

(b)

Figure 6.12. Simulated and Gaussian-approximated BERs versus E_b/N_0 in a single user system. $2\beta = 100$ is used. Simulated results are plotted as solid lines and analytical results are plotted as dotted lines for (a) ECDSK-M1 system ($G = 1$); (b) ECDSK-M2 system ($G = 1/2$).

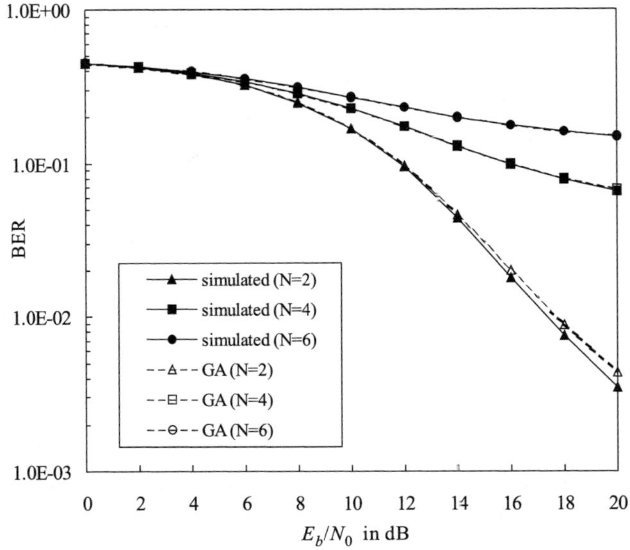

(a)

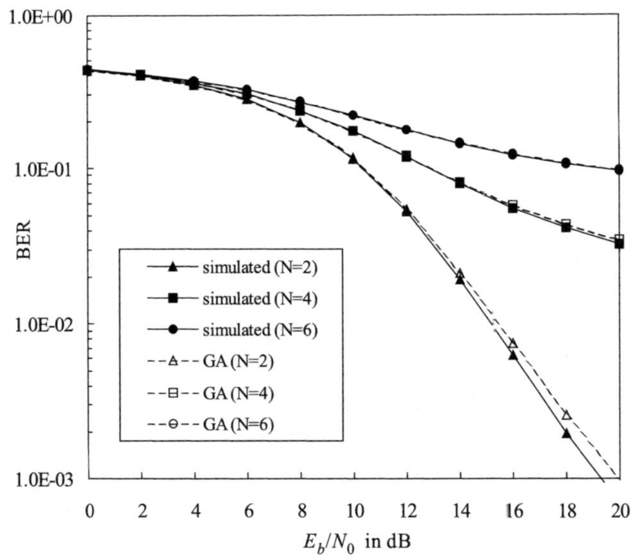

(b)

Figure 6.13. Simulated and Gaussian-approximated BERs versus E_b/N_0 n 2-, 4- and 6-user systems. $2\beta = 100$ is used. Simulated results are plotted as solid lines and analytical results are plotted as dotted lines using (a) PMA-ECDSK-M1 scheme ($G = 1$); (b) PMA-ECDSK-M2 scheme ($G = 1/2$).

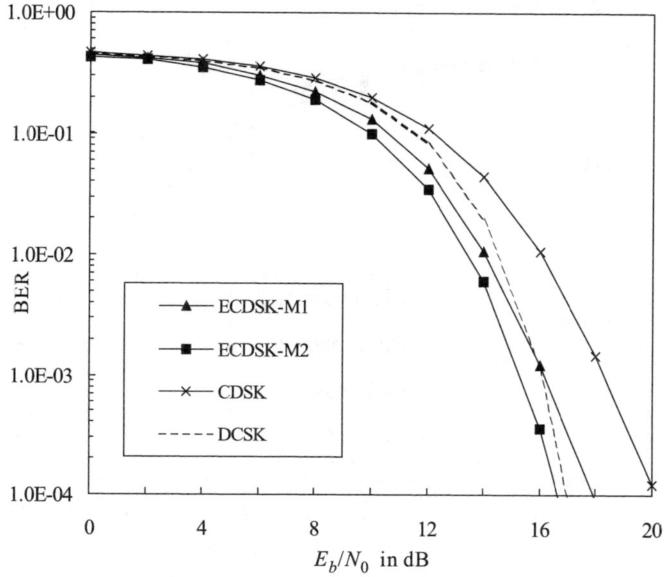

Figure 6.14. Simulated BER versus E_b/N_0 for the single-user ECDSK-M1 ($G = 1$), ECDSK-M2 ($G = 1/2$), CDSK and DCSK systems. $2\beta = 100$ is used for all systems except DCSK where $2\beta = 200$ is used.

is small for the ECDSK-M2 scheme, whereas it is large for the CDSK scheme. It is also observed that the performance of the ECDSK-M1 system is close to that of the DCSK system. Moreover, due to the structure of the transmitted signal, the ECDSK-M1, ECDSK-M2 and CDSK systems can achieve twice the data rate compared to the DCSK system.

Figure 6.15 compares the simulated BERs of the PMA-ECDSK-M1, PMA-ECDSK-M2, PMA-CDSK, and PMA-DCSK systems. The number of users $N = 2, 4$ and 6 are considered. When the value of E_b/N_0 is below about 16 dB, the PMA-ECDSK-M2 scheme can provide the best performance. For higher E_b/N_0 values, the PMA-DCSK scheme can achieve the lowest BER. In all cases, it is seen that the PMA-CDSK scheme has the worst performance.

6.5 SUMMARY

In this chapter, we have looked at two enhanced methods based on the correlation-delay-shift-keying (CDSK) scheme, namely, ECDSK-M1 and ECDSK-M2 schemes. The multiple-access schemes for the ECDSK-M1, ECDSK-M2 and CDSK systems have also been studied. Moreover, using the Gaussian approxi-

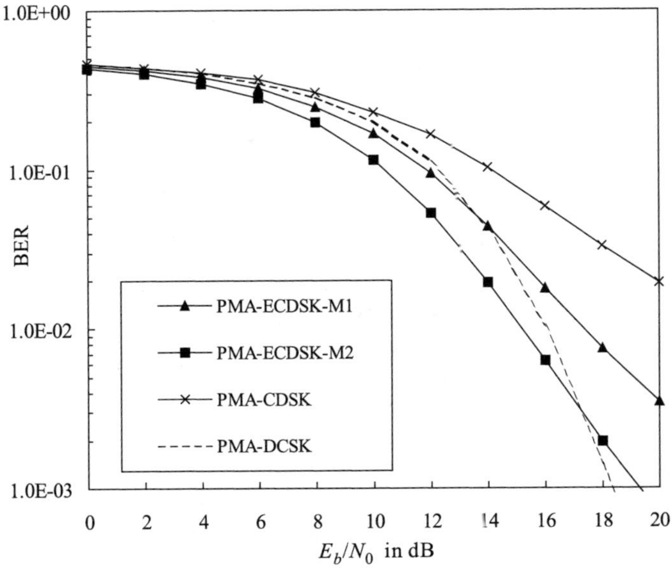

(a)

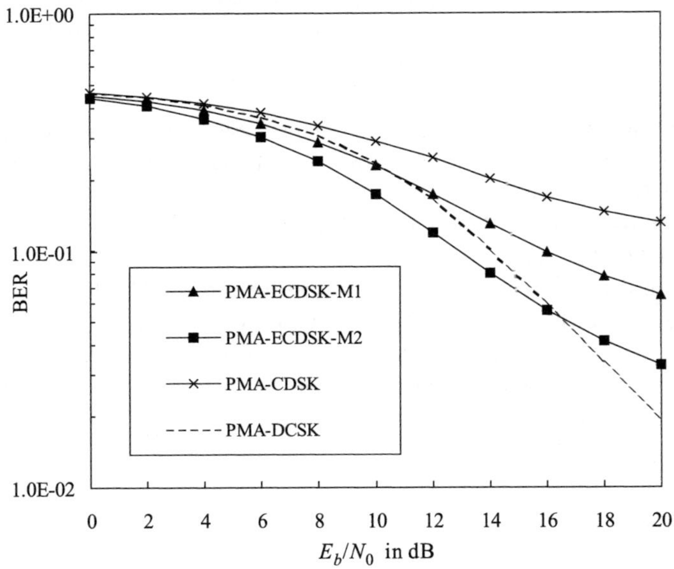

(b)

Figure 6.15. Simulated BER versus E_b/N_0 for the PMA-ECDSK-M1 ($G = 1$), PMA-ECDSK-M2 ($G = 1/2$), PMA-CDSK and PMA-DCSK systems. $2\beta = 100$ is used for all systems except DCSK where $2\beta = 200$ is used. (a) 2-user system; (b) 4-user system; (c) 6-user system.

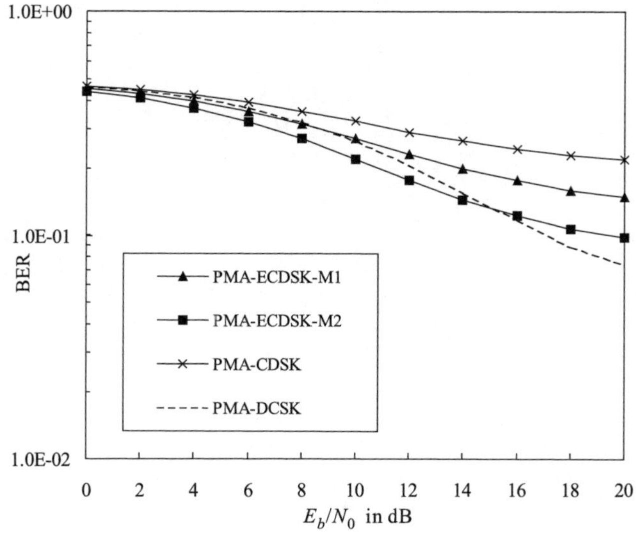

(c)

Figure 6.15. (Continued)

mation method, we have illustrated the technique that derives the analytical BER performance of these systems. Finally, the BERs for the ECDSK-M1, ECDSK-M2 and CDSK systems are compared to that for the DCSK system in both single-user and multi-user environments. Results show that when E_b/N_0 is below about 16 dB, the ECDSK-M2 scheme can achieve the best performance.

In the previous chapter and in this chapter, we have taken an in-depth study of the CDSK system and its derivatives. While some improvements have been made on the CDSK system, the performances of CDSK-based communication systems are still limited by the intra-signal interference as well as the inter-user interference in a multiple-access environment. In the next two chapters, we will shift our focus to other types of noncoherent communication techniques that can provide better bit error performances.

7

MULTIPLE-ACCESS SCHEME UTILIZING TRANSMITTED REFERENCES AND TRAINING SEQUENCES

In Chapters 5 and 6, we focus our investigation on the CDSK scheme and its derivatives. In this chapter, we continue with our study on multiple-access noncoherent chaos-based digital communication systems by looking into an alternative multiple-access scheme utilizing transmitted references [Tam *et al.*, 2004b]. In this scheme, each user transmits reference chaotic signals modulated by a training sequence to the receiver periodically. Also, users are differentiated by their chaotic signals and distinct training sequences. Based on the incoming reference signals, each receiver adjusts its internal parameters according to the user's pre-assigned training sequence. After the training period is completed, the receiver will start demodulating the information-bearing chaotic signals.

We organize this chapter as follows. We begin with the descriptions of the transmitter structure and the transmitted signal in Section 7.1. In Section 7.2, two receiver structures will be presented. The working principles of each receiver will be explained, together with a discussion on the performance bounds. Finally, we present the simulations results in Section 7.3.

7.1 TRANSMITTER STRUCTURE

Shown in Fig. 7.1 is a multiple-access chaos-based communication system under a noisy environment. Figure 7.2 shows the transmitter structure of the ith user. The transmitter consists of a chaos generator, a number of delay blocks, a switch and a multiplier. The transmitted signal is organized into frames which are further sub-divided into a number of time slots, as shown in Fig. 7.3.

The first K_1 slots are used to send the reference chaotic samples for training, while the information-bearing chaotic samples are sent in the remaining K_2 slots. Without loss of generality, we consider the transmitted signal for a single frame period. At the output of the chaos generator, the chaotic signal is first biased before transmission such that its mean value is zero. This avoids transmission of the dc component which carries no useful information but increases transmission power. Suppose β chaotic samples are sent within one slot. For each frame, the same β chaotic samples will be used in each slot. We denote the chaotic samples used by

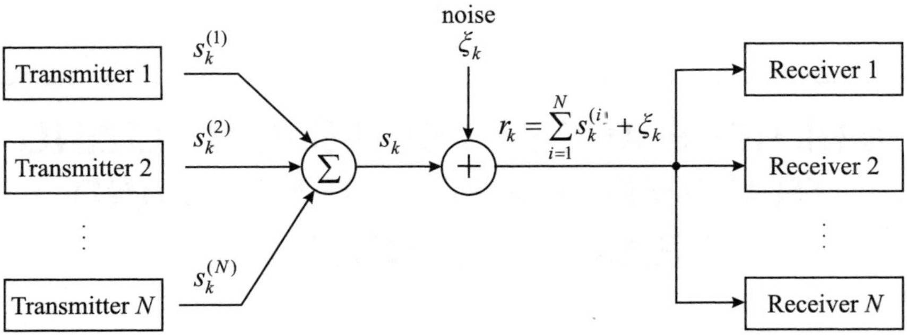

Figure 7.1. Multiple-access chaos-based communication system under an additive white Gaussian noise environment.

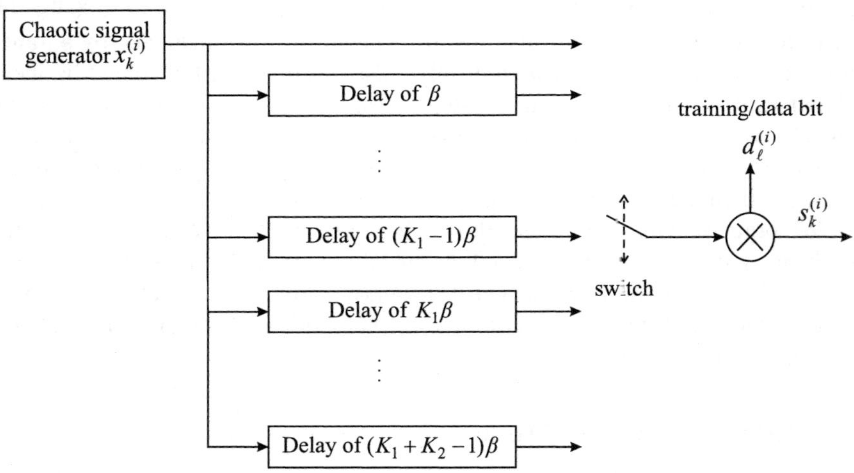

Figure 7.2. Transmitter structure of the multiple-access chaos-based communication system under study.

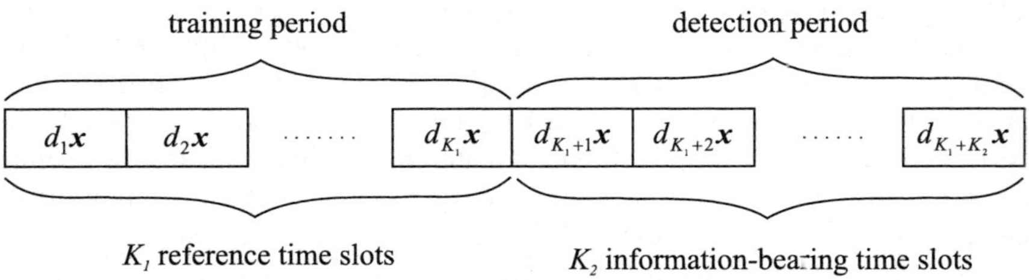

Figure 7.3. Transmitted frame structure of the ith user.

the ith user within this frame by $\{x_k^{(i)} : k = 1, 2, \ldots, \beta\}$. For algebraic brevity, we also define a chaotic-sample vector $\boldsymbol{x}^{(i)}$ as

$$\boldsymbol{x}^{(i)} = \begin{bmatrix} x_1^{(i)} & x_2^{(i)} & \cdots & x_\beta^{(i)} \end{bmatrix}^{\mathrm{T}} \tag{7.1.1}$$

where T represents the matrix transposition.

Moreover, the chaotic-sample vector will be modulated by the training bits or the data bits before transmission. For the ℓth time slot, the sample vector sent by the ith user, denoted by $\boldsymbol{s}_\ell^{(i)}$, is given by

$$\boldsymbol{s}_\ell^{(i)} = \begin{bmatrix} s_{(\ell-1)\beta+1}^{(i)} & s_{(\ell-1)\beta+2}^{(i)} & \cdots & s_{\ell\beta}^{(i)} \end{bmatrix}^{\mathrm{T}} = d_\ell^{(i)} \boldsymbol{x}^{(i)} \tag{7.1.2}$$

where $d_\ell^{(i)}$ equals "+1" or "−1". When $1 \leqslant \ell \leqslant K_1$, $d_\ell^{(i)}$ represents the training bit, and when $K_1 + 1 \leqslant \ell \leqslant K_1 + K_2$, $d_\ell^{(i)}$ denotes the information bit. In other words, if the training/data bit is "+1", the transmitted samples are the same as the chaotic samples. If the training/data bit equals "−1", the sign of the chaotic samples will be inverted and then transmitted. A typical frame structure for the ith user is shown in Fig. 7.3.

Consider the N-user system shown in Fig. 7.1. We assume that the frame sizes of all users are identical and the frames transmitted for all users are synchronized. Also, the number of slots per frame and the number of chaotic samples transmitted per slot are identical for all users. We further assume that $K_1 = K_2 = K$ such that on the average, two slots are required to send one data bit. Thus the average number of chaotic samples transmitted per bit (spreading factor) equals 2β. Assuming that all transmitters are switched on, the overall transmitted sample vector of the system, denoted by $\boldsymbol{s}_\ell$, is obtained by summing the outputs from all users, i.e.,

$$\boldsymbol{s}_\ell = \begin{bmatrix} s_{(\ell-1)\beta+1} & s_{(\ell-1)\beta+2} & \cdots & s_{\ell\beta} \end{bmatrix}^{\mathrm{T}} = \sum_{i=1}^{N} \boldsymbol{s}_\ell^{(i)}. \tag{7.1.3}$$

7.2 RECEIVER STRUCTURES

We make the usual assumption that the channel is additive white Gaussian. Thus, during the ℓth time slot, the received sample vector, $\boldsymbol{r}_\ell$, is

$$\boldsymbol{r}_\ell = \begin{bmatrix} r_{(\ell-1)\beta+1} & r_{(\ell-1)\beta+2} & \cdots & r_{\ell\beta} \end{bmatrix}^{\mathrm{T}} = \boldsymbol{s}_\ell + \boldsymbol{\Phi}_\ell, \tag{7.2.1}$$

where

$$\boldsymbol{\Phi}_\ell = \begin{bmatrix} \xi_{(\ell-1)\beta+1} & \xi_{(\ell-1)\beta+2} & \cdots & \xi_{\ell\beta} \end{bmatrix}^{\mathrm{T}}, \tag{7.2.2}$$

and ξ_k represents the kth noise sample, the mean and variance (power spectral density) of which are zero and $N_0/2$, respectively. The first K received sample vectors, i.e., $\{\boldsymbol{r}_\ell : \ell = 1, 2, \ldots, K\}$, are the training signals. Based on these training signals, the receiver needs to update its internal parameters before decoding the information-bearing vectors that follow. Two receiver structures will be investigated. The first one is based on an adaptive transversal filter, and the second one estimates the chaotic-sample vector from the training signals.

7.2.1 Adaptive Transversal Filter

7.2.1.1 *Receiver Structure* Figure 7.4 shows the structure of an adaptive transversal filter, which makes use of the first K received sample vectors for updating the tap weights. The tap weights are set to zero for all users at the beginning of each training process. During the training period, the estimated training bit corresponding to the ℓth slot of the jth user, denoted by $\tilde{d}_\ell^{(j)}$, is first computed using

$$\tilde{d}_\ell^{(j)} = \boldsymbol{r}_\ell^{\mathrm{T}} \boldsymbol{w}_{\ell-1}^{(j)} \tag{7.2.3}$$

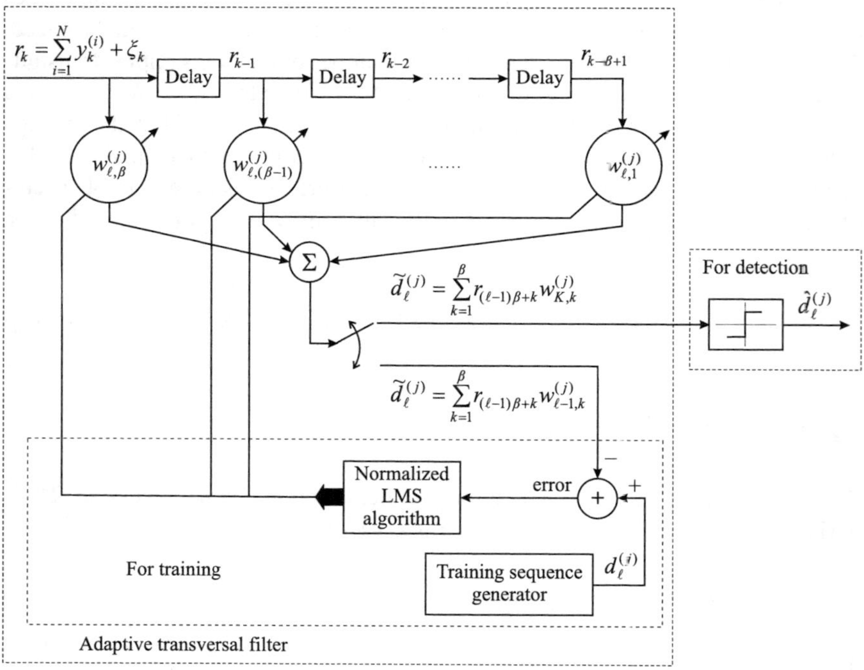

Figure 7.4. Receiver based on an adaptive transversal filter.

where

$$\boldsymbol{w}_\ell^{(j)} = \left[w_{\ell,1}^{(j)} \; w_{\ell,2}^{(j)} \cdots w_{\ell,\beta}^{(j)} \right]^{\mathrm{T}} \tag{7.2.4}$$

is a vector containing the tap weights of the adaptive filter after the ℓth ($\ell = 1, 2, \ldots, K$) iteration (time slot). Then the estimated training bit is compared with the desired symbol $d_\ell^{(j)}$, which is different for different users (receivers). Based on the error between the desired symbol and the estimated symbol, i.e., $e_\ell^{(j)} = d_\ell^{(j)} - \tilde{d}_\ell^{(j)}$, the tap weights in each receiver are updated at the end of each time slot using the normalized least-mean-square (LMS) algorithm. The whole iterative process is summarized in the following steps [Haykin, 1996]:

$$\boldsymbol{w}_0^{(j)} = 0, \tag{7.2.5}$$

$$e_\ell^{(j)} = d_\ell^{(j)} - \tilde{d}_\ell^{(j)} = d_\ell^{(j)} - \boldsymbol{r}_\ell^{\mathrm{T}} \boldsymbol{w}_{\ell-1}^{(j)}, \tag{7.2.6}$$

$$\boldsymbol{w}_\ell^{(j)} = \boldsymbol{w}_{\ell-1}^{(j)} + \frac{\tilde{\mu}}{\epsilon + \|\boldsymbol{r}_\ell\|^2} e_\ell^{(j)} \boldsymbol{r}_\ell, \quad 0 < \tilde{\mu} < 2, \; \epsilon \geqslant 0, \tag{7.2.7}$$

where $\|\boldsymbol{r}_\ell\|$ is the Euclidean norm of the input vector $\boldsymbol{r}_\ell$ defined as

$$\|\boldsymbol{r}_\ell\| = \sqrt{\sum_{k=1}^{\beta} [r_{(\ell-1)\beta+k}]^2}. \tag{7.2.8}$$

At the end of the training period, i.e., after K iterations, we will obtain the tap-weight vector $\boldsymbol{w}_K$ which can then be used to estimate the data symbols embedded in the remaining time slots of the frame. The decoded data symbol, i.e., $\hat{d}_\ell^{(j)}$, $\ell = K+1, K+2, \ldots, 2K$, is then determined according to the following rule:

$$\hat{d}_\ell^{(j)} = \begin{cases} +1 & \text{if } \tilde{d}_\ell^{(j)} = \boldsymbol{r}_\ell^{\mathrm{T}} \boldsymbol{w}_K^{(j)} > 0, \; \ell = (K+1), \ldots, 2K, \\ -1 & \text{if } \tilde{d}_\ell^{(j)} = \boldsymbol{r}_\ell^{\mathrm{T}} \boldsymbol{w}_K^{(j)} \leqslant 0, \; \ell = (K+1), \ldots, 2K. \end{cases} \tag{7.2.9}$$

In other words, for the remaining K time slots of each frame, if the estimated data symbol $\tilde{d}_\ell^{(j)}$ is larger than zero, then "+1" is detected. Otherwise, "−1" is decoded.

7.2.1.2 *Theoretical Performance Bound* Assume that the channel is noiseless during the training period. When the training period is long enough, the error between the desired data and the estimated data will approach zero for all users. Hence, we assume that the error $e_\ell^{(j)}$ equals zero for all $\ell > \vartheta$, where ϑ is a sufficiently large integer. In other words,

$$e_\ell^{(j)} = 0, \quad j = 1, 2, \ldots, N. \tag{7.2.10}$$

Based on (7.2.7), it can be further deduced that for all $\ell > \vartheta$,

$$\boldsymbol{w}_\ell^{(j)} = \boldsymbol{w}_\vartheta^{(j)} \tag{7.2.11}$$

for all $j = 1, 2, \ldots, N$. Moreover, substituting (7.2.10) and (7.2.11) into (7.2.6), and making use of (7.1.1) through (7.2.1), we obtain

$$\begin{aligned}
& d_\ell^{(j)} = \boldsymbol{r}_\ell^{\mathrm{T}} \boldsymbol{w}_\vartheta^{(j)} \\
& \Rightarrow d_\ell^{(j)} = \boldsymbol{s}_\ell^{\mathrm{T}} \boldsymbol{w}_\vartheta^{(j)} \\
& \Rightarrow d_\ell^{(j)} = \sum_{i=1}^{N} \left(d_\ell^{(i)} \left[\boldsymbol{x}^{(i)}\right]^{\mathrm{T}} \boldsymbol{w}_\vartheta^{(j)} \right) \\
& \Rightarrow d_\ell^{(j)} = \sum_{i=1}^{N} d_\ell^{(i)} \left(\sum_{k=1}^{\beta} x_k^{(i)} w_{\vartheta,k}^{(j)} \right), \qquad j = 1, 2, \ldots, N \\
& \Rightarrow \sum_{k=1}^{\beta} x_k^{(i)} w_{\vartheta,k}^{(j)} = \delta_{ij}, \qquad i, j = 1, 2, \ldots, N,
\end{aligned} \tag{7.2.12}$$

where δ_{ij} denotes the Kronecker's delta defined as

$$\delta_{ij} = \begin{cases} 1 & \text{when } i = j, \\ 0 & \text{when } i \neq j. \end{cases} \tag{7.2.13}$$

The equality in (7.2.12) shows that the correlation between the chaotic-sample vector of the ith user and the tap-weight vector of the jth transversal filter (receiver) equals zero and unity, respectively, for $i \neq j$ and $i = j$. In other words, after the training process has been completed, the inter-user interference can be eliminated totally using the transversal filter, while the required signal can still be recognized and produces unity output at the transversal filter. Therefore, under such conditions, the cross-correlation estimation problem can also be minimized [Kolumbán *et al.*, 2002b].

When the transversal filter is used to decode the information-bearing chaotic signal, which is now assumed to be corrupted by noise, the system behaves like a coherent single-user antipodal chaos-shift-keying (ACSK) system because the inter-user interference has been removed under a noiseless condition. The bit error rate (BER) of the aforementioned coherent CSK system is approximated by that of the conventional coherent binary phase-shift-keying (BPSK) communication scheme [Kolumbán and Kennedy, 2000], and is given by

$$\mathrm{BER} = \frac{1}{2}\,\mathrm{erfc}\left(\sqrt{\frac{\hat{E}_b}{N_0}}\right), \tag{7.2.14}$$

where $\hat{E}_b$ denotes the energy of each bit in the demodulation process. In the non-coherent system under study, we assume that the number of training time slots and the number of information-bearing time slots are the same within each frame. As a consequence, if we denote the average bit energy of the jth user by $E_b^{(j)}$, which equals

$$E_b^{(j)} = 2\beta P_s^{(j)}, \tag{7.2.15}$$

where

$$P_s^{(j)} = \mathrm{E}\left[\left(x_k^{(j)}\right)^2\right] \tag{7.2.16}$$

denotes the average transmission power of the jth user, then $E_b^{(j)}$ will double the bit energy spent in the demodulation process, i.e.,

$$E_b^{(j)} = 2\hat{E}_b. \tag{7.2.17}$$

Substituting (7.2.17) into (7.2.14), the performance bound of the receiver for the jth user, denoted by $\mathrm{BER}^{(j)}$, equals

$$\mathrm{BER}^{(j)} = \frac{1}{2}\,\mathrm{erfc}\left(\sqrt{\frac{E_b^{(j)}}{2N_0}}\right). \tag{7.2.18}$$

In practice, the BER performance is expected to be worse than that given in (7.2.18) because the tap-weight vector cannot be estimated perfectly under the influence of noise and with a finite length of training sequence.

7.2.1.3 *Choice of Training Sequences* In the modulation scheme under study, the chaotic samples used to carry the data vary in every frame. At the beginning of each frame, the tap weights of the adaptive filter need to be reset and then updated so as to track the next set of chaotic samples. Hence the training period should not be too long. To shorten the convergence time and to avoid getting the same tap weights for more than one users at the end of the training period, specific sets of orthogonal codes are assigned to the users as the training sequences.

Suppose $K = 2^{\kappa}$ where κ is a positive integer. We can construct K orthogonal Walsh functions based on the Hadamard matrix $\tilde{\boldsymbol{H}}_K$, which can be generated using the following recursive procedure [Proakis and Salehi, 1994].

$$\tilde{\boldsymbol{H}}_1 = [-1] \quad \text{for } \kappa = 0 \tag{7.2.19}$$

$$\tilde{\boldsymbol{H}}_2 = \begin{bmatrix} \tilde{\boldsymbol{H}}_1 & \tilde{\boldsymbol{H}}_1 \\ \tilde{\boldsymbol{H}}_1 & -\tilde{\boldsymbol{H}}_1 \end{bmatrix} = \begin{bmatrix} -1 & -1 \\ -1 & +1 \end{bmatrix} \quad \text{for } \kappa = 1, \tag{7.2.20}$$

$$\tilde{\boldsymbol{H}}_4 = \begin{bmatrix} \tilde{\boldsymbol{H}}_2 & \tilde{\boldsymbol{H}}_2 \\ \tilde{\boldsymbol{H}}_2 & -\tilde{\boldsymbol{H}}_2 \end{bmatrix} = \begin{bmatrix} -1 & -1 & -1 & -1 \\ -1 & +1 & -1 & +1 \\ -1 & -1 & +1 & +1 \\ -1 & +1 & +1 & -1 \end{bmatrix} \quad \text{for } \kappa = 2, \tag{7.2.21}$$

$$\tilde{\boldsymbol{H}}_8 = \begin{bmatrix} \tilde{\boldsymbol{H}}_4 & \tilde{\boldsymbol{H}}_4 \\ \tilde{\boldsymbol{H}}_4 & -\tilde{\boldsymbol{H}}_4 \end{bmatrix}$$

$$= \begin{bmatrix} -1 & -1 & -1 & -1 & -1 & -1 & -1 & -1 \\ -1 & +1 & -1 & +1 & -1 & +1 & -1 & +1 \\ -1 & -1 & +1 & +1 & -1 & -1 & +1 & +1 \\ -1 & +1 & +1 & -1 & -1 & +1 & +1 & -1 \\ -1 & -1 & -1 & -1 & +1 & +1 & +1 & +1 \\ -1 & +1 & -1 & +1 & +1 & -1 & +1 & -1 \\ -1 & -1 & +1 & +1 & +1 & +1 & -1 & -1 \\ -1 & +1 & +1 & -1 & +1 & -1 & -1 & +1 \end{bmatrix} \quad \text{for } \kappa = 3, \tag{7.2.22}$$

$$\vdots$$

$$\tilde{\boldsymbol{H}}_K = \begin{bmatrix} \tilde{\boldsymbol{H}}_{K/2} & \tilde{\boldsymbol{H}}_{K/2} \\ \tilde{\boldsymbol{H}}_{K/2} & -\tilde{\boldsymbol{H}}_{K/2} \end{bmatrix} \quad \text{for } \kappa = \log_2 K. \tag{7.2.23}$$

Each row in the Hadamard matrix represents a Walsh code. Denote the ith row of the $K \times K$ Hadamard matrix by $\boldsymbol{W}^{(i)}_{K\times K}$, $i = 1, 2, \ldots, K$. It is readily shown that the Walsh codes are orthogonal, i.e.,

$$\boldsymbol{W}^{(i)}_{K\times K}\left[\boldsymbol{W}^{(j)}_{K\times K}\right]^{\mathrm{T}} = \begin{cases} 0 & \text{if } i \neq j, \\ K & \text{if } i = j. \end{cases} \tag{7.2.24}$$

The choice of K depends on the number of users in the system. For example, $\tilde{\boldsymbol{H}}_4$ gives four orthogonal Walsh codes, and hence up to four users can be supported.

7.2.2 Inverse-and-Average Receiver

7.2.2.1 *Receiver Structure* The structure of the inverse-and-average receiver is shown in Fig. 7.5. The roles of the receiver are to estimate the chaotic-sample vector of the required user during the training period, and to decode the information-bearing chaotic signal by correlating it with the estimated sample vector. To estimate the chaotic-sample vector, the modulation due to the training bit at the transmitter is first removed at the receiving end by multiplying the received signal in each time slot by the corresponding training bit again. Since the training bit equals ± 1, multiplying the chaotic-sample vector of the jth user $\boldsymbol{x}^{(j)}$ twice (one at transmitter and another at receiver) by the same training bit produces no effect on the vector. Subsequently, the resultant vectors in all the training time slots are averaged to produce an estimation of the chaotic-sample vector which is represented by

$$\hat{\boldsymbol{x}}^{(j)} = \begin{bmatrix} \hat{x}_1^{(j)} & \hat{x}_2^{(j)} & \cdots & \hat{x}_\beta^{(j)} \end{bmatrix}^{\mathrm{T}}. \tag{7.2.25}$$

When the training process is finished, the chaotic signals in the information-bearing time slots will correlate with the estimated chaotic-sample vector. When the output of the jth receiver, denoted by $y_\ell^{(j)}$, is larger than zero, a "+1" is decoded. Otherwise, a "−1" will be detected.

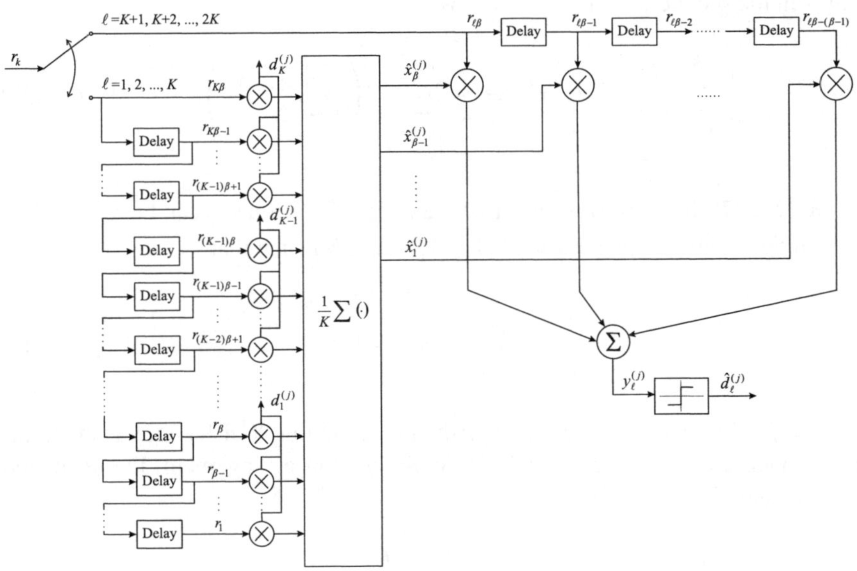

Figure 7.5. Structure of the inverse-and-average receiver.

7.2.2.2 *Theoretical Performance Bound* Using the same notations defined in Section 7.1, for the jth user, the recovered chaotic-sample vector is given by

$$\begin{aligned}\hat{\boldsymbol{x}}^{(j)} &= \frac{1}{K}\sum_{\ell=1}^{K} d_\ell^{(j)}\boldsymbol{r}_\ell \\ &= \frac{1}{K}\sum_{\ell=1}^{K} d_\ell^{(j)}\left(\sum_{i=1}^{N}\boldsymbol{s}_\ell^{(i)} + \boldsymbol{\Phi}_\ell\right) \\ &= \frac{1}{K}\sum_{\ell=1}^{K} d_\ell^{(j)}\left(\sum_{i=1}^{N} d_\ell^{(i)}\boldsymbol{x}^{(i)} + \boldsymbol{\Phi}_\ell\right) \\ &= \boldsymbol{x}^{(j)} + \frac{1}{K}\sum_{\ell=1}^{K}\sum_{\substack{i=1\\ i\neq j}}^{N} d_\ell^{(j)} d_\ell^{(i)}\boldsymbol{x}^{(i)} + \frac{1}{K}\sum_{\ell=1}^{K} d_\ell^{(j)}\boldsymbol{\Phi}_\ell. \end{aligned} \tag{7.2.26}$$

In (7.2.26), the first term is the required chaotic-sample vector. Both the second and third terms are unwanted signals and should be removed. The third term, being derived from the noise, cannot be eliminated. The second term, however, is introduced by the transmitted signals of other users and can be removed by assigning appropriate training sequences to the users. Re-arranging the summation signs in the second term in (7.2.26) gives

$$\frac{1}{K}\sum_{\ell=1}^{K}\sum_{\substack{i=1\\ i\neq j}}^{N} d_\ell^{(j)} d_\ell^{(i)}\boldsymbol{x}^{(i)} = \frac{1}{K}\sum_{\substack{i=1\\ i\neq j}}^{N}\boldsymbol{x}^{(i)}\left(\sum_{\ell=1}^{K} d_\ell^{(j)} d_\ell^{(i)}\right). \tag{7.2.27}$$

From (7.2.27), it can be observed that the effect of other users can be removed by assigning training sequences $\{d_\ell^{(i)} : i = 1, 2, \ldots, N\}$ such that

$$\sum_{\ell=1}^{K} d_\ell^{(j)} d_\ell^{(i)} = 0 \tag{7.2.28}$$

for $i \neq j$. This can be easily accomplished by assigning Walsh codes as the training sequences, as in Section 7.2.1.3. With such an arrangement, the estimated chaotic-sample vector becomes

$$\hat{\boldsymbol{x}}^{(j)} = \boldsymbol{x}^{(j)} + \frac{1}{K}\sum_{\ell=1}^{K} d_\ell^{(j)}\boldsymbol{\Phi}_\ell. \tag{7.2.29}$$

The estimated chaotic-sample vector $\hat{\boldsymbol{x}}^{(j)}$ will then correlate with the chaotic signals transmitted in the information-bearing time slots. Denote the input to the detector by $y_\ell^{(j)}$ for the ℓth time slot of the jth user. Then,

$$y_\ell^{(j)} = \left[\hat{\boldsymbol{x}}^{(j)}\right]^{\mathrm{T}} \boldsymbol{r}_\ell. \tag{7.2.30}$$

The received symbol, denoted by $\hat{d}_\ell^{(j)}$, is decoded according to the following rule:

$$\hat{d}_\ell^{(j)} = \begin{cases} +1 & \text{if } y_\ell^{(j)} > 0,\ \ell = (K+1), \ldots, 2K, \\ -1 & \text{if } y_\ell^{(j)} \leqslant 0,\ \ell = (K+1), \ldots, 2K. \end{cases} \tag{7.2.31}$$

To estimate the performance bound, we further assume that the number of reference time slots K is large enough such that the term due to noise can be ignored in (7.2.29). Therefore, (7.2.29) is simplified to

$$\hat{\boldsymbol{x}}^{(j)} \approx \boldsymbol{x}^{(j)}, \tag{7.2.32}$$

implying that a clean chaotic-sample vector can be recovered at the receiver. Under such circumstances, during the decoding of the information-bearing time slots, the communication system is equivalent to a multi-user coherent antipodal CSK system. Therefore, the performance of the inverse-and-average receiver is bounded by that of the multi-user coherent antipodal CSK system, which will be found by computer simulations in the following section.

Comparing the complexity of the two types of receivers mentioned in this section, it can be seen that the receiver based on an adaptive transversal filter has a slightly more complicated structure (mainly because of the LMS algorithm) over the inverse-and-average receiver.

7.3 RESULTS AND DISCUSSIONS

In our simulations, all users use the same map, each with a different initial condition, to generate the chaotic samples. The form of the map is given by

$$x_{k+1} = 4x_k^3 - 3x_k. \tag{7.3.1}$$

With this map, the average transmission power of each user P_s is readily shown equal to 0.5 (see Appendix 3A). It can be substituted into (7.2.15) in computing the performance bound equation of the adaptive-transversal-filter (ATF) receiver, as given in (7.2.18). Also, for the ATF receivers, the parameters $\tilde{\mu}$ and ϵ in the LMS algorithm are set to 0.5 and 10^{-20}, respectively. The bit error rates of the communication system are then simulated to reveal the effects of varying the following parameters:

- bit-energy-to-noise-power-spectral-density ratio (E_b/N_0) which is given by $2\beta P_s/N_0$;
- length of training sequence K;
- number of users N; and
- spreading factor 2β.

For comparison, we will show the BERs of the typical time-delay-based multiple-access approach, such as the multiple-access differential chaos-shift-keying (MA-DCSK) system proposed by Kolumbán *et al.* [1997a] and studied by Lau *et al.* [2002b]. The performance bounds of the ATF receiver (by (7.2.18)) and inverse-and-average (IA) receiver (by simulations) will also be shown as references.

First, we evaluate how the length of the training sequence affects the training result of the ATF receiver under a noiseless environment. Figure 7.6 plots the correlation of $\boldsymbol{x}^{(i)}$ and $\boldsymbol{w}_K^{(j)}$, i.e., $\sum_{k=1}^{\beta} x_k^{(i)} w_{K,k}^{(j)}$, versus K. As shown in the figure, when the length of the training sequence increases, the correlation value approaches unity and zero, respectively, for $i = j$ and $i \neq j$. The results verify

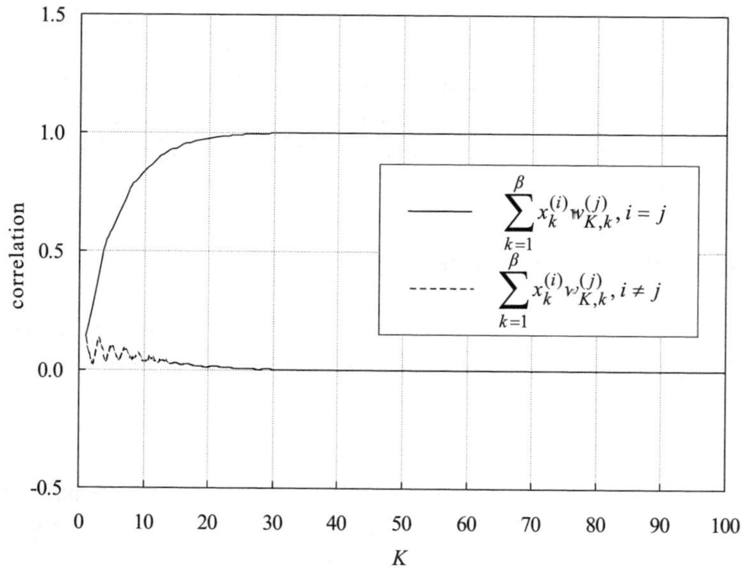

Figure 7.6. Correlation of $\boldsymbol{x}^{(i)}$ and $\boldsymbol{w}_K^{(j)}$ versus K for an adaptive-transversal-filter receiver under a noiseless environment. Number of users = 4. Spreading factor = 200.

that the ATF receiver has the capability of recognizing the required signal and removing the inter-user interference after the training process.

In the next set of simulations, we study a single-user system and vary the length of the training sequence. The corresponding BER curves are plotted in Fig. 7.7. For the case with $K = 1$, the structure of the transmitted signal is the same as that of a single-user DCSK system [Kolumbán *et al.*, 1998], the BER performance of which is also shown in the figure for comparison. Also, the structure of the IA receiver is the same as the noncoherent DCSK receiver when $K = 1$. From the results, it can be observed that for $K = 1$, the ATF receiver achieves similar BERs as the DCSK scheme (and consequently the IA receiver) in a single-user environment. For $K = 4, 8$ and 16, the ATF and IA receivers have similar performance. Also, when the length of the training sequence K increases, the BERs of both types of receivers have improved because a larger number of training signals can reduce the effect of noise during the training process.

Figure 7.8 plots the BERs versus E_b/N_0 for a 4-user system. The length of the training sequence K is equal to 4, 8 and 16. Spreading factors of 20, 40, 100 and 200 are employed. The BER curve for the MA-DCSK system is also shown here for comparison [Tam *et al.*, 2003]. Here, we see clearly that the ATF receiver significantly outperforms the IA receiver and the MA-DCSK scheme in all cases.

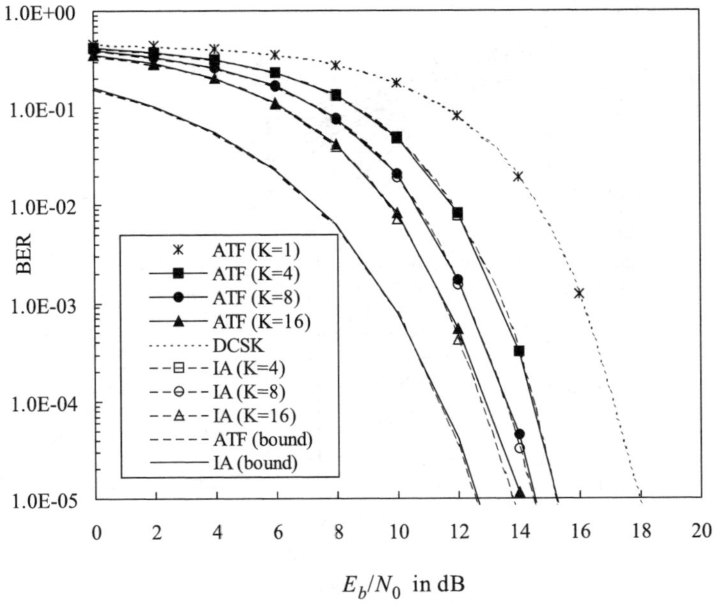

Figure 7.7. BER versus E_b/N_0 for the chaos-based communication system with one user. $K = 1, 4,$ 8 and 16. Spreading factor is 200.

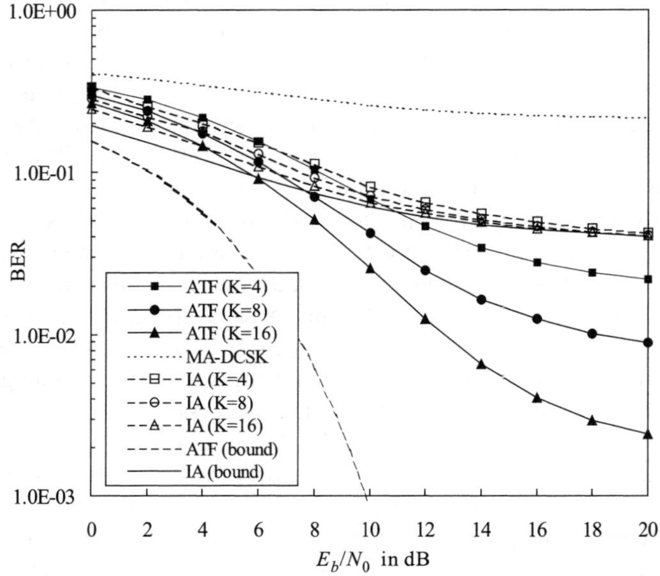

(a)

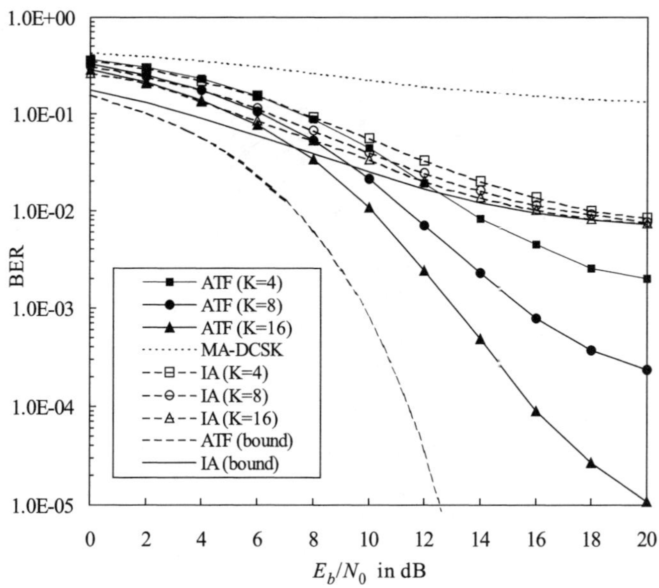

(b)

Figure 7.8. BER versus E_b/N_0 for the chaos-based communication system with 4 users. Spreading factor is (a) 20; (b) 40; (c) 100; and (d) 200.

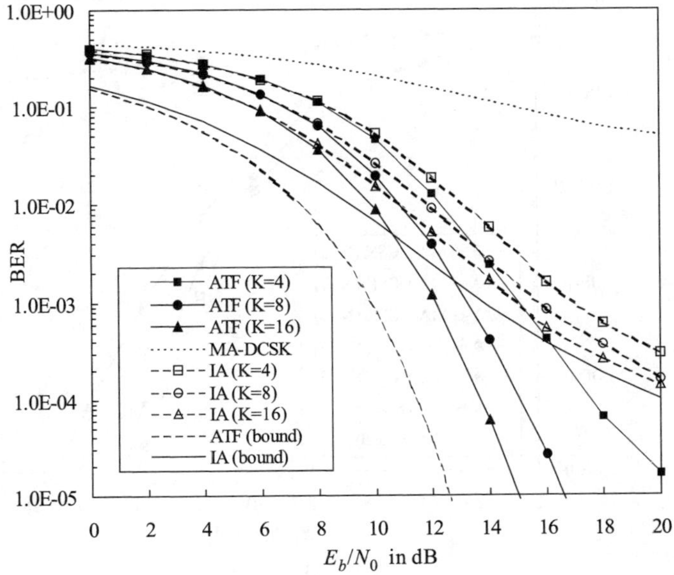

(c)

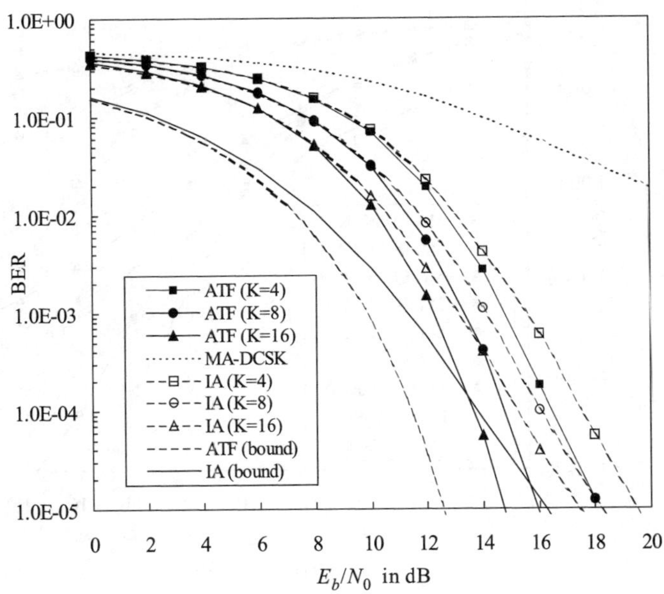

(d)

Figure 7.8. (Continued)

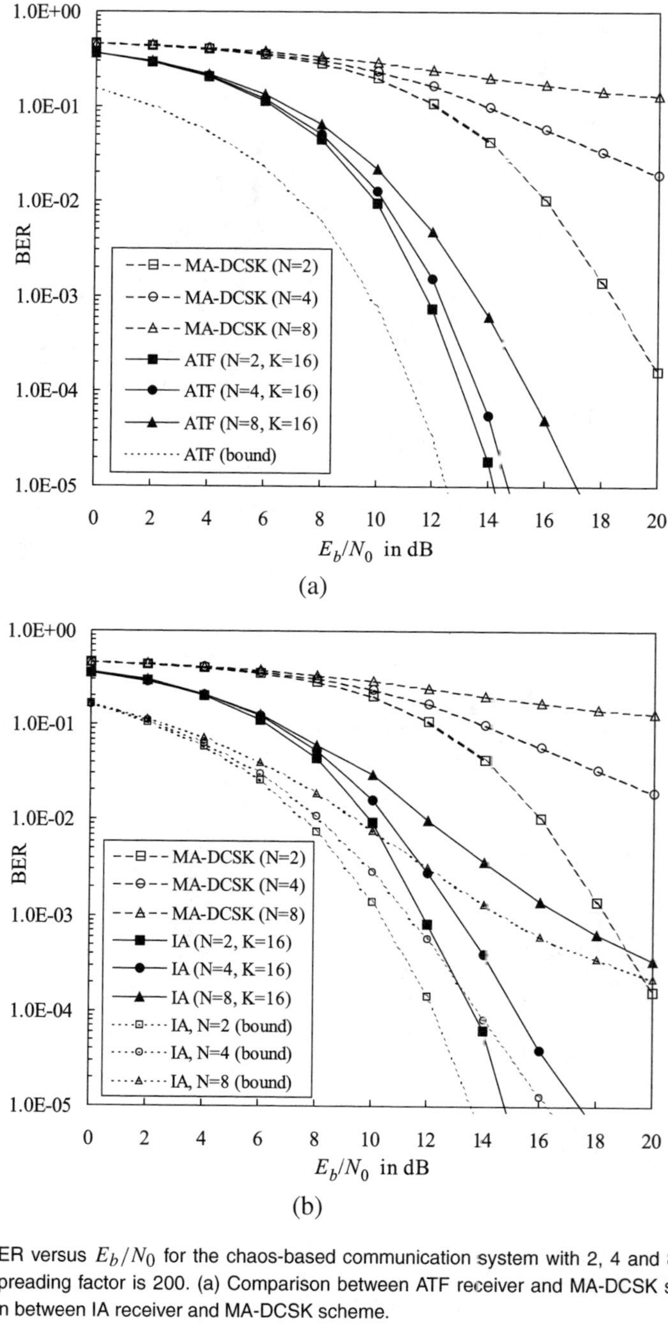

Figure 7.9. BER versus E_b/N_0 for the chaos-based communication system with 2, 4 and 8 users. K = 16 and spreading factor is 200. (a) Comparison between ATF receiver and MA-DCSK scheme. (b) Comparison between IA receiver and MA-DCSK scheme.

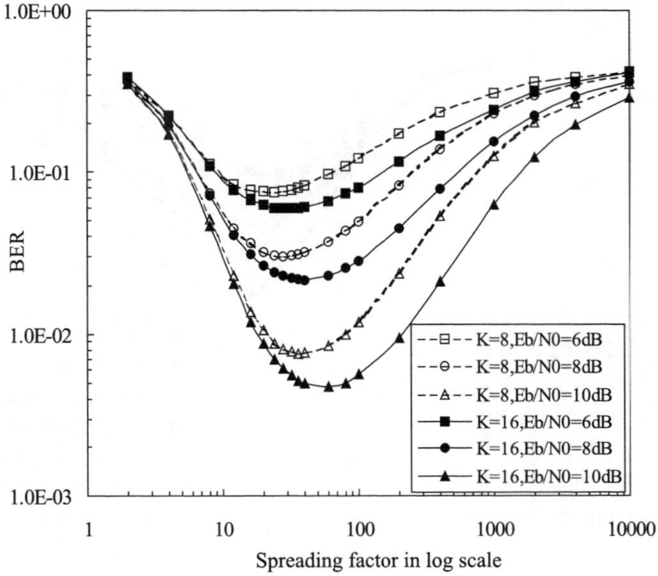

(a)

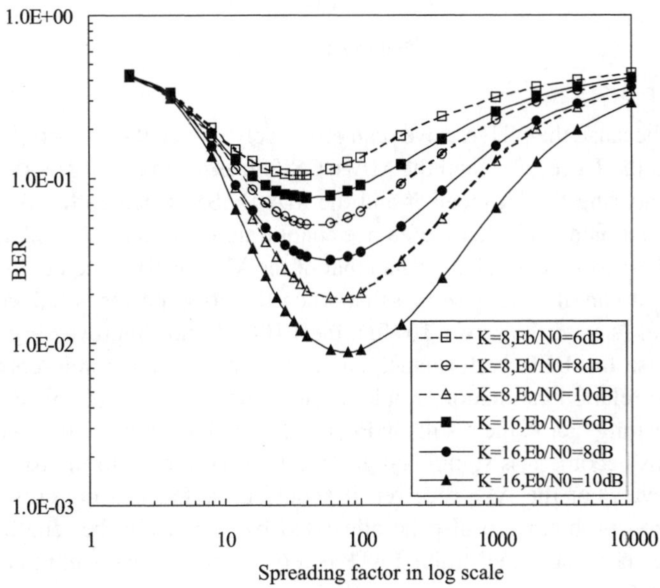

(b)

Figure 7.10. BER versus spreading factor for the chaos-based communication system employing ATF receivers. The number of users is (a) $N = 2$; (b) $N = 4$; and (c) $N = 8$.

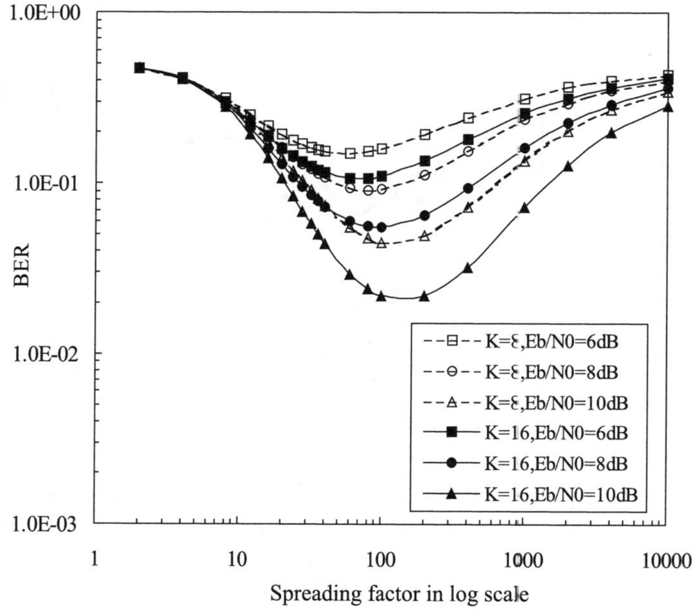

(c)

Figure 7.10. (Continued)

This is because the ATF receiver can effectively reduce the inter-user interference whereas the IA receiver and the MA-DCSK scheme do not possess such capability. Comparing the IA receiver and the MA-DCSK scheme, the IA receiver can always estimate a "cleaner" reference chaotic signal used for demodulation. Thus, its BER performance is better than that of the MA-DCSK scheme. In Figs. 7.8(a) and (b), it can also be observed that when the noise power is reduced to a very low level (say, E_b/N_0 over 16 dB), the BERs do not improve very much. This is because the BER performance has been limited by the auto-correlation and cross-correlation estimation problems [Kolumbán *et al.*, 2002b]. By increasing the spreading factor, the results in Figs. 7.8(c) and (d) indicate that the estimation problems become less significant and the BER continues to improve with lower noise level. For the ATF receiver, it should be noted that the cross-correlation estimation problem can also be alleviated by increasing the length of training sequence because a well-trained ATF receiver can effectively eliminate the inter-user interference.

Figure 7.9 plots the BER curves versus E_b/N_0 for 2-, 4- and 8-user systems. The length of the training sequence equals 16 and the spreading factor used is 200. The corresponding curves for the MA-DCSK system are also shown in the

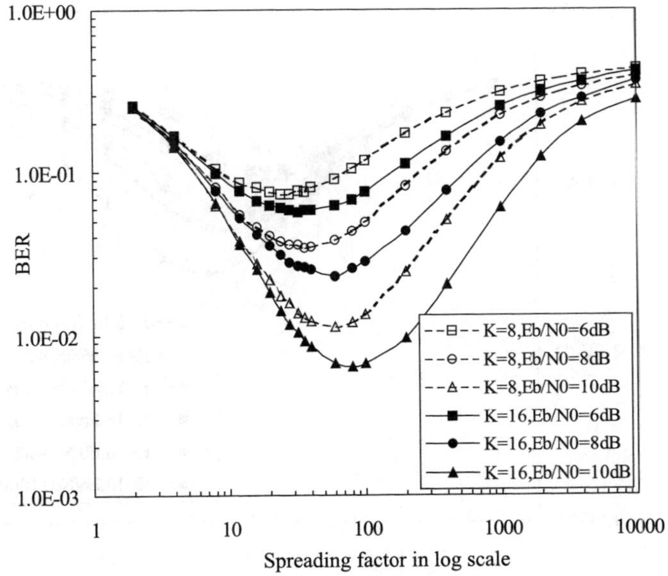

(a)

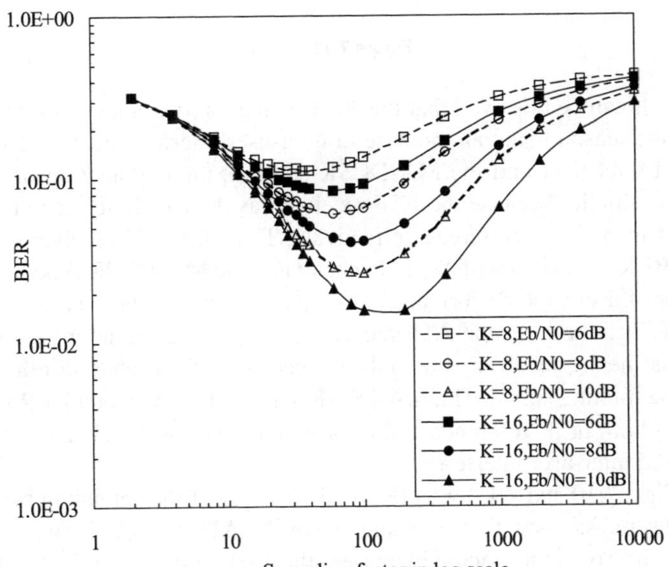

(b)

Figure 7.11. BER versus spreading factor for the chaos-based communication system employing IA receivers. The number of users is (a) $N = 2$; (b) $N = 4$; and (c) $N = 8$.

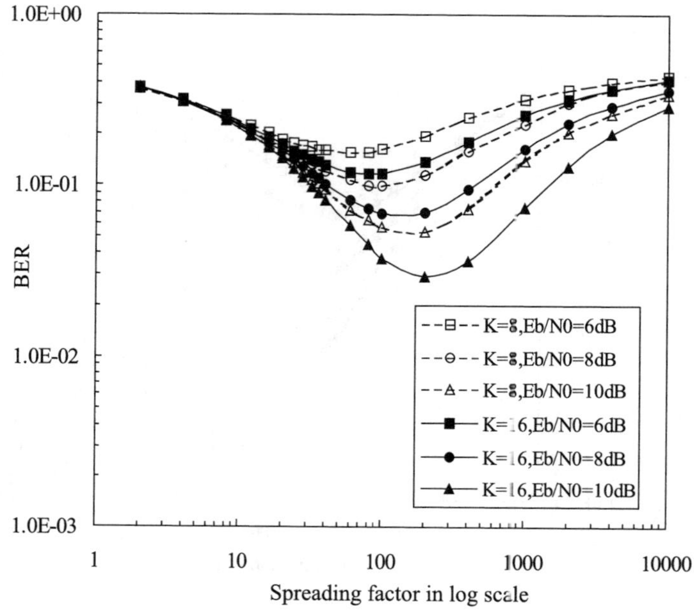

(c)

Figure 7.11. (Continued)

figures. It can be observed that the BER increases (degrades) with the number of users, the reason being an increase in inter-user interference. Compared to those of the IA receiver and the MA-DCSK scheme, the degradation of the ATF receiver is smaller because the ATF receiver has the capability of reducing the effects of inter-user interference effectively. Thus, the ATF receiver can achieve a better BER compared with the IA receiver and the MA-DCSK scheme. For a large number of users, the performance discrepancy between the ATF and IA receivers is large. It is because the ATF receiver can alleviate the inter-user interference whereas the IA receiver aims only to recover a clean chaotic reference signal from the training signal. The MA-DCSK scheme has the worst BER performance because both of its reference and information-bearing signals are suffering from noise and inter-user interference.

In Figs. 7.10 and 7.11, we plot the BERs against the spreading factor in 2-, 4- and 8-user systems with $K = 8$ and 16 for the ATF and IA receivers. It is shown that as the spreading factor increases, the BER improves initially, but reaches an optimum value at a certain spreading factor beyond which performance starts to degrade. The decrease in the BER is due to the reduced effect of the auto-correlation and cross-correlation estimation problems, while the increase in the BER is caused by the large noise level.

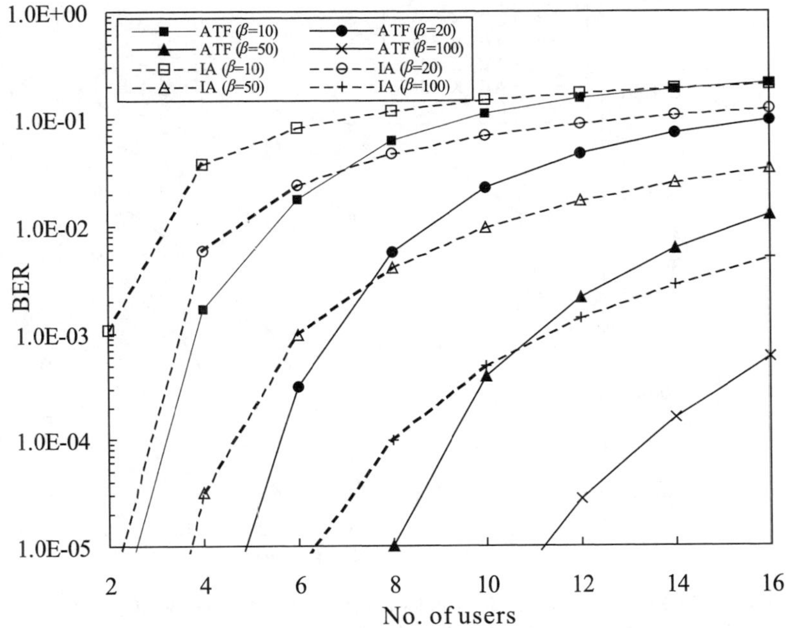

Figure 7.12. BER versus number of users for the chaos-based communication system for different values of spreading factor in a noiseless environment. $K = 16$.

Finally, we investigate the non-coherent multiple-access system under a noiseless environment. In Fig. 7.12, the BERs of the ATF and IA receivers are plotted against the number of users for different spreading factors. As expected, the BERs increase with the number of users because of the increase in inter-user interference. Also, the BERs improve with the spreading factor because the chaotic signals sent by different users can be more easily distinguished from one another when the spreading factor is large. Moreover, it can be observed that the ATF receiver performs better than the IA receiver in all cases. In Fig. 7.13, the BERs are plotted against the number of users for different lengths of the training sequence. The performance of both ATF and IA receivers degrades with the number of users. For the ATF receiver, the BER reduces as K increases because a longer training sequence enables the ATF receiver to remove the inter-user interference more effectively. Under a noiseless condition, the performance of the IA receiver is independent of K because a clean reference chaotic signal can always be recovered. Therefore, only one curve is shown for the IA receiver. As explained earlier, since the IA receiver cannot eliminate the inter-user interference, its performance is always worse than that of the ATF receiver.

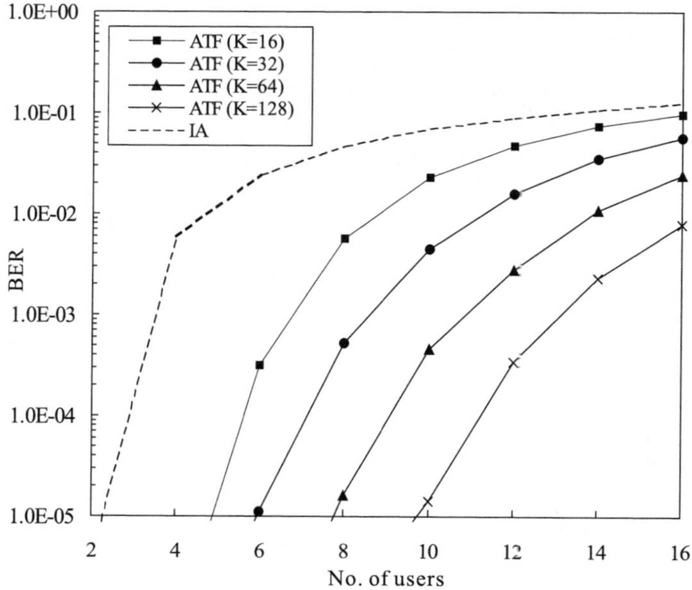

Figure 7.13. BER versus number of users for the chaos-based communication system for different values of K in a noiseless environment. Spreading factor is 40.

7.4 SUMMARY

In this chapter, we have investigated a non-coherent multiple-access chaos-based communication system utilizing transmitted references and training sequences. The transmission scheme is simple and easy to implement. Essentially, a series of training chaotic signals are sent to train the receivers at the receiving side. Two types of receivers have been studied. The first one is based on an adaptive transversal filter (ATF). By using the normalized least-mean-square algorithm to update the tap weights of the filter, the mean-square-error between the incoming training symbols and the expected symbols is reduced, which in turn mitigates the interference between users when decoding the information-bearing signal. The second type of receiver, namely the inverse-and-average (IA) receiver, aims at recovering clean reference chaotic signals by multiplying the signals in the reference time slots by the corresponding training bits, followed by averaging. Comparing the complexity of the two receivers, the ATF receiver has a slightly more complicated training algorithm over the IA receiver. However, results show that the ATF receiver outperforms the IA receiver and other time-delay-based multiple-access system.

8

NEAR-OPTIMUM COMMUNICATION SCHEMES

In Chapters 5 to 7, several noncoherent communication schemes are studied. The major drawback of noncoherent systems is the relatively poor performance, compared to their coherent counterparts. In this final chapter, we introduce another method for improving the performance of noncoherent chaos-based communication systems. This method involves transforming chaotic signals into pre-defined patterns according to the relative magnitudes of the signal samples. First, for large spreading factors, the transformation leads to a large correlation between two different chaotic signals having the same probability density function (pdf). Such a property can be used to recover the information in a noncoherent system. This is because if the chaotic map and the transformation function are known, we can generate another chaotic signal at the receiver and construct a reference signal for use in the correlator. In this scheme, both synchronization circuit and reference-transmission are not required. This scheme is referred to as near-optimum noncoherent (NON) scheme. Second, for small spreading factors where high autocorrelation of the different chaotic signals cannot be achieved, we make use of a near-optimum noncoherent scheme with transmitted reference (NON-TR). For comparison, the corresponding performance bound using coherent detection is also given. This reference scheme is called near-optimum coherent (NOC) scheme. It can be shown that the NON scheme can achieve a similar performance as the CSK system.

We organize this chapter as follows. In Section 8.1, we will describe the implementation and gives the performance bound for the NOC and NON systems. Section 8.2 introduces the NON-TR scheme. Multiple-access techniques are also discussed for the NOC, NON and NON-TR systems. Finally, in Section 8.3, the simulated results are reported and compared with those of the conventional coherent CSK and noncoherent differential chaos-shift-keying (DCSK) schemes.

8.1 NEAR-OPTIMUM COHERENT (NOC) AND NONCOHERENT (NON) SYSTEMS

8.1.1 Single-User System

Figure 8.1 shows the transmitter and receiver structures of a single-user near-optimum coherent (NOC)/noncoherent (NON) system. A chaotic signal, denoted by $\{x_k\}$, is first generated in the transmitter. A transformation block is added after the chaotic signal generator. We denote the spreading factor by 2β, and consider the lth transmitted symbol, i.e., $k = 2(l-1)\beta+1, 2(l-1)\beta+2, \ldots, 2l\beta$. Let

$$\boldsymbol{x}_l = [\, x_{2(l-1)\beta+1} \quad x_{2(l-1)\beta+2} \quad \cdots \quad x_{2l\beta}\,]. \tag{8.1.1}$$

We introduce a transformation function, $F(\cdot)$, which transforms vector $\boldsymbol{x}_l$ to another vector $\tilde{\boldsymbol{x}}_l$, i.e.,

$$\tilde{\boldsymbol{x}}_l = F(\boldsymbol{x}_l), \tag{8.1.2}$$

where

$$\tilde{\boldsymbol{x}}_l = [\, \tilde{x}_{2(l-1)\beta+1} \quad \tilde{x}_{2(l-1)\beta+2} \quad \cdots \quad \tilde{x}_{2l\beta}\,]. \tag{8.1.3}$$

In the transformation, the bit period can be divided into one or several slots. The elements of the vector $\boldsymbol{x}_l$ within each slot are then rearranged in a pre-defined order according to the magnitudes of the signal samples.

The resulting re-ordered waveform will be used to represent symbol "+1". Multiplying the transformed chaotic signal by -1, we obtain the waveform that represents "-1". For simplicity, we denote the lth transmitted symbol by $d_l \in \{-1, +1\}$, and we assume that the symbols "+1" and "-1" are transmitted with identical probabilities. During the lth symbol duration, the transmitted signal, denoted by s_k, is given by

$$s_k = d_l \tilde{x}_k. \tag{8.1.4}$$

Assuming that the transmitted signal is passed through an additional white Gaussian noise (AWGN) channel with a two-sided power spectral density $N_0/2$,

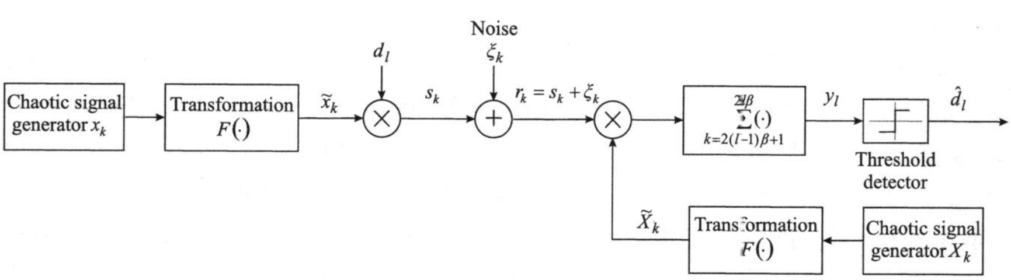

Figure 8.1. Transmitter and receiver structures of the near-optimum system.

the received signal, denoted by r_k, is given by

$$r_k = s_k + \xi_k, \tag{8.1.5}$$

where ξ_k represents the AWGN signal, the mean and variance of which are zero and $N_0/2$, respectively.

At the receiving side, depending on whether the NOC or NON system is being considered, the same chaotic signal or another chaotic signal with the same pdf, denoted by X_k, is first generated. In other words,

$$X_k \begin{cases} = x_k & \text{for near-optimum coherent (NOC) system,} \\ \neq x_k & \text{for near-optimum noncoherent (NON) system.} \end{cases} \tag{8.1.6}$$

Define

$$\boldsymbol{X}_l = [\, X_{2(l-1)\beta+1} \quad X_{2(l-1)\beta+2} \quad \cdots \quad X_{2l\beta}\,]. \tag{8.1.7}$$

The same transformation $F(\cdot)$ is performed over $\boldsymbol{X}_l$, yielding

$$\tilde{\boldsymbol{X}}_l = F(\tilde{\boldsymbol{X}}_l), \tag{8.1.8}$$

where

$$\tilde{\boldsymbol{X}}_l = [\, \tilde{X}_{2(l-1)\beta+1} \quad \tilde{X}_{2(l-1)\beta+2} \quad \cdots \quad \tilde{X}_{2l\beta}\,]. \tag{8.1.9}$$

Then, $\tilde{X}_k$ will be used to correlate with the received signal. The output of the correlator at the end of the lth symbol duration, denoted by y_l, becomes

$$\begin{aligned} y_l &= \sum_{k=2(l-1)\beta+1}^{2l\beta} r_k \tilde{X}_k \\ &= \sum_{k=2(l-1)\beta+1}^{2l\beta} (d_l \tilde{x}_k + \xi_k)\tilde{X}_k \\ &= \sum_{k=2(l-1)\beta+1}^{2l\beta} \Bigg(\underbrace{d_l \tilde{x}_k \tilde{X}_k}_{\text{required signal}} + \underbrace{\tilde{X}_k \xi_k}_{\text{noise component}} \Bigg). \end{aligned} \tag{8.1.10}$$

For the NOC system, y_l is given by

$$y_l = \sum_{k=2(l-1)\beta+1}^{2l\beta} \Bigg(\underbrace{d_l x_k^2}_{\text{required signal}} + \underbrace{x_k \xi_k}_{\text{noise component}} \Bigg). \tag{8.1.11}$$

The symbol can be detected according to the sign of the correlator output, which is described by the following equation

$$\hat{d}_l = \begin{cases} +1 & \text{if } y_l > 0, \\ -1 & \text{if } y_l \leqslant 0. \end{cases} \tag{8.1.12}$$

For the NON scheme, assume that 2β is large enough such that the transformed chaotic waveforms for all bit durations are similar. Then, the term $\sum_{k=2(l-1)\beta+1}^{2l\beta} d_l \tilde{x}_k \tilde{X}_k$ can be approximated by $\sum_{k=2(l-1)\beta+1}^{2l\beta} d_l x_k^2$. In this case, it is readily shown that the performance of the near-optimum noncoherent system is similar to those of the NOC system and the antipodal CSK system. Assuming that the conditional correlator output follows a Gaussian distribution, the analytical solution for the bit-error-rate (BER) can be shown equal to (see also (3.3.16))

$$\begin{aligned} \mathrm{BER}_{\mathrm{NON}} &\approx \mathrm{BER}_{\mathrm{CSK}} (= \mathrm{BER}_{\mathrm{NOC}}) \\ &= \frac{1}{2} \operatorname{erfc}\left(\left[\frac{\Psi}{\beta} + \left(\frac{E_b}{N_0} \right)^{-1} \right]^{-1/2} \right), \end{aligned} \tag{8.1.13}$$

where

$$P_s = \mathrm{E}\left[x_k^2\right] \tag{8.1.14}$$

and

$$\Psi = \frac{\mathrm{var}[x_k^2]}{P_s^2}. \tag{8.1.15}$$

8.1.2 Multiple-User System

Figure 8.2 depicts the transmitter and receiver structures of a multiple-access near-optimum coherent/noncoherent system with N users. Here, the transformations are different for different users. To further minimize the interference, we can make use of the Walsh codes when designing the transformations. For example, in a 4-user system, we can split the bit duration into four equal time intervals and define $\boldsymbol{x}_{l,m}^{(i)}$ $(m = 1, 2, 3, 4)$ for the ith user as

$$\boldsymbol{x}_{l,m}^{(i)} = \left[x^{(i)}_{2(l-1)\beta + \frac{2(m-1)\beta}{4} + 1} \quad x^{(i)}_{2(l-1)\beta + \frac{2(m-1)\beta}{4} + 2} \quad \cdots \quad x^{(i)}_{2(l-1)\beta + \frac{2m\beta}{4}} \right]. \tag{8.1.16}$$

Denote

$$\boldsymbol{x}_l^{(i)} = \left[x^{(i)}_{2(l-1)\beta+1} \quad x^{(i)}_{2(l-1)\beta+2} \quad \cdots \quad x^{(i)}_{2l\beta} \right] \tag{8.1.17}$$

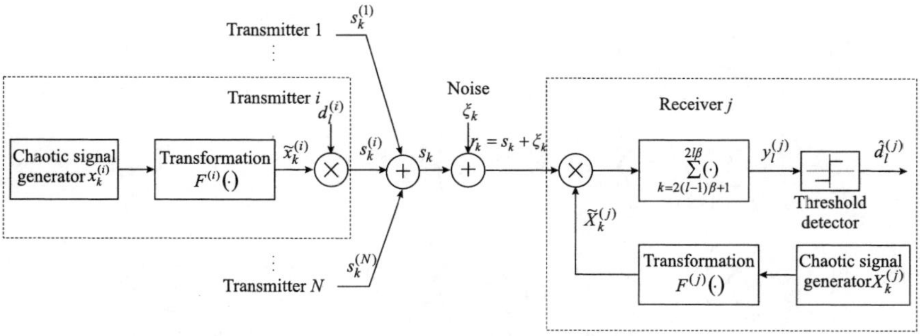

Figure 8.2. Transmitter and receiver structures of a multiple-access near-optimum system.

and

$$\tilde{x}_l^{(i)} = \left[\tilde{x}_{2(l-1)\beta+1}^{(i)} \quad \tilde{x}_{2(l-1)\beta+2}^{(i)} \quad \cdots \quad \tilde{x}_{2l\beta}^{(i)}\right]. \tag{8.1.18}$$

For the ith user, the chaotic signal vector after transformation, denoted by $\tilde{\boldsymbol{x}}_l^{(i)}$, is given by

$$\begin{aligned}\tilde{\boldsymbol{x}}_l^{(i)} &= F^{(i)}\left(\boldsymbol{x}_l^{(i)}\right)\\ &= \left[\boldsymbol{H}_{i,1}G\left(\boldsymbol{x}_{l,1}^{(i)}\right) \; \boldsymbol{H}_{i,2}G\left(\boldsymbol{x}_{l,2}^{(i)}\right) \; \boldsymbol{H}_{i,3}G\left(\boldsymbol{x}_{l,3}^{(i)}\right) \; \boldsymbol{H}_{i,4}G\left(\boldsymbol{x}_{l,4}^{(i)}\right)\right],\end{aligned} \tag{8.1.19}$$

where $G(\cdot)$ is a common transformation and $\boldsymbol{H}_{i,m}$ represents the (i,m)th element of a 4×4 Hadamard matrix (see (7.2.21)). Then the transmitted signal for ith user is given by

$$s_k^{(i)} = d_l^{(i)}\tilde{x}_k^{(i)}. \tag{8.1.20}$$

At the receiver of the jth user, the received signal will correlate with a transformed chaotic signal, denoted by $\tilde{X}_k^{(j)}$, created by the receiver. Denote

$$\boldsymbol{X}_l^{(j)} = \left[X_{2(l-1)\beta+1}^{(j)} \quad X_{2(l-1)\beta+2}^{(j)} \quad \cdots \quad X_{2l\beta}^{(j)}\right] \tag{8.1.21}$$

and

$$\begin{aligned}\tilde{\boldsymbol{X}}_l^{(j)} &= \left[\tilde{X}_{2(l-1)\beta+1}^{(j)} \quad \tilde{X}_{2(l-1)\beta+2}^{(j)} \quad \cdots \quad \tilde{X}_{2l\beta}^{(j)}\right]\\ &= F^{(j)}\left(\boldsymbol{X}_l^{(j)}\right).\end{aligned} \tag{8.1.22}$$

The output of the correlator is then represented by

$$
\begin{aligned}
y_l^{(j)} &= \sum_{k=2(l-1)\beta+1}^{2l\beta} r_k \tilde{X}_k^{(j)} \\
&= \sum_{k=2(l-1)\beta+1}^{2l\beta} \left(\sum_{i=1}^{N} d_l^{(i)} \tilde{x}_k^{(i)} + \xi_k \right) \tilde{X}_k^{(j)} \\
&= \sum_{k=2(l-1)\beta+1}^{2l\beta} \Bigg[\underbrace{d_l^{(j)} \tilde{x}_k^{(j)} \tilde{X}_k^{(j)}}_{\text{required signal}} + \underbrace{\sum_{i=1, i\neq j}^{N} d_l^{(i)} \tilde{x}_k^{(i)} \tilde{X}_k^{(j)}}_{\text{interference}} + \underbrace{\tilde{X}_k^{(j)} \xi_k}_{\text{noise component}} \Bigg].
\end{aligned}
\tag{8.1.23}
$$

When the spreading factor is large, the waveforms given by $G(\boldsymbol{x}_{l,m}^{(i)})$ are similar even for different m and i. Thus, the cross-correlation between $\tilde{x}_k^{(i)}$ and $\tilde{X}_k^{(j)}$ is close to zero when $i \neq j$. Moreover, the cross-correlation between $\tilde{x}_k^{(j)}$ and $\tilde{X}_k^{(j)}$ is close to the autocorrelation of $x_k^{(j)}$. Hence, the output of the correlator can be approximated by

$$
y_l^{(j)} \approx \sum_{k=2(l-1)\beta+1}^{2l\beta} \Bigg[\underbrace{d_l^{(j)} \left(x_k^{(j)}\right)^2}_{\text{required signal}} + \underbrace{\tilde{X}_k^{(j)} \xi_k}_{\text{noise component}} \Bigg]. \tag{8.1.24}
$$

Comparing (8.1.24) to (8.1.10) and (8.1.11), we conclude that under the aforementioned conditions, the performance of the multiuser system is similar to that of a single-user system.

8.2 NEAR-OPTIMUM NONCOHERENT SYSTEM WITH TRANSMITTED REFERENCE

8.2.1 Single-User System

In the NON system, when the spreading factor is small, the variance of the term $\sum_{k=2(l-1)\beta+1}^{2l\beta} \tilde{x}_k \tilde{X}_k$ is large and the system performance becomes worse. In this situation, a transmitted reference will be needed, as in the case of the DCSK system where the transmitted bit duration is divided into reference and information-bearing time slots.

Figure 8.3 shows the transmitter and receiver structures of a near-optimum noncoherent system with transmitted reference (NON-TR). Suppose β chaotic samples are sent in each time slot (spreading factor = 2β). The reference part of

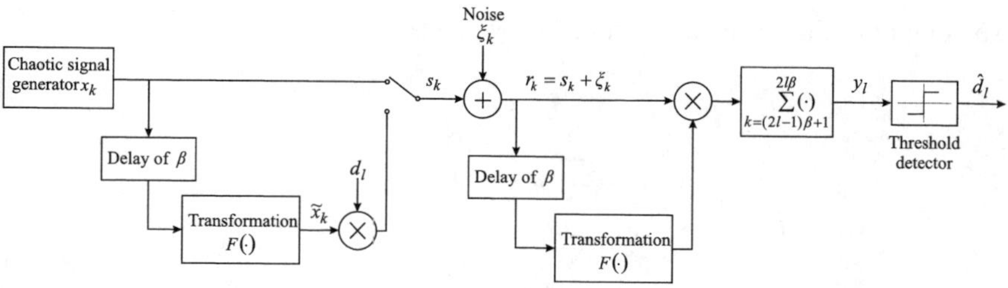

Figure 8.3. Transmitter and receiver structures of the near-optimum noncoherent system with transmitted reference.

the transmitted signal during the $(2l-1)$th time slot, i.e., for time $k = (2l-2)\beta + 1, (2l-2)\beta + 2, \ldots, (2l-1)\beta$, is given by

$$s_k = x_k. \tag{8.2.1}$$

For the information-bearing part, the chaotic signal generated in the last slot is first transformed, i.e.,

$$\tilde{\boldsymbol{x}}_{2l} = F(\boldsymbol{x}_{2l-1}), \tag{8.2.2}$$

where

$$\boldsymbol{x}_{2l-1} = [x_{(2l-2)\beta+1} \quad x_{(2l-2)\beta+2} \quad \cdots \quad x_{(2l-1)\beta}], \tag{8.2.3}$$

$$\tilde{\boldsymbol{x}}_{2l} = [\tilde{x}_{(2l-1)\beta+1} \quad \tilde{x}_{(2l-1)\beta+2} \quad \cdots \quad \tilde{x}_{2l\beta}]. \tag{8.2.4}$$

Then, the transformed signal is multiplied by the transmitted symbol (either $+1$ or -1) and the transmitted signal in the information-bearing part is given by

$$s_k = d_l \tilde{x}_k. \tag{8.2.5}$$

Assuming an AWGN channel, the signal arriving at the receiver is given by (8.1.5).

At the receiver, the received signal in the reference slot (the $(2l-1)$th slot) will undergo the same transformation $F(\cdot)$, and the correlation between this transformed signal and the received signal in the information-bearing slot (the $2l$ th slot) is computed. For brevity, we define

$$\boldsymbol{r}_{2l-1} = [r_{(2l-2)\beta+1} \quad r_{(2l-2)\beta+2} \quad \cdots \quad r_{(2l-1)\beta}] \tag{8.2.6}$$

and

$$\begin{aligned}\tilde{\boldsymbol{r}}_{2l} &= [\tilde{r}_{(2l-1)\beta+1} \quad \tilde{r}_{(2l-1)\beta+2} \quad \cdots \quad \tilde{r}_{2l\beta}] \\ &= F(\boldsymbol{r}_{2l-1}).\end{aligned} \tag{8.2.7}$$

The output of the correlator, denoted by y_l, equals

$$
\begin{aligned}
y_l &= \sum_{k=(2l-1)\beta+1}^{2l\beta} r_k \tilde{r}_k \\
&= \sum_{k=(2l-1)\beta+1}^{2l\beta} (d_l \tilde{x}_k + \xi_k)\, \tilde{r}_k.
\end{aligned} \tag{8.2.8}
$$

The optimum performance of the receiver occurs when $\tilde{x}_k$ and $\tilde{r}_k$ are having a similar pdf and the above equation can be rewritten as

$$
\begin{aligned}
y_l &\approx \sum_{k=(2l-1)\beta+1}^{2l\beta} (d_l \tilde{x}_k + \xi_k)\, C\tilde{x}_k \\
&= \sum_{k=(2l-1)\beta+1}^{2l\beta} \Bigg[\underbrace{C d_l \tilde{x}_k^2}_{\text{required signal}} + \underbrace{C\tilde{x}_k \xi_k}_{\text{noise component}} \Bigg],
\end{aligned} \tag{8.2.9}
$$

where C is a proportionality constant. The best BER performance is therefore given by

$$
\text{BER} = \frac{1}{2}\,\text{erfc}\left(\left[\frac{2\Psi}{\beta} + 2\left(\frac{E_b}{N_0}\right)^{-1} \right]^{-1/2} \right). \tag{8.2.10}
$$

8.2.2 Multiple-User System

Shown in Fig. 8.4 is a multiple-access near-optimum noncoherent system with transmitted reference. Consider the ith user. During the reference time slot, the transmitted signal is given by

$$
s_k^{(i)} = x_k^{(i)}. \tag{8.2.11}
$$

For the information-bearing time slot (the $2l$th slot), the chaotic signal transmitted in the reference slot (the $(2l-1)$th slot) is first transformed and we obtain

$$
\begin{aligned}
\tilde{\boldsymbol{x}}_{2l}^{(i)} &= \left[\tilde{x}_{(2l-1)\beta+1}^{(i)} \quad \tilde{x}_{(2l-1)\beta+2}^{(i)} \quad \cdots \quad \tilde{x}_{2l\beta}^{(i)} \right] \\
&= F^{(i)}\left(\boldsymbol{x}_{2l-1}^{(i)}\right),
\end{aligned} \tag{8.2.12}
$$

where

$$
\boldsymbol{x}_{2l-1}^{(i)} = \left[x_{(2l-2)\beta+1}^{(i)} \quad x_{(2l-2)\beta+2}^{(i)} \quad \cdots \quad x_{(2l-1)\beta}^{(i)} \right]. \tag{8.2.13}
$$

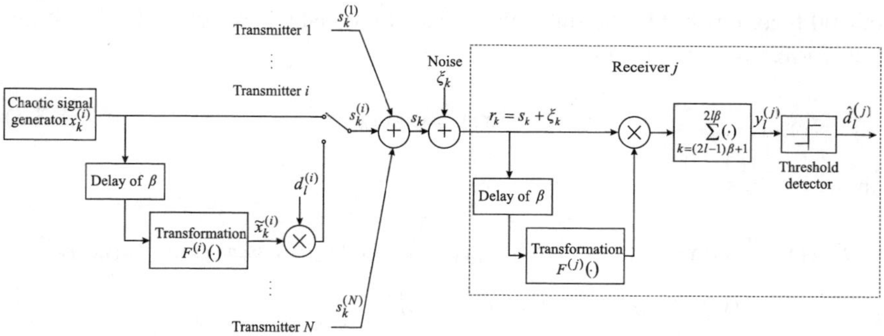

Figure 8.4. Transmitter and receiver structures of a multiple-access near-optimum noncoherent system with transmitted reference.

The transmitted signal during the information-bearing time slot is then given by

$$s_k^{(i)} = d_l^{(i)} \tilde{x}_k^{(i)}. \tag{8.2.14}$$

At the receiver of the jth user, the received signal in the reference time slot is first transformed before correlating with the signal in the information-bearing time slot. Because of the interference introduced by other users, the BER performance will degrade.

8.3 RESULTS AND DISCUSSIONS

In our study, the Chebyshev map of degree 2 and the Hénon map are used to generate the chaotic signals. These maps are given by

- Chebyshev map of degree 2:

$$x_{k+1} = 2x_k^2 - 1; \tag{8.3.1}$$

- Hénon map:

$$\begin{aligned} x_{k+1} &= 1 + z_k - 1.4x_k^2, \\ z_{k+1} &= 0.3x_k. \end{aligned} \tag{8.3.2}$$

8.3.1 Effects of the Choice of Transformation Function $F(\cdot)$ in the NOC System

First, we study the effect of the choice of transformation function $F(\cdot)$ on the performance of the near-optimum coherent (NOC) system. A spreading factor of 100 is assumed. On the transmitter side, we consider two cases in which the transformations are given by

$$F(\boldsymbol{x}_l) = G(\boldsymbol{x}_l) \tag{8.3.3}$$

and

$$F(\boldsymbol{x}_l) = \Big[G\big([x_{2(l-1)\beta+1} \cdots x_{2(l-1)\beta+33}]\big) \quad G\big([x_{2(l-1)\beta+34} \cdots x_{2(l-1)\beta+66}]\big) \\ G\big([x_{2(l-1)\beta+67} \cdots x_{2(l-1)\beta+100}]\big)\Big]. \tag{8.3.4}$$

Here, we assume that $G(\cdot)$ is a simple function that sorts the magnitudes of the input vector according to a pre-defined order.

In the receiver, we assume that the exact chaotic signal is regenerated. It is then sent to five different transformation blocks before correlating with the incoming signal. The transformations are

$$F_1(\boldsymbol{x}_l) = G(\boldsymbol{x}_l), \tag{8.3.5}$$

$$F_2(\boldsymbol{x}_l) = \Big[G\big([x_{2(l-1)\beta+1} \cdots x_{2(l-1)\beta+33}]\big) \quad G\big([x_{2(l-1)\beta+34} \cdots x_{2(l-1)\beta+66}]\big) \\ G\big([x_{2(l-1)\beta+67} \cdots x_{2(l-1)\beta+100}]\big)\Big], \tag{8.3.6}$$

$$F_3(\boldsymbol{x}_l) = \Big[G\big([x_{2(l-1)\beta+1} \cdots x_{2(l-1)\beta+20}]\big) \quad G\big([x_{2(l-1)\beta+21} \cdots x_{2(l-1)\beta+40}]\big) \\ G\big([x_{2(l-1)\beta+41} \cdots x_{2(l-1)\beta+60}]\big) \quad G\big([x_{2(l-1)\beta+61} \cdots x_{2(l-1)\beta+80}]\big) \\ G\big([x_{2(l-1)\beta+81} \cdots x_{2(l-1)\beta+100}]\big)\Big], \tag{8.3.7}$$

$$F_4(\boldsymbol{x}_l) = \Big[G\big([x_{2(l-1)\beta+1} \cdots x_{2(l-1)\beta+10}]\big) \quad G\big([x_{2(l-1)\beta-11} \cdots x_{2(l-1)\beta+20}]\big) \\ G\big([x_{2(l-1)\beta+21} \cdots x_{2(l-1)\beta+30}]\big) \quad G\big([x_{2(l-1)\beta-31} \cdots x_{2(l-1)\beta+40}]\big) \\ G\big([x_{2(l-1)\beta+41} \cdots x_{2(l-1)\beta+50}]\big) \quad G\big([x_{2(l-1)\beta-51} \cdots x_{2(l-1)\beta+60}]\big) \\ G\big([x_{2(l-1)\beta+61} \cdots x_{2(l-1)\beta+70}]\big) \quad G\big([x_{2(l-1)\beta-71} \cdots x_{2(l-1)\beta+80}]\big) \\ G\big([x_{2(l-1)\beta+81} \cdots x_{2(l-1)\beta+90}]\big) \quad G\big([x_{2(l-1)\beta+91} \cdots x_{2(l-1)\beta+100}]\big)\Big], \tag{8.3.8}$$

$$F_5(\boldsymbol{x}_l) = \Big[G\big([x_{2(l-1)\beta+1} \cdots x_{2(l-1)\beta+50}]\big) \quad G\big([x_{2(l-1)\beta+51} \cdots x_{2(l-1)\beta+100}]\big)\Big]. \tag{8.3.9}$$

Figure 8.5 plots the simulated BERs versus E_b/N_0 when different transformation functions are used in the detection. It is clearly seen that when the same transformation function is used in both the transmitter and receiver (corresponding to the NOC system), the BER is the lowest. Comparing Fig. 8.5(a) and Fig. 8.5(b),

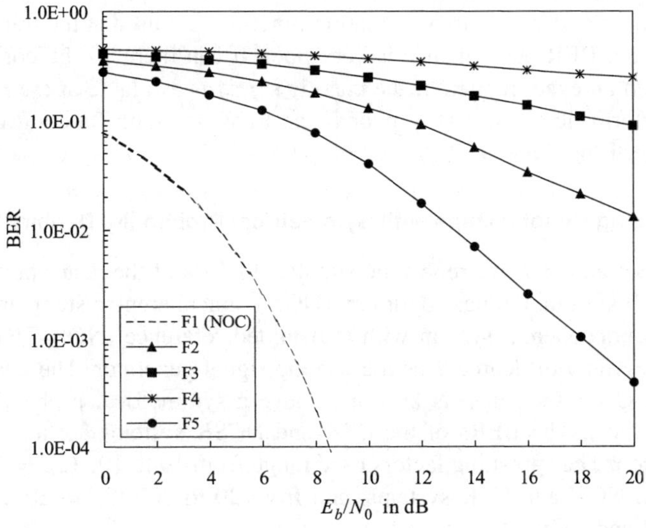

(a)

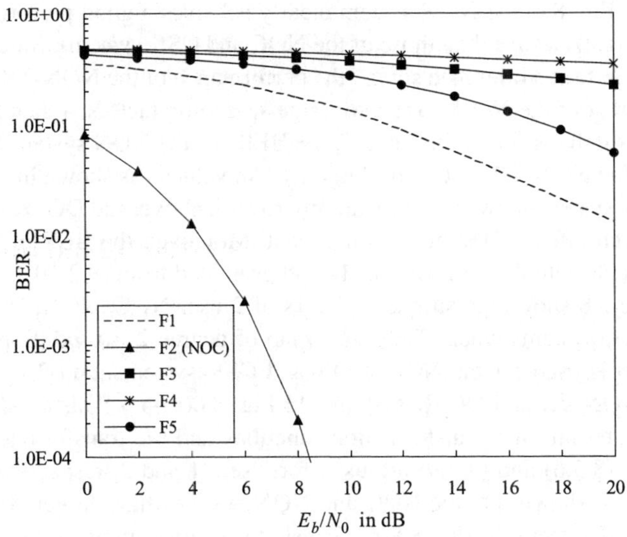

(b)

Figure 8.5. BER of the near-optimum coherent (NOC) system when different transformation functions are used to detect the signal. $2\beta = 100$ is used. Transformation function used at the transmitter: (a) $F(\boldsymbol{x}_l) = G(\boldsymbol{x}_l)$; (b) $F(\boldsymbol{x}_l) = [G([x_{(l-1)\beta+1} \cdots x_{(l-1)\beta+33}]) \;\; G([x_{(l-1)\beta+34} \cdots x_{(l-1)\beta+66}]) \;\; G([x_{(l-1)\beta+67} \cdots x_{(l-1)\beta+100}])]$.

moreover, the BER curves corresponding to the two NOC cases are very similar. However, if different transformation functions are used at the transmitter and receiver, the BER performance is very poor. It can therefore be concluded that even when an exact replica of the chaotic signal is available at the receiver, the BER performance is still very poor if the transformation function used by the transmitter is not known.

8.3.2 Using Chaotic Maps with Symmetrical Probability Density Function

In Figs. 8.6 and 8.7, we report the simulated BERs of the near-optimum coherent (NOC) system, the near-optimum (NON) noncoherent system, and the near-optimum noncoherent system with transmitted reference (NON-TR), using the Chebyshev map of degree 2 as the chaotic signal generator. The chaotic signal from the Chebyshev map is known to have a symmetrical probability density function (pdf). The BERs of the CSK and DCSK systems are also plotted for comparison. The spreading factors used range from 10 to 100 ($2\beta = 10$–100) for the NOC, NON and CSK systems, and from 20 to 200 ($2\beta = 20$–200) for the NON-TR and DCSK systems.

In theory, the NOC system makes use of the high autocorrelation of the chaotic signals to achieve comparable performance as the antipodal CSK system. As shown in Fig. 8.6, the NON system clearly achieves a good performance, which is only slightly worse than those of the NOC and CSK systems. Since a half-bit is used to transmit a reference signal, the performance of the NON-TR system is inferior to that of the NON system for large spreading factors. When the spreading factor is small, as shown in Fig. 8.7, the BERs of the NON system is higher than those of the NON-TR system for large E_b/N_0 values. As shown in all figures, the NON-TR system shows a significant improvement over the DCSK system due to the reduced effect of the noise component. Moreover, the BER of the NON-TR system is close to the performance bound evaluated from (8.2.10).

Figure 8.8 shows the simulated BERs of 2-user NOC, NON, NON-TR, CSK and DCSK systems when Chebyshev map of degree 2 is used. Spreading factor $2\beta = 100$ is used for the NOC, NON and CSK systems, and $2\beta = 200$ is used for the NON-TR and DCSK systems. In Fig. 8.8(a), no Walsh codes have been incorporated into the transformation functions and the transformation functions shown in (8.3.6) and (8.3.9) are used for Users 1 and 2, respectively. The BER curves have shown that the NOC and NON systems have larger BERs than that of the CSK system. In Fig. 8.8(b), Walsh codes are combined with the transformation (8.3.9) to form different transformation functions for the users and the transformed chaotic signals for the two users are nearly orthogonal. When coherent detection is performed, the BER of the NOC system is lower than that of the CSK system. Since the NOC system can give a much higher autocorrelation, the performance of the NOC system is also superior to that of the NON system.

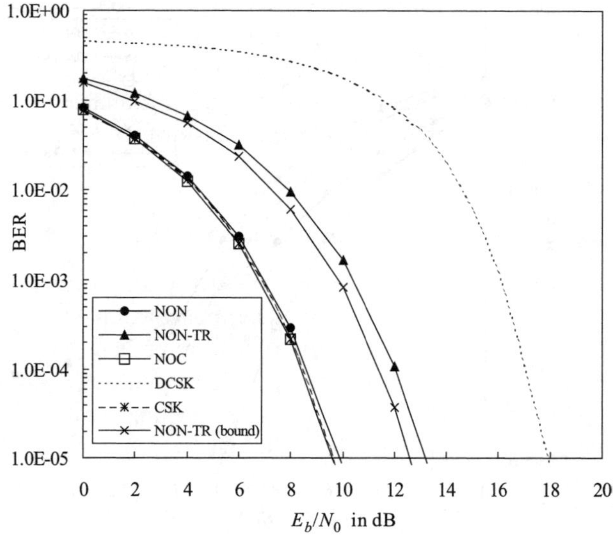

(a)

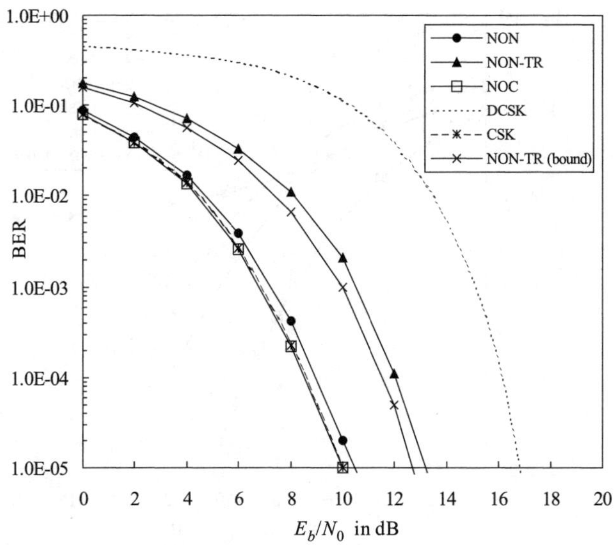

(b)

Figure 8.6. Simulated BERs of the NOC, NON, NON-TR, CSK and DCSK systems when the Chebyshev map of degree 2 is used. (a) $2\beta = 100$ is used for the NOC, NON and CSK systems, and $2\beta = 200$ is used for the NON-TR and DCSK systems; (b) $2\beta = 50$ is used for the NOC, NON and CSK systems, and $2\beta = 100$ is used for the NON-TR and DCSK systems.

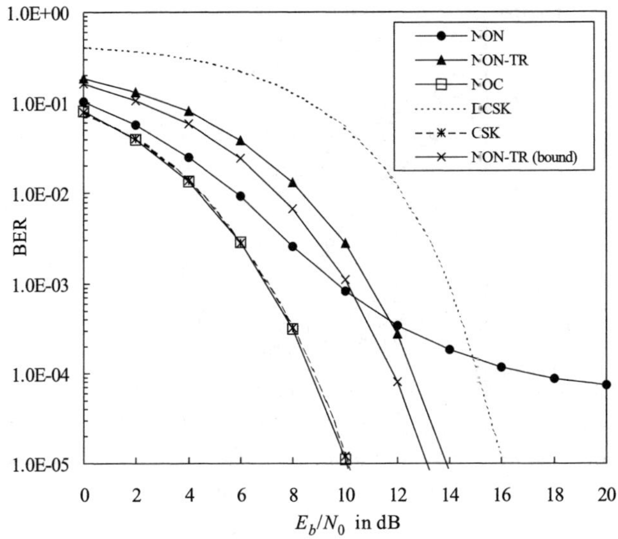

(a)

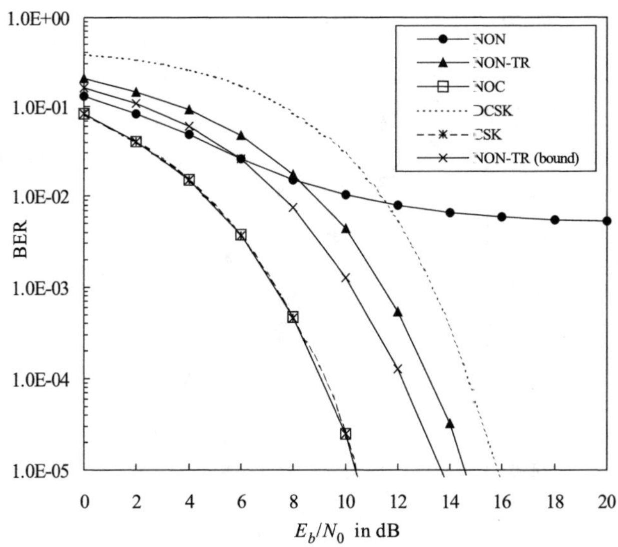

(b)

Figure 8.7. Simulated BERs of the NOC, NON, NON-TR, CSK and DCSK systems when the Chebyshev map of degree 2 is used. (a) $2\beta = 20$ is used for the NOC, NON and CSK systems, and $2\beta = 40$ is used for the NON-TR and DCSK systems; (b) $2\beta = 10$ is used for the NOC, NON and CSK systems, and $2\beta = 20$ is used for the NON-TR and DCSK systems.

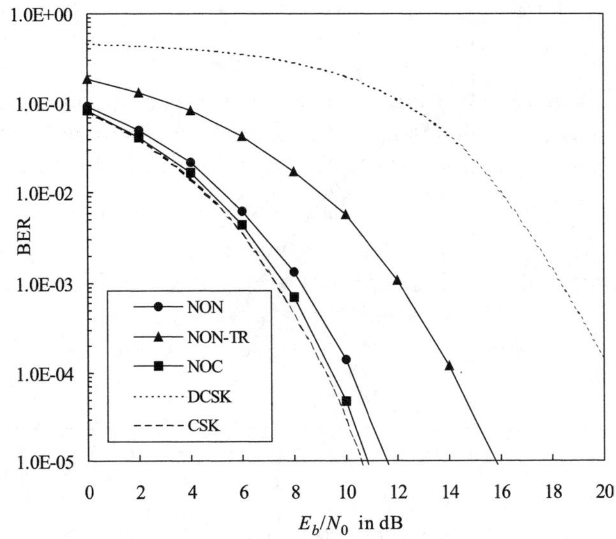

(a)

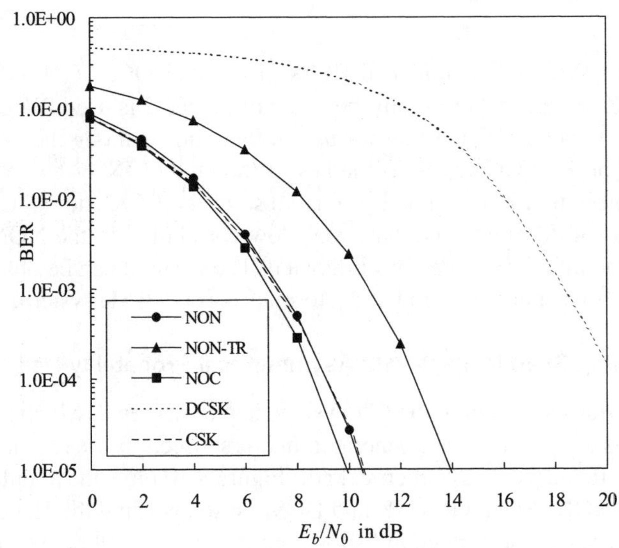

(b)

Figure 8.8. Simulated BERs of 2-user NOC, NON, NON-TR, CSK and DCSK systems when Chebyshev map of degree 2 is used. A spreading factor of $2\beta = 100$ is used for the NOC, NON and CSK systems, and $2\beta = 200$ is used for the NON-TR and DCSK systems. (a) Walsh codes are not used in the transformations. (b) Walsh codes are used in the transformations.

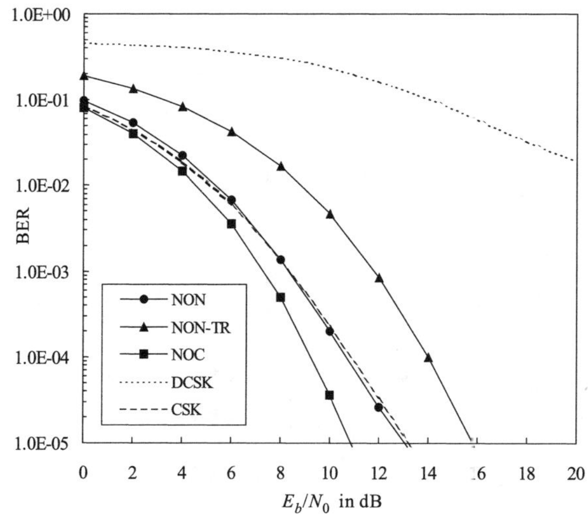

Figure 8.9. Simulated BERs of 4-user NOC, NON, NON-TR, CSK and DCSK systems when Chebyshev map of degree 2 is used. Walsh codes have been used in designing the transformation functions. Spreading factor $2\beta = 100$ is used for the NOC, NON and CSK systems, $2\beta = 200$ is used for the NON-TR and DCSK systems.

Figure 8.9 plots the simulated BERs of 4-user NOC, NON, NON-TR, CSK and DCSK systems when Chebyshev map of degree 2 is used. Walsh codes have been used in designing the transformation functions. Among the curves, the performance of the NOC system is the best. While the NON system cannot achieve an autocorrelation function as large as those in the CSK and NOC systems, the waveforms of different users have very low correlation in the NON system and much of the interference can be eliminated. Therefore, it can be observed that the NON system can obtain a similar performance as the CSK system.

8.3.3 Using Chaotic Maps with Asymmetrical Probability Density Function

In the previous subsection, the Chebyshev map of degree 2, which has a symmetrical pdf, is used as the chaos generator. In this subsection, we repeat the study, but using chaotic maps of asymmetrical pdf. Figure 8.10 plots the simulated BERs of the NOC, NON, NON-TR, CSK and DCSK systems using the Hénon map as the chaos generator. The spreading factor used is 100 (i.e., $2\beta = 100$) for the NOC, NON and CSK systems, and 200 (i.e., $2\beta = 200$) for the NON-TR and DCSK systems. By comparing to Fig. 8.6(a), it can be seen that the chaotic map has little effect on the BER performance of the systems.

Shown in Figs. 8.11 and 8.12 are the simulated BERs of 2-user and 4-user NOC, NON, NON-TR, CSK and DCSK systems when Hénon map is used. The

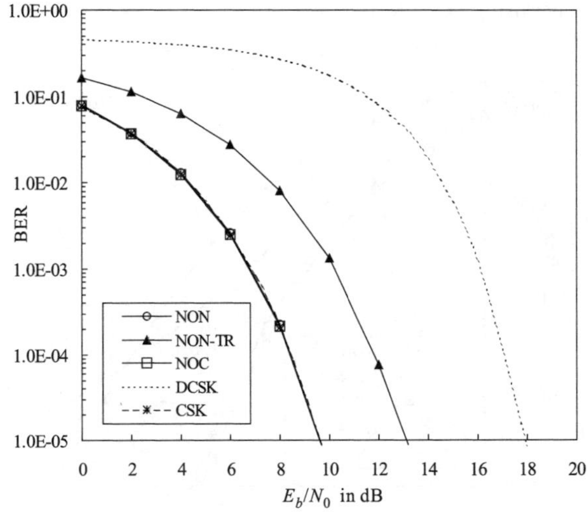

Figure 8.10. Simulated BERs of NOC, NON, NON-TR, CSK and DCSK systems when the Hénon map is used as chaos generator. A spreading factor of $2\beta = 100$ is used for the NOC, NON and CSK systems, and $2\beta = 200$ is used for the NON-TR and DCSK systems.

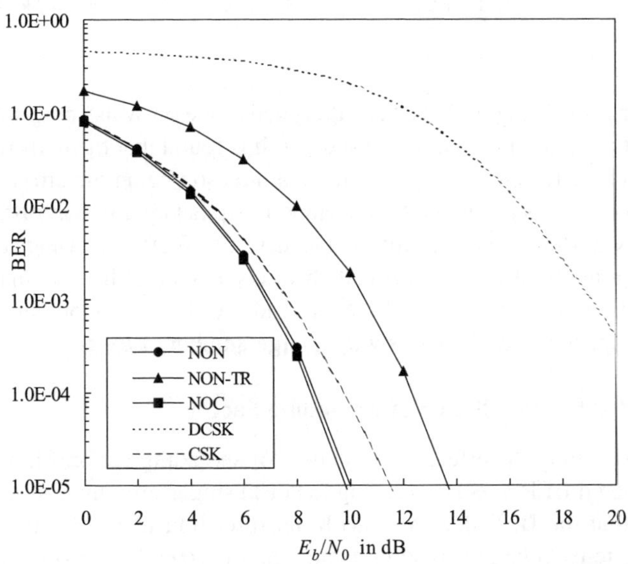

Figure 8.11. Simulated BERs of 2-user NOC, NON, NON-TR, CSK and DCSK systems when Hénon map is used. The transformation functions have been designed based on Walsh codes. Spreading factor $2\beta = 100$ is used for the NOC, NON and CSK systems, $2\beta = 200$ is used for the NON-TR and DCSK systems.

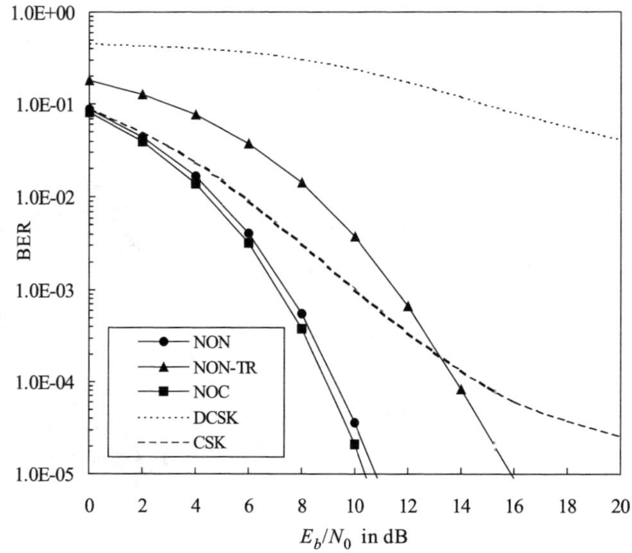

Figure 8.12. Simulated BERs of 4-user NOC, NON, NON-TR, CSK and DCSK systems when Hénon map is used. The transformation functions have been designed based on Walsh codes. Spreading factor $2\beta = 100$ is used for the NOC, NON and CSK systems, $2\beta = 200$ is used for the NON-TR and DCSK systems.

transformation functions have been designed based on Walsh codes. We compared the results to those in Figs. 8.8(b) and 8.9. It is found that the performances of the CSK and DCSK systems degrade, the main reason being the cross-correlation of the chaotic signals generated by the Hénon map is large. In contrast, for the NOC and NON systems, there is little change in the BER performance because the cross-correlation of the transformed chaotic signals is still very small. Therefore, the results indicate that both the NOC and NON systems outperform the CSK system, especially for the cases with a large number of users.

8.3.4 Effect of the Choice of Spreading Factor

Finally, we study the effect of the choice of spreading factor. Figure 8.13 shows the simulated BER versus spreading factor in single-user and 2-user systems. We observe that the BER decreases with the spreading factor. Unlike other noncoherent systems, a large spreading factor can improve the performance. When the spreading factor increases, the performance of the NON system gets closer to that of the NOC system. But when the spreading factor is very small, e.g., $2\beta < 10$, the NON system becomes unreliable. In that case, the NON-TR system offers a better performance than the NON system.

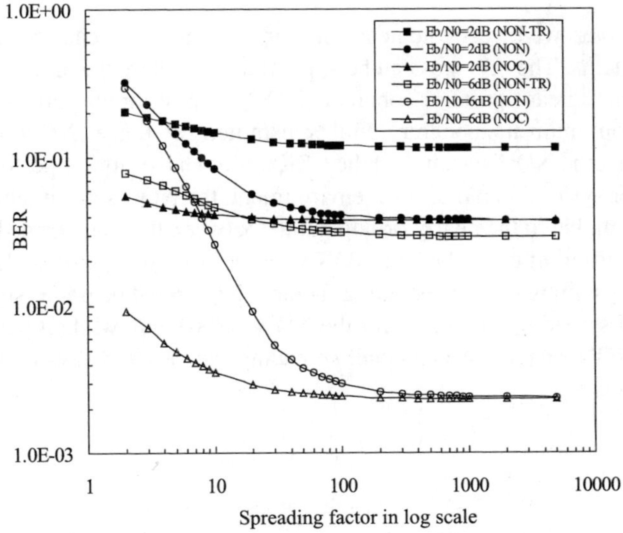

(a)

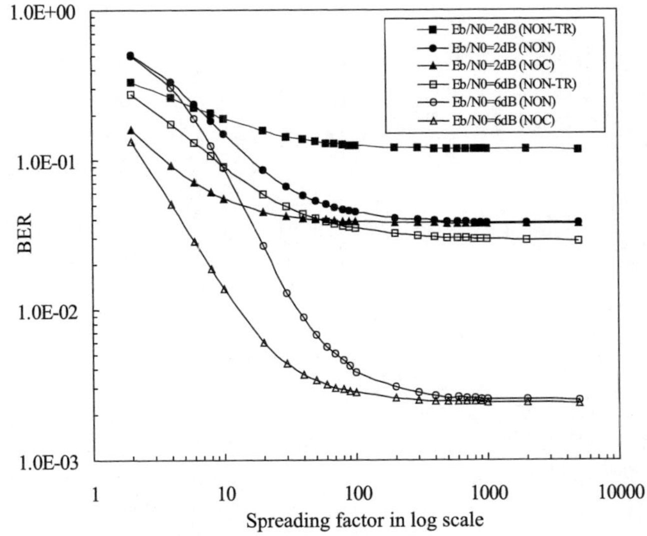

(b)

Figure 8.13. Simulated BER versus spreading factor for the NOC, NON, NON-TR systems when Chebyshev map of degree 2 is used. (a) Single-user system; (b) 2-user system.

8.4 SUMMARY

In this chapter, we investigate near-optimum schemes for chaos-based communication systems. The schemes can be applied to the coherent and noncoherent systems, namely near-optimum coherent (NOC), near-optimum noncoherent NON and near-optimum noncoherent with transmitted reference (NON-TR) systems. We find that the NOC system and the CSK system have similar performances in a single-user system. In a multiuser environment, the NOC system outperforms the CSK system due to the nearly orthogonality between the transformed chaotic signals from different users. For the NON system, its performance is close to that of the NOC system when the spreading factor is large but it degrades significantly at very small spreading factor. As for the NON-TR system, while it performs better than the NON system only at small spreading factors, it always outperforms the DCSK system.

BIBLIOGRAPHY

Alligood K.T., Sauer T.D. and Yorke J.A. [1997], *Chaos: An Introduction to Dynamical Systems*, Springer, New York.

Argüello F., Bugallo M. and Amor M. [2002], "Multi-user receivers for spread spectrum communications based on chaotic sequences", *International Journal of Bifurcation and Chaos*, **12**(4), pp. 847–853.

Briggs W., Bryan G.H. and Walker G. [1985], *The Tutorial Algebra*, Vol. 1, Hong Kong University Press, Hong Kong.

Brown R. [1996], "Clarifying chaos: Examples and counterexamples", *International Journal of Bifurcation and Chaos*, **6**(2), pp. 219–249.

Carroll T.L. and Pecora L.M. [1999], "Using multiple attractor chaotic systems for communication", *Chaos*, **9**, pp. 445–451.

Chong C.C., Yong S.K. and Lee S.S. [2005a], "UWB direct chaotic communication technology", *IEEE Antennas and Wireless Propagation Letters*, **4**, pp. 316–319.

Chong C.C., Yong S.K., Kim Y.H., Kim J.H., Lee S.S., Dmitriev A.S., Panas A.I., Starkov S.O., Andreyev Y.V., Efremova E.V., Kuzmin L.V., Kim H. and Cha J. [2005b], "Samsung electronics (SAIT) CFP presentation for IEEE 802.15.4a alternative PHY", P802.15-05/330r2-SG4a, www.ieee802.org/15/pub/05/15-05-0030-00-004a-samsung-electronics-sait-cfp-presentation.pdf.

Chow T.W.S., Feng. J.C. and Ng K.T. [2000], "An adaptive demodulator for the chaotic modulation communication system with RBF neural network", *IEEE Transactions on Circuits and Systems Part I*, **47**(6), pp. 902–909.

Cuomo K.M. and Oppenheim A.V. [1993], "Circuit implementation of synchronized chaos with applications to communications", *Physical Review Letters*, **71**(1), pp. 65–68.

Dedieu H., Kennedy M.P. and Hasler M. [1993], "Chaos shift keying: Modulation and demodulation of a chaotic carrier using self-synchronizing Chua's circuit", *IEEE Transactions on Circuits and Systems Part II*, **40**(10), pp. 634–642.

Dmitriev A.S., Panas A.I. and Starkov S.O. [2005], "Ultra-wide band communications using chaos", *Proceedings, Second International Workshop Networking with Ultra Wideband Workshop on Ultra Wideband for Sensor Networks*, Rome, Italy, pp. 25–29.

Elaydi S. [2000], *Discrete Chaos*, Chapman and Hall/CRC, Boca Raton, FL.

Elmirghani J.M.H. and Cryan R.A. [1994], "New chaotic based communication technique with multiuser provision", *Electronics Letters*, **30**, pp. 1206–1207.

Feng J. and Tse C.K. [2001], "An on-line adaptive chaotic demodulator based on radial-basis-function neural networks", *Physical Review E*, **63**(2), 026202-1-10.

Galias Z. and Maggio G.M. [2001], "Quadrature chaos-shift keying: Theory and performance analysis", *IEEE Transactions on Circuits and Systems Part I*, **48**(12), pp. 1510–1519.

Geisel T. and Fairen V. [1984], "Statistical properties of chaos in Chebyshev maps", *Physics Letters*, **105A**(6), pp. 263–266.

Hasler M. and Schimming T. [2000], "Chaos communication over noisy channels", *International Journal of Bifurcation and Chaos*, **10**, pp. 719–735.

Haykin S. [1996], *Adaptive Filter Theory*, Prentice-Hall, Englewood Cliffs.

Heidari-Bateni G. and McGillem C.D. [1994], "A chaotic direct-sequence spread-spectrum communication system", *IEEE Transactions on Communications*, **42**(2/3/4), pp. 1524–1527.

Heidari-Bateni G., McGillem C.D. and Tenorio M.F. [1992], "A novel multiple-address digital communictaion system using chaotic signals", in: *Proc. International Communication Conference*, Chicago, USA, pp. 1232–1236.

Hilborn R.C. [2000], *Chaos and Nonlinear Dynamics: An Introduction for Scientists and Engineers*, Oxford University Press, New York.

Itoh M. and Murakami H. [1995], "New communication systems via chaotic synchronization", *IEICE Transactions Fundamentals*, **E78-A**(3), pp. 285–290.

Itoh M., Murakami H. and Chua L.O. [1994], "Communication systems via chaotic modulations", *IEICE Transactions on Fundamentals*, **E77-A**(3), pp. 1000–1006.

Jáko Z. [1998], "Performance improvement of DCSK modulation", in: *Proc. International Workshop on Nonlinear Dynamics of Electronic Systems (NDES'98)*, Budapest, Hungary, pp. 119–122.

Jákó Z., Kis G. and Kolumbán G. [2000], "Multiple access capability of the FM-DCSK chaotic communications system", in: *Proc. International Workshop on Nonlinear Dynamics of Electronic Systems (NDES'2000)*, Catania, Italy, pp. 52–55.

Kennedy M.P. [1998], "Chaotic communications: From chaotic synchronization to FM-DCSK", in: *Proc. International Workshop on Nonlinear Dynamics of Electronic Systems (NDES'98)*, Budapest, Hungary, pp. 31–40.

Kennedy M.P. and Kolumbán G. [2000], "Digital communications using chaos", in: Chen G., *Controlling Chaos and Bifurcations in Engineering Systems*, CRC Press, New York, pp. 477–500.

Kennedy M.P., Kolumbán G., Kis G. and Jáko Z. [1998], "Recent advances in communicating with chaos", *Proc. IEEE International Symposium on Circuits and Systems (ISCAS'98)*, Montery, California, USA, pp. 461–464.

Kis G., Jáko Z., Kennedy M.P. and Kolumbán G. [1998], "Chaotic communications without synchronization", in: *Proc. 6th IEE Conference on Telecommunications*, Edinburgh, UK, pp. 49–53.

Kisel A., Dedieu H. and Schimming T. [2001], "Maximum likelihood approaches for noncoherent communication with chaotic carriers", *IEEE Transactions on Circuits and Systems Part I*, **48**, pp. 533–542.

Kocarev L., Halle K.S., Eckert K., Chua L.O. and Parlitz U. [1992], "Experimental demonstration of secure communications via chaotic synchronization", *International Journal of Bifurcation and Chaos*, **2**, pp. 709–713.

Kohda T. and Tsuneda A. [1994], "Even- and odd-correlation functions of chaotic Chebyshev bit sequences for CDMA", in: *Proc. IEEE International Symposium on Spread Spectrum Technology and Applications*, Oulu, Finland, pp. 391–395.

Kolumbán G. [1998], "Performance evaluation of chaotic communications systems: Determination of low-pass equivalent model", in: *Proc. International Workshop on Nonlinear Dynamics of Electronic Systems (NDES'98)*, Budapest, Hungary, pp. 41–51.

Kolumbán G. [2000a], "Basis function description of chaotic modulation schemes", in: *Proc. International Workshop on Nonlinear Dynamics of Electronic Systems (NDES'00)*, Catania, Italy, pp. 165–169.

Kolumbán G. [2000b], "Theoretical noise performance of correlator-based chaotic communications schemes", *IEEE Transactions on Circuits and Systems Part I*, **47**(12), pp. 1692–1701.

Kolumbán G. [2002], "The theory and implementation of a robust chaotic digital communications system", *http://rfic.ucsd.edu/chaos/ws2002/Kolumban.pdf.*

Kolumbán G. and Kennedy M.P. [2000], "The role of synchronization in digital communications using chaos — Part III: Performance bounds for correlation receivers", *IEEE Transactions on Circuits and Systems Part I*, **47**(12), pp. 1673–1683.

Kolumbán G., Vizvari G.K., Schwarz W. and Abel A. [1996], "Differential chaos shift keying: A robust coding for chaos communication", in: *Proc. International Workshop on Nonlinear Dynamics of Electronic Systems,* Seville, Spain, pp. 87–92.

Kolumbán G., Kennedy M.P. and Kis G. [1997a], "Multilevel differential chaos shift keying", in: *Proc. International Workshop on Nonlinear Dynamics of Electronic Systems (NDES'97)*, Moscow, Russia, pp. 191–196.

Kolumbán G., Kis G., Kennedy M.P. and Jáko Z. [1997b], "FM-DCSK: A new and robust solution to chaos communications", in: *Proc. International Symposium on Nonlinear Theory and Its Applications (NOLTA'97)*, Hawaii, pp. 117–120.

Kolumbán G., Kennedy M.P. and Chua L.O. [1998], "The role of synchronization in digital communications using chaos — Part II: Chaotic modulation and chaotic synchronization", *IEEE Transactions on Circuits and Systems Part I*, **45**(4), pp. 1129–1140.

Kolumbán G., Jáko Z. and Kennedy M.P. [1999], "Enhanced versions of DCSK and FM-DCSK data transmission systems", in: *Proc. IEEE International Symposium on Circuits and Systems (ISCAS'99)*, Orlando, USA, pp. 475–478.

Kolumbán G., Jáko Z. and Kis G. [2002a], "Enhance versions of the DCSK systems", in: *Proc. International Symposium on Nonlinear Theory and Its Applications (NOLTA'02)*, Xi'an, China, pp. 95–98.

Kolumbán G., Kennedy M.P., Jáko Z. and Kis G [2002b], "Chaotic communications with correlator receiver: Theory and performance limit", *Proceedings of the IEEE*, **90**(5), pp. 711–732.

Lau F.C.M. and Tse C.K. [2002], "Optimum correlator-type receiver design for CSK communication systems", *International Journal of Bifurcation and Chaos*, **12**(5), pp. 1029–1038.

Lau F.C.M. and Tse C.K. [2003a], *Chaos-Based Digital Communication Systems*, Springer-Verlag, Heidelberg.

Lau F.C.M. and Tse C.K. [2003b], "On optimal detection of noncoherent chaos-shift-keying signals in a noisy environment", *International Journal of Bifurcation and Chaos*, **13**(6), pp. 1587–1597.

Lau F.C.M. and Tse C.K. [2003c], "Approximate-optimal detection for chaos communication systems", *International Journal of Bifurcation and Chaos*, **13**(5), pp. 1329–1335.

Lau F.C.M., Cheong K.Y. and Tse C.K. [2002a], "A permutation-based multiple access DCSK system", in: *Proc. International Symposium on Nonlinear Theory and Its Applications (NOLTA'02)*, Xi'an, China, pp. 511–514.

Lau F.C.M., Yip M.M., Tse C.K. and Hau S.F. [2002b], "A multiple access technique for differential chaos shift keying", *IEEE Transactions on Circuits and Systems Part I*, **49**(1), pp. 96–104.

Lau F.C.M., Cheong K.Y. and Tse C.K. [2003], "Permutation-based DCSK and multiple access DCSK systems", *IEEE Transactions on Circuits and Systems Part I*, **50**(6), pp. 733–742.

Lawrance A.J. and Ohama G. [2002], "Exact and analytically approximate bit error rates in chaos communication", in: *Proc. International Symposium on Nonlinear Theory and Its Applications (NOLTA'02)*, Xi'an, China, pp. 99–102.

Leung H. and Lam J. [1997], "Design of demodulator for the chaotic modulation communication system", *IEEE Transactions on Circuits and Systems Part I*, **44**(3), pp. 262–267.

Lupas R. and Verdú S. [1989], "Linear multiuser detectors for synchronous code-division multiple-access channels", *IEEE Transactions on Information Theory*, **35**(1), pp. 123–136.

Lupas R. and Verdú S. [1990], "Near-far resistance of multiuser detectors in asynchronous channels", *IEEE Transactions on Communications*, **38**(4), pp. 496–508.

Madsen A.H. and Cho K.S. [1999], "MMSE/PIC multiuser detection for DS/CDMA systems with inter- and intra-cell interference", *IEEE Transactions on Communications*, **47**(2), pp. 291–299.

Mazzini G., Setti G. and Rovatti R. [1997], "Chaotic complex spreading sequences for asynchronous DS-CDMA Part I: System modeling and results", *IEEE Transactions on Circuits and Systems Part I*, **44**(10), pp. 937–947.

Mazzini G., Setti G. and Rovatti R. [1998], "Chaotic complex spreading sequences for asynchronous DS-CDMA Part II: Some theoretical performance bounds", *IEEE Transactions on Circuits and Systems Part I*, **45**(4), pp. 496–506.

Miller S.H., Elmirghani J.M.H. and Cryan R.A. [1995], "Efficient chaotic-driven echo path modelling", *Electronics Letters*, **31**, pp. 429–430.

Parlitz U. and Ergezinger S. [1994], "Robust communication based on chaotic spreading sequences", *Physics Letters*, **A188**, pp. 146–150.

Parlitz U., Chua L.O., Kocarev L., Halle K.S. and Shang A. [1992], "Transmission of digital signals by chaotic synchronization", *International Journal of Bifurcation and Chaos*, **2**, pp. 973–977.

Peterson R.L., Ziemer R.E. and Borth D.F. [1995], *Introduction to Spread Spectrum Communications*, Prentice Hall, Englewood Cliffs.

Proakis J.G. [1995], *Digital Communications*, McGraw–Hill, Singapore.

Proakis J.G. and Salehi M. [1994], *Communications Systems Engineering*. Prentice Hall, Englewood Cliffs.

Schimming T. and Hasler M. [2000], "Optimal detection of differential chaos shift keying", *IEEE Transactions on Circuits and Systems Part I*, **47**(12), pp. 1712–1719.

Simon M.K., Omura J.K., Scholtz R.A. and Levitt B.K. [2002], *Spread Spectrum Communications Handbook*, McGraw–Hill, New York.

Skaug R. and Hjelmstad J.F. [1985], *Spread Spectrum in Communication*, P. Peregrinus on behalf of the Institution of Electrical Engineers, London.

Sushchik M., Tsimring L.S. and Volkovskii A.R. [2000], "Performance analysis of correlation-based communication schemes utilizing chaos", *IEEE Transactions on Circuits and Systems Part I*, **47**(12), pp. 1684–1691.

Tam W.M., Lau F.C.M., Tse C.K. and Yip M.M. [2002a], "An approach to calculating the bit error rate of a coherent chaos-shift-keying digital communication system under a noisy multiuser environment", *IEEE Transactions on Circuits and Systems Part I*, **49**(2), pp. 210–223.

Tam W.M., Lau F.C.M. and Tse C.K. [2002b], "Multi-user detection techniques for multiple access chaos-based digital communication systems", in: *Proc. International Symposium on Nonlinear Theory and Its Applications (NOLTA'02)*, Xi'an, China, pp. 503–506.

Tam W.M., Lau F.C.M. and Tse C.K. [2003], "Analysis of bit error rates for multiple access CSK and DCSK communication systems", *IEEE Transactions on Circuits and Systems Part I*, **50**(5), pp. 702–707.

Tam W.M., Lau F.C.M., Tse C.K. and Lawrance A.J. [2004a], "Exact analytical bit error rates for multiple access chaos-based communication systems," *IEEE Transactions on Circuits and Systems Part II*, **51**(9), pp. 473–481.

Tam W.M., Lau F.C.M. and Tse C.K. [2004b], "A multiple access scheme for chaos-based digital communication systems utilizing transmitted reference", *IEEE Transactions on Circuits and Systems Part I*, **51**(9), pp. 1868–1878.

Tam W.M., Lau F.C.M. and Tse C.K. [2004c], "Performance analysis of multiple access chaotic-sequence spread-spectrum communication systems using parallel interference cancellation receivers", *International Journal of Bifurcation and Chaos*, **14**(10), pp. 3633–3646.

Tam W.M., Lau F.C.M. and Tse C.K. [2004d], "An approach to calculating the bit error rates of multiple access chaotic-sequence spread-spectrum communication systems employing multi-user detectors", *International Journal of Bifurcation and Chaos*, **14**(1), pp. 183–206.

Tam W.M., Lau F.C.M. and Tse C.K. [2006], "Generalized correlation-delay-shift-keying scheme for noncoherent chaos-based communication systems", *IEEE Transactions on Circuits and Systems Part I*, **53**(3), pp. 712–721.

Tse C.K., Cheong K.Y., Lau F.C.M. and Hau S.F. [2001], "An approach for CSK detection based on return maps", *Proc. International Symposium on Nonlinear Theory and Its Applications (NOLTA'01)*, Miyagi, Japan, pp. 637–640.

Tse C.K., Cheong K.Y., Lau F.C.M. and Hau S.F. [2002], "Return-map-based approaches for noncoherent detection in chaotic digital communications", *IEEE Transactions on Circuits and Systems Part I*, **49**(10), pp. 1495–1499.

Tsimring L.S, Volkovskii A.R., Young S.C. and Rulkov N.F. [2001], "Multi-user communication using chaotic frequency modulation", in: *Proc. International Symposium on Nonlinear Theory and Its Applications (NOLTA'01)*, Miyagi, Japan, pp. 561–564.

Verdú S. [1986], "Minimum probability of error for asynchronous Gaussian multiple-access channels", *IEEE Transactions on Information Theory*, **32**(1), pp. 85–96.

Xie Z., Short R.T. and Rushforth C.K. [1990], "A family of suboptimum detectors for coherent multiuser communications", *IEEE Journal on Selected Areas in Communications*, **8**(4), pp. 683–690.

Yang T. and Chua L.O. [1997], "Chaotic digital code-division multiple access (CDMA) communication systems", *International Journal of Bifurcation and Chaos*, **7**(12), pp. 2789–2805.

Yang T. and Chua L.O. [1998], "Error performance of chaotic digital code-division multiple access (CDMA) systems", *International Journal of Bifurcation and Chaos*, **8**(10), pp. 2047–2059.

Yoon Y.C., Kohno R. and Imai H. [1993], "A spread-spectrum multiaccess system with cochannel interference cancellation for multipath fading channels", *IEEE Journal on Selected Areas in Communications*, **11**(7), pp. 1067–1075.

Yoon S. and Ness Y.B. [2002], "Performance analysis of linear multiuser detectors for randomly spread CDMA using Gaussian approximation", *IEEE Journal on Selected Areas in Communications*, **20**(2), pp. 409–418.

INDEX

a posteriori probability, 19

ACSK, *see* antipodal chaos shift keying
adaptive transversal filter, 190
additive noise, 13
additive white Gaussian noise, 35, 58, 112, 159
adjoint, 66
anti-jamming, 2
antipodal chaos shift keying, 12, 34
ATF, *see* adaptive transversal filter
autocorrelation, 51, 94
autovariance, 50
average bit energy, 22, 38, 65, 88, 119, 164, 193
average power, 38, 64, 88, 193
AWGN, *see* additive white Gaussian noise

bandpass system, 9
bandwidth, 9
basis function, 6, 26, 27
Bayes' rule, 19
BER, *see* bit error rate
BF simulation, *see* brute-force simulation
binary phase shift keying, 193
bit duration, 7–9, 12, 19, 26, 31, 34, 113, 115, 126, 158, 164, 169, 212, 214
bit error rate, 86–88, 193, 197
– BF simulation, 73
– exact, 38–42
– GA, 36–38, 63, 65, 66, 69, 70, 72, 117, 118, 129, 130, 141, 147, 148, 156, 162, 164, 168, 169, 175, 176, 212, 216
– MAS, 63, 67, 70, 73
bit-energy estimator, 13, 15, 19
bit-energy-to-noise-power-spectral-density ratio, 13, 38, 73, 119, 198
BPSK, *see* binary phase shift keying
brute-force simulation, 42, 73, 89

CDMA, *see* code-division-multiple-access
CDSK, *see* correlation delay shift keying
CFM, *see* chaotic frequency modulation
channel, 22, 25, 35
chaos, 1
– application to communications, 5
chaos shift keying, 6, 9, 12, 15, 17, 34
– antipodal, 12
– with two chaos generators, 15
chaotic dynamical system, 1
– continuous-time, 1
– discrete-time, 1
chaotic frequency modulation, 30
chaotic masking, 5
chaotic modulation, 5
chaotic on-off-keying, 13
chaotic signal, 1
– autocorrelation, 1
– cross-correlation, 1
– power spectrum, 1
chaotic system, 1
Chebyshev map, 38, 41, 49, 50
– of degree 2, 49, 117, 217
– of degree 3, 49, 52, 53, 63, 197
– statistical properties, 49
code-division-multiple-access, 26
communication systems
– chaos-based, 11
– conventional, 11
– digital, 11
– spread-spectrum, 26
– UWB, 7
– wireless, 7
complementary error function, 36, 63, 119, 148, 164, 193, 212
COOK, *see* chaotic on-off-keying
correlation, 1, 17, 35
correlation delay shift keying, 24, 112, 169
correlation matrix, 58, 61, 65
correlator-type, 17
covariance, 37, 38, 62, 72, 101–104
CSK, *see* chaos shift keying
cubic map, 2, 49, 53, 63, 73, 197

dc components, 57, 85
DCSK, *see* differential chaos shift keying
DD, *see* decorrelating detector
decorrelating detector, 59, 61
demodulation/detection
– coherent
– – CFM, 31
– – COOK, 13
– – CSK, 13, 16, 17
– – MA-ACSK, 35
– – MA-NOC, 213
– – NOC, 211
– – synchronization-error-based, 16
– correlator-type, 17, 116
– noncoherent
– – approximate-optimal, 7
– – bit-energy based, 13, 19
– – CDSK, 25
– – COOK, 13
– – CSK, 17
– – DCSK, 21
– – ECDSK-M1, 159
– – ECDSK-M2, 165
– – FM-DCSK, 22
– – GCDSK, 116
– – MA-CDSK, 127
– – MA-GCDSK, 127
– – MA-NON, 213
– – MA-NON-TR, 217
– – NON, 211
– – NON-TR, 215
– – optimal, 7
– – optimal classifier, 19
– – PMA-CDSK, 171
– – PMA-ECDSK-M1, 171
– – PMA-ECDSK-M2, 171
– – return-map based, 19
determinant operator, 110
differential chaos shift keying, 6, 20, 22, 26
direct-sequence code division multiple access, 55
direct-sequence spread-spectrum, 8
DS-CDMA, *see* direct-sequence code division multiple access
DS-SS, *see* direct-sequence spread-spectrum

ECDSK-M1, *see* enhanced correlation delay shift keying with Method 1
ECDSK-M2, *see* enhanced correlation delay shift keying with Method 2
enhanced correlation delay shift keying with Method 1, 158
enhanced correlation delay shift keying with Method 2, 164
erfc, *see* complementary error function
Euclidean norm, 191
exact method, 42

FCC, *see* Federal Communications Commission
Federal Communications Commission, 7
FM-DCSK, *see* frequency-modulated DCSK
frame, 187, 189, 191, 193
frequency-modulated DCSK, 7, 22, 26, 27

GA, *see* Gaussian-approximation-based
Gaussian-approximation-based, 36, 117, 129, 141, 148, 162, 167, 172
GCDSK, *see* generalized correlation delay shift keying
generalized correlation delay shift keying, 115

Hénon map, 217
Hadamard matrix, 194, 213

IA, *see* inverse and average
independent variables, 37, 39, 40, 57, 85
information-bearing signal, 6, 20, 22, 24–26, 192, 195
initial condition, 1, 38
– sensitive dependence, 1
interference
– co-channel, 27, 33, 84
– inter-symbol, 117, 129
– inter-user, 129, 192
– intra-signal, 114, 117
– intra-user, 129
inverse and average, 195

Kronecker's delta, 71

least mean square, 191
LMS, *see* least mean square
low probability of detection, 2
low-pass system, 9

MA-ACSK, *see* multiple access ACSK
MA-CDSK, *see* multiple access CDSK
MA-DCSK, *see* multiple access time-delay-based (permutation-based) DCSK

MA-GCDSK, *see* multiple access GCDSK
MA-NOC, *see* multiple access NOC
MA-NON, *see* multiple access NON
MA-NON-TR, *see* multiple access NON-TR
MAI, *see* multiple-access interference
MAS, *see* mixed analysis–simulation
mean, 34, 36–39, 41, 50–53, 57, 62–66, 68, 70, 72, 85, 88, 94–102, 105, 118, 129, 142, 148, 162, 167, 175
minimum mean-square-error, 60
mixed analysis–simulation, 63, 67, 70, 73
MMSE, *see* minimum mean-square-error
modulation
– CDSK, 24
– CFM, 30
– COOK, 13
– CSK, 12, 15
– DCSK, 20
– ECDSK-M1, 158
– ECDSK-M2, 164
– FM-DCSK, 22
– GCDSK, 115
– MA-ACSK, 34
– MA-CDSK, 126
– MA-GCDSK, 126
– MA-NOC, 212
– MA-NON, 212
– MA-NON-TR, 216
– NOC, 210
– NON, 210
– NON-TR, 214
– PMA-CDSK, 169
– PMA-ECDSK-M1, 169
– PMA-ECDSK-M2, 169
multipath fadings, 2
multipath signals, 4
multiple access, 5, 26
– ACSK, 34
– CDSK, 126
– CFM, 30
– DCSK, 26
– FM-DCSK, 26, 27
– GCDSK, 126
– NOC, 212
– NON, 212
– NON-TR, 216
– permutation-based
– – CDSK, 169
– – DCSK, 169
– – ECDSK-M1, 169
– – ECDSK-M2, 169
– time-delay-based
– – DCSK, 198
multiple-access interference, 55
multiuser detection, 55
– linear, 55, 56, 60
– – DD, 59, 61
– – MMSE, 60, 67
– nonlinear, 56, 83
– – conventional/PIC, 88
– – DD/PIC, 88
– – MMSE/PIC, 88
– – PIC, 83

narrowband signal, 2
narrowband system, 4
near-optimum coherent, 210
near-optimum noncoherent, 210
near-optimum noncoherent system with transmitted reference, 214
NOC, *see* near-optimum coherent
noise, 2
– additive noise, 13
– additive white Gaussian noise, 35, 58, 112, 159
NON, *see* near-optimum noncoherent
NON-TR, *see* near-optimum noncoherent system with transmitted reference
Nyquist sampling rate, 9

optimal classifier, 19

P-DCSK, *see* permutation-based DCSK
parallel interference cancellation, 83
pdf, *see* probability density function
permutation, 170
permutation-based DCSK, 7
phase-lock loop, 31
PIC, *see* parallel interference cancellation
PLL, *see* phase-lock loop
PMA-CDSK, *see* multiple access permutation-based CDSK
PMA-DCSK, *see* multiple access permutation-based DCSK
PMA-ECDSK-M1, *see* multiple access permutation-based ECDSK-M1
PMA-ECDSK-M2, *see* multiple access permutation-based ECDSK-M2
power spectral density, 2, 35, 112, 159, 190, 210

power spectrum, 1
probability density function, 13, 17, 19, 22, 41, 50, 63, 211, 216, 220, 224
– N-dimensional normal, 59, 110
– Chebyshev map, 49
– joint, 40
– two-dimensional (bivariate) normal, 108
psd, *see* power spectral density

QCSK, *see* Quadrature chaos shift keying
Quadrature chaos shift keying, 7

radial-basis function, 5
RAKE receiver, 4
RBF, *see* radial-basis function
reference signal, 20, 24, 26
regression, 20
return map, 20

sampling frequency, 9
Schwarz's inequality, 62
single-user detector, 59
spread-spectrum communications, 4
– advantage, 4
– application, 5
– direct-sequence, 8
spreading factor, 9, 34, 57, 113, 159, 189, 210
spreading sequence, 8
– *m* sequence, 8
– Gold sequence, 8
– quantized chaotic sequence, 8
synchronization, 5, 16, 30

tap weight, 191
threshold detector, 15, 17
time slot, 20, 26, 27, 187, 189, 191, 195, 214, 216
time-delay-based MA-DCSK, *see* multiple access time-delay-based DCSK
training sequence 193
transformation, 210–213, 218
– permutation, 7, 169, 170

ultra-wideband, 7
UWB, *see* ultra-wideband

variance, 36–41, 50, 62–64, 66, 68, 70, 72, 85, 87, 88, 94, 95, 100–102, 142, 147, 148, 155, 156, 162, 168, 175
VCO, *see* voltage controlled oscillator
voltage controlled oscillator, 31

Walsh code, 26, 194, 212